Exploring
drafting
basic fundamentals

by

John R. Walker
Charlottesville, Virginia

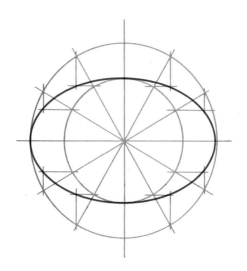

South Holland, Illinois
THE GOODHEART-WILLCOX COMPANY, INC.
Publishers

Library of Congress Catalog Card Number 86-31968
International Standard Book Number 0-87006-620-X

23456789-87-54321098

Library of Congress Cataloging in Publication Data
Walker, John R.,   Exploring drafting.
1. Mechanical drawing.   I. Title. T353.W22   1987   604.2   86-31968 ISBN 0-87006-620-X

INTRODUCTION

A course in drafting should be a part of your education. It will help you to develop the capacity to plan in an orderly fashion, to interpret the ideas of others, and to express yourself in an understandable manner.

EXPLORING DRAFTING is a first course which teaches drafting fundamentals and basic constructions. As you proceed with the course, you will become familiar with the drafting methods and processes used by industry. You will develop and practice drafting skills and techniques.

Conventional drawings are constructed in step-by-step fashion in this text to demonstrate drawing principles. Other more challenging drawings are also provided to help develop your originality and creativity. The Test Your Knowledge section at the end of each unit will allow you to check your comprehension of the material covered. The problems at the end of the unit allow you an opportunity to practice your drafting skills.

EXPLORING DRAFTING is current with the latest ANSI (American National Standards Institute) practices. You will learn how to use symbols to communicate in the international language.

EXPLORING DRAFTING provides you with the basic understanding necessary to allow you to progress into CADD (Computer Aided Design and Drafting). The knowledge learned in this text will give you the background to draw and design on computer screens and plotters.

Drafting, the "language of industry," offers many career opportunities for men and women. Some of these exciting and rewarding futures are described in the unit on Careers in Drafting and Design. You will find that the ability to draw and to understand drafting will be of benefit throughout your lifetime and your career.

John R. Walker

CONTENTS

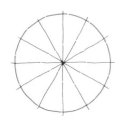

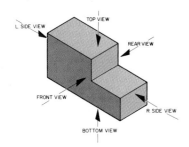

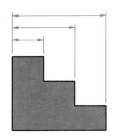

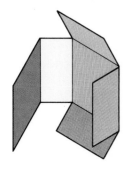

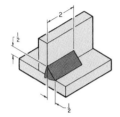

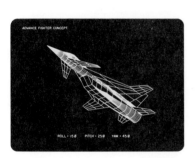

(Lockheed-California Co.)

Drafting is the graphic communication used to move ideas into production. The skills and techniques of the drafter are used to communicate the designs and drawings used in our technology based society. (Pontiac Motor Div., GMC)

Unit 1

WHY STUDY DRAFTING

This unit will introduce you to the many fields of drafting. After studying it, you will discover why drafting is called a "universal language." You will understand why drawings are often the best way to explain or show our ideas. You will be able to explain why drafting is so important to modern industry. You will recognize the tremendous technological changes that affect the ways drawings are made and stored.

DRAFTING is a form of graphic communication. It is concerned with the preparation of the drawings needed to develop and manufacture a product, Fig. 1-1. It is a very important part of modern industry because drawings are often the only way to explain or show our ideas, Fig. 1-2.

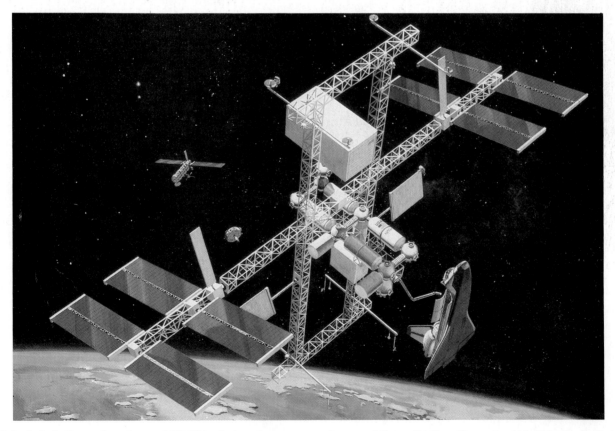

Fig. 1-1. Artist's illustration of one of NASA's designs for a proposed manned Space Station. Computer Aided Design (called CAD) indicated that sections or modules of the Space Station (shown above the Space Shuttle) be attached to a boom-like structure resembling a radio antenna tower. Large solar arrays for generating electrical power are attached to the opposite end of the tower (nicknamed ''power tower''). This arrangement that separates the module from the solar panels causes the Space Station to act as a pendulum as it orbits the earth giving balance to the entire structure. (Rockwell International)

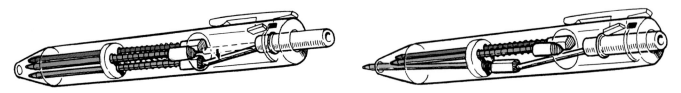

Fig. 1-2. Drawings are often the best way to explain or show ideas. Think how difficult it would be to explain how this simple three-color pen operates using only the written word.

Drafting is frequently called a "universal language." Like other languages, symbols (lines and figures) that have specific meaning are used. The symbols accurately describe the shape, size, material, finish, and fabrication or assembly of a product. These symbols have been standardized over most of the world. This makes it possible to interpret or understand drawings made in other countries, Fig. 1-3.

Drafting is also the "language of industry." Industry uses this precise language because the drawings must communicate the information the designer had in mind to those who produce the product.

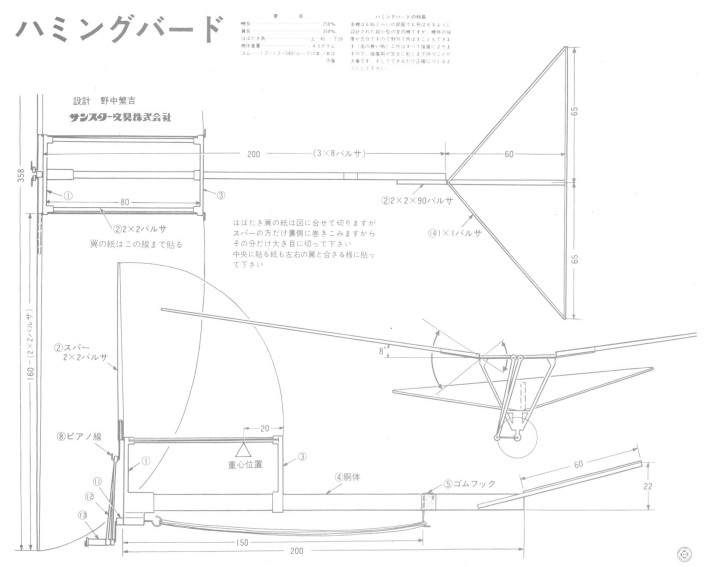

Fig. 1-3. Although the instructions are written in Japanese and the dimensions in the metric system, it would be possible for you to use these plans to construct this model ornithopter (it flies by flapping its wings like a bird).

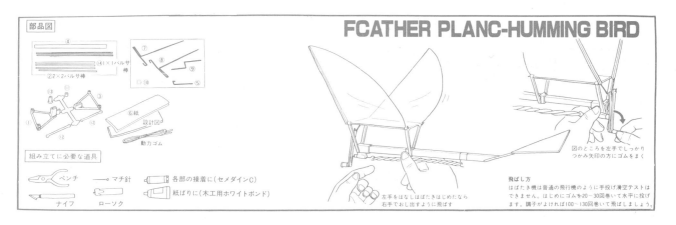

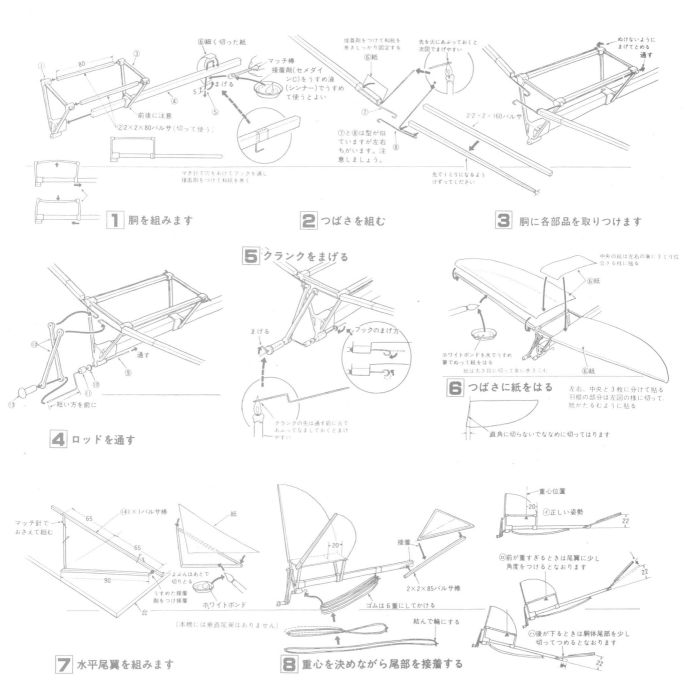

Fig. 1-3 continued.

Drawings can be made MANUALLY (by hand) on paper or film using drafting tools such as drawing board, pencil or pen, angles, compass, etc., Fig. 1-4. Drawings are also produced by COMPUTER GRAPHICS, Fig. 1-5, using *Computer Aided Design and Drafting (CADD)* system. When the design is completed, high speed PLOTTERS or AUTOMATED DRAFTING SYSTEMS turn out hard (paper) copy showing the part or design, Fig. 1-6. (Computer graphics is explained in more detail in Unit 23.)

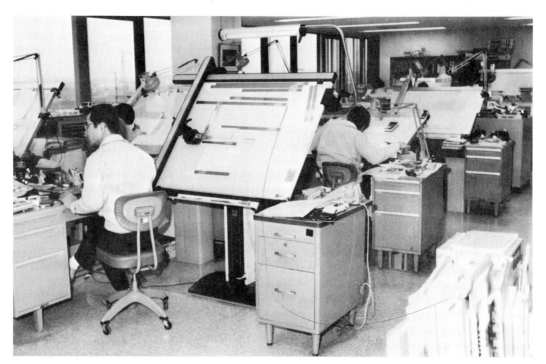

Fig. 1-4. Example from industry showing where drawings are made manually on paper or film using drafting tools. (MRC/Tamiya)

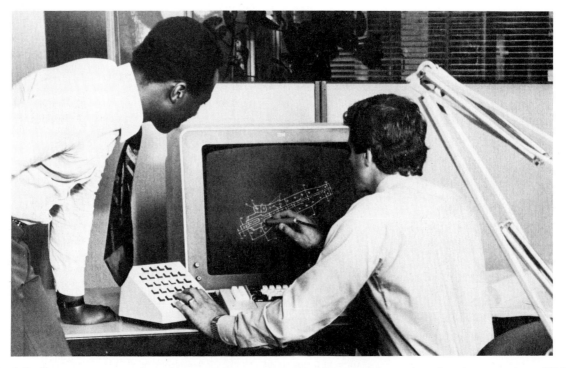

Fig. 1-5. Computer generated graphics uses the computer to aid in designing and engineering a product. (IBM)

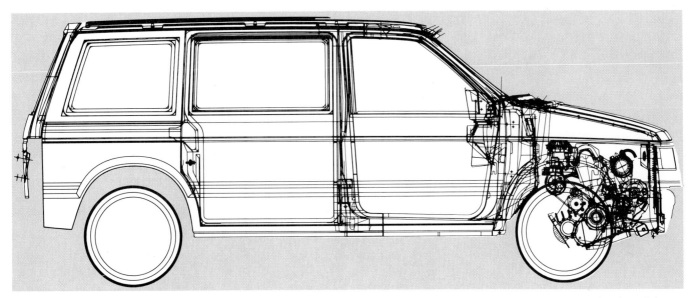

Fig. 1-6. A computer design can be converted into hard (paper) copy using a high speed plotter. This computer- developed drawing is a design study to determine the best position of the engine in a small passenger van. (Chrysler Corp.)

It would be difficult in today's world to name an occupation that does not require the ability to read and understand graphic information such as drawings, charts, or diagrams. YOU use drawings when you construct a model or electronic kit, Fig. 1-7.

Many specialized fields of drafting have been developed—aerospace, architectural, automotive, electrical and electronic, printed circuitry design, and topographical.

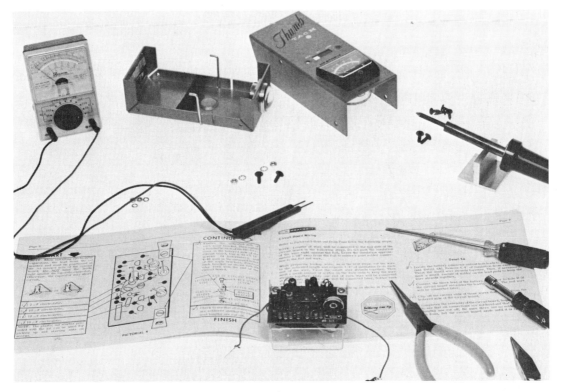

Fig. 1-7. You use drawings when you construct a model or an electronic kit like the optical tachometer shown here. A tachometer measures the RPM (revolutions per minute) of rotating objects.

Basically, they all use the same drafting equipment and employ similar drafting techniques. (Computer graphics may require specialized programs.) However, the finished product of each field varies greatly, Figs. 1-8, 1-9, and 1-10. You will learn the skills and techniques of the drafter as you study this book.

Fig. 1-8. An aerospace drafter must have the technical knowledge to carry out the ideas of the aerospace engineer. (Grumman Aerospace Corp.)

Fig. 1-9. The drafter conveys or communicates with drawings all the information required to construct this bridge. (Prestressed Concrete Institute)

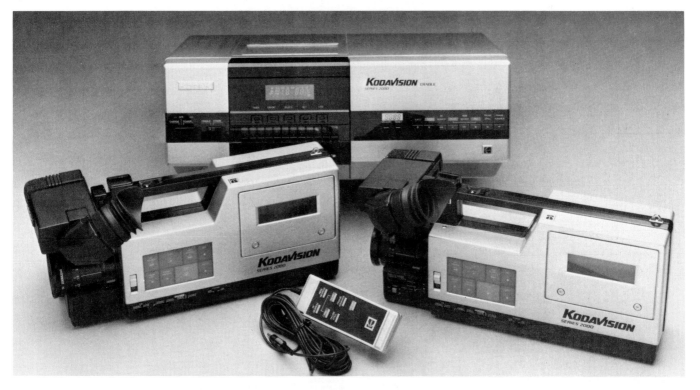

Fig. 1-10. A drafter must often have a working knowledge of many technical areas. The design of this video equipment involved optics, printed circuits, plastics, metals, and manufacturing processes. While engineers furnished the basic information, drafters had to put their ideas on paper. (Eastman Kodak Company)

DRAFTING VOCABULARY

Accurate, Aerospace, Architectural, Automated, Communicate, Computer aided drafting and design, Computer graphics, Diagram, Difficult, Drafting, Fabrication, Graphic communications, Interpret, Manually, Manufacture, Plotters, Preparation, Standardize, Symbols, Techniques, Topographical, Understand.

TEST YOUR KNOWLEDGE—UNIT 1

Please do not write in the book. Place your answers on another sheet of paper.

1. What does the term drafting mean?
2. Drawings are often used because: (List the correct answer/s.)
   a. They are easy to make.
   b. They are the best means available to explain or show many ideas.
   c. People who cannot read can understand drawings.
3. Drafting is called: (List the correct answer/s.)
   a. A picture language.
   b. The language of industry.
   c. A universal language.
4. There are very few occupations that do not require the ability to read and understand graphic information. Name five (5) occupations that require this skill.
5. What does the term CADD mean?
6. When a CADD designed problem is completed, it is converted into hard (paper) copy on a _____ or _____.
7. List five (5) specialized fields of drafting.

OUTSIDE ACTIVITIES

1. Prepare a list with two columns. In the first column write in the names of as many occupations that you can think of that require the ability to read and understand graphic information. In the second column, place the names of occupations that do not require this skill.
2. Secure samples of drawings used by the following industries:
   a. Aerospace.
   b. Building construction.
   c. Structural.
   d. Manufacturing.
   e. Map making.
   f. Electrical and electronic.
3. Make a collection of pictures (magazine clippings, photographs, etc.) that show products made by industries listed in activity 2. (DO NOT cut them from library books and magazines.)
4. Visit a local architect who designs residences. After discussing a project in work, prepare a report on the steps normally followed when designing a home for a client. Prepare your questions carefully before you make your visit.
5. Visit a local surveyor and make a report on the work he or she does. Borrow samples of completed work for a bulletin board display.
6. Obtain and display copies of drawings made in a foreign country.

Unit 2

CAREERS IN DRAFTING AND DESIGN

After studying this unit, you will be able to identify many of the career possibilities related to the fields of drafting and design. You will discover the characteristics and skills needed to prepare for various careers. You will comprehend the role of leadership in society and how it relates to drafting technology. You will be able to identify the traits of good leadership. Perhaps you will be encouraged to seek employment in these challenging fields where women and men compete on equal terms for jobs.

It would be difficult to name an industry that does not use drawings. These drawings may be in the form of conventional production drawings, Fig. 2-1, instructional booklets, charts, graphs, or maps. It typically takes more than 27,000 drawings to manufacture an automobile. The field of drafting provides employment for over one million men and women. The work of other millions requires them to be able to read and interpret drawings. See Fig. 2-2.

Fig. 2-1. Skilled drafters are always in demand. This photo was taken at a National VICA (Vocational Industrial Clubs of America, Inc.) Student Graphic Communication Skills Olympics. Competition was keen because the students were highly skilled in drafting procedures and techniques. How do you think you would fare in such a competition? (VICA)

Fig. 2-2. Skilled workers like this quality control specialist must be able to read and understand drawings. Here, special measuring tools are used to determine whether this auto body assembly meets production standards. (American Motors Corp.)

DRAFTING OCCUPATIONS

Job titles and duties will vary somewhat from one company to another. However, the following drafting occupations are typical of those found in industry.

DRAFTERS make working plans and detailed drawings. They prepare them from specifications and information received verbally, from sketches, and from notes.

The drafter usually starts out as a TRAINEE DRAFTER where he or she redraws or repairs damaged drawings. The trainee may revise engineering drawings or make simple detail drawings under the direct supervision of a senior drafter.

During the training period, trainee drafters are often enrolled in formal classes such as computer graphics, as shown in Fig. 2-3, mathematics, electronics, or manufacturing processes. These classes may be held within the company or at a local technical school or community college.

Upon completion of the training period, the trainee drafter usually advances to JUNIOR DRAFTER. Similar positions are known as: DETAILER, DETAIL DRAFTER, and ASSISTANT DRAFTER.

This position calls for the preparation of detail and working drawings of machine parts, electrical/electronic devices, or structures, from rough design drawings. It may also require preparation of simple assembly drawings, charts, or graphs. The junior drafter must be able to prepare simple calculations made according to established drafting room procedures.

From junior drafter, the next step forward is DRAFTER. The drafter applies independent judgement in the preparation of original layouts with intricate details. He or she must have an understanding of machine shop practices, the proper use of materials, and be able to make extensive use of reference books and handbooks.

Fig. 2-3. Today's drafter must be familiar with computer aided design and drafting techniques. (California Computer Products, Inc.)

16

Fig. 2-4. Many drafters were needed to prepare production drawings for this helicopter. Much of the design and drafting was done using computer aided design and drafting techniques. (United Technologies Sikorsky Div.)

With experience, the drafter will become a SENIOR DRAFTER and be expected to do complex original work.

In time, the senior drafter can become a LEAD DRAFTER or CHIEF DRAFTER. Such a person is responsible for all work done by the department.

Most drafters specialize in a particular field of technical drawing: aerospace, architectural, structural, etc. Regardless of the field of specialization, drafters should be able to draw rapidly, with accuracy and neatness. They will also need a working knowledge of computer graphics.

Firms employing computer aided design and drafting usually establish job applications for computer graphics drafters/specialists. A few job titles include CAD DRAFTER, CAD/CAM SPECIALIST, COMPUTER GRAPHICS SPECIALIST, and COMPUTER GRAPHICS TECHNICIAN. These specialists seldom prepare drawings manually on "the board." Instead, their "drawings" are computer generated with hard (paper) copy produced on a plotter.

ALL drafters must have a thorough understanding of mathematics, science, materials, and manufacturing processes in their areas of specialization. The manufacturing industries, Fig. 2-4, employ large numbers of drafters. Others are employed by architectural and engineering firms and local, state, and federal government.

INDUSTRIAL DESIGNERS

The work of the INDUSTRIAL DESIGNER influences most every item used in our daily living, whether it is the design of a small tape player or a giant jet plane. Many designers are on the staffs of major automobile manufacturers, Fig. 2-5.

In general, the chief function of the industrial designer is to simplify and improve the operation and appearance of industrial products. Design simplification usually means fewer parts to wear or malfunction. Appearance plays an important role in the sale of a product. The industrial designer must be aware of changing customs and tastes. He or she must know why people buy and use different products.

Fig. 2-5. A thorough knowledge of advanced technical practices is required to place a present day automobile into production. (General Motors Corp.)

It is recommended that prospective designers have an engineering degree and a working knowledge of engineering and manufacturing techniques and materials. Artistic ability is necessary.

TOOL DESIGNERS

Tools and devices needed to manufacture industrial products are designed by the TOOL DESIGNER. The tool designer originates the designs for cutting tools, special holding devices (fixtures), jigs, dies, and machine attachments that are necessary to manufacture the product. He or she must be familiar with machine shop practices, be an accomplished drafter, and have a working knowledge of algebra, geometry, and trigonometry. It is also important to be familiar with computer aided design, and computer aided manufacturing (CAD/CAM) and robotics, Fig. 2-6.

As industrial technology expands and more automated machinery is utilized, there will be a constantly increasing demand for competent tool designers.

TEACHERS

TEACHING is a satisfying and challenging profession, Fig. 2-7. Teachers of industrial

Fig. 2-6. Tool designers engineer and develop the tools and devices needed to manufacture industrial products. At this manufacturing facility, robots play an important part in spot welding various body panels together.
(Oldsmobile Div., General Motors Corp.)

Fig. 2-7. Teaching is a challenging and rewarding profession that offers considerable freedom. (Harford County Maryland Vo/tech High School)

technology, vocational, and technical education are in a fortunate position because they have a freedom not found in many other professions.

Four years of college training are needed and some industrial experience is highly recommended.

ENGINEERS

The ENGINEER usually specializes in one of the recognized engineering specialties — aeronautical, industrial, marine, structural, civil, electrical/electronic, and metallurgy to name but a few. See Fig. 2-8. Engineers provide technical and, in many instances, managerial leadership in industry and government. Depending upon the area of specialization, engineers may be responsible for the design and development of new products and processes, plan structures and highways, or work out new ways of transforming raw materials into salable products.

Laws provide for licensing engineers whose work may affect life, health, and property. A professional engineering license usually requires graduation from an approved engineering college,

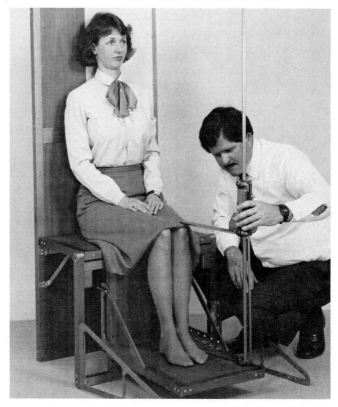

Fig. 2-8. Human factors engineers demonstrate a body measuring device used by Ford Motor Company to record key body measurements of people of various sizes. This information is critical in helping designers and engineers design car interiors that maximize driver and passenger comfort and convenience. (Ford Motor Company)

four years of experience, and passing an examination. Some states will accept experience in place of a college degree.

ARCHITECTS

In general, ARCHITECTS plan and design all kinds of structures. However, they may specialize in specific fields of architecture such as private homes, industrial buildings, schools, or commercial buildings.

When planning a structure, an architect first consults the client on the purpose of the building, size, location, cost range, and other requirements. Upon completion and approval of preliminary design drawings, detailed working drawings, and specification sheets are prepared. As construction progresses, the architect usually makes periodic inspections to determine whether the plans are being followed and construction details are to specifications.

Most architects are licensed. This requires graduation from an approved college program, several years of experience working for an architectural firm (similar to the internship of a physician), and passing a special examination. Several years of experience plus certain college courses are acceptable in a few states in place of graduation from an approved architectural program.

MODELMAKERS

Industry makes extensive use of models, mockups, and prototypes for engineering and planning purposes. Their preparation is often the responsibility of the engineering drafting group although professional MODELMAKERS are frequently used. See Fig. 2-9. Modelmakers must be able to read and understand drawings. They must also have skills in working metal, wood, and plastics.

TECHNICAL ILLUSTRATORS

Technical illustration is a process of preparing art work for industry. The TECHNICAL ILLUSTRATOR prepares pictorial matter for engineer-

Fig. 2-9. In order to interpret the designer's or engineer's plans accurately, modelmakers must be able to read and understand drawings. (General Motors Corp.)

ing, marketing, or educational purposes. See Fig. 2-10. Technical illustrators must have both technical and artistic abilities. The technical ability is necessary to understand the mechanical aspects of the job. The artistic ability will show the building or product in three dimensions. A technical illustrator must be able to make accurate sketches and finished drawings according to industry standards.

LEADERSHIP IN DRAFTING TECHNOLOGY

So far in your study of this unit, you have learned that drafting technology requires highly skilled people. However, the nature of industry is that strong and dynamic LEADERSHIP is required at all levels if it is to be competitive with other technologically oriented countries.

LEADERSHIP is the ability to be a leader. A LEADER is a person who is in charge or command. The quality of leadership usually determines whether an organization will be successful or be a failure, Fig. 2-11.

WHAT MAKES A GOOD LEADER

Most good leaders usually have the following traits (qualities) in common:

20

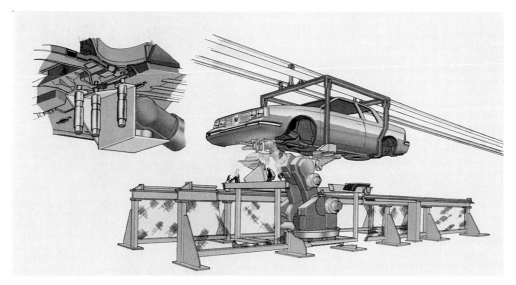

Fig. 2-10. This drawing showing how underbody bolt securing is done using a vision guided robot is the work of a technical illustrator. It was prepared for engineering and training purposes. (Oldsmobile Div., General Motors Corp.)

VISION. Knows what must be done. Continually looks for positive ways to reach these goals.

COMMUNICATION. Can communicate in such a way that will encourage the assistance and cooperation of others.

PERSISTENCE. Willing to work long hours to achieve success.

ORGANIZATIONAL QUALITIES. Knows how to organize and direct the activities of the group.

RESPONSIBILITY. Accepts responsibility for actions but readily gives credit to others when they deserve or warrant it.

DELEGATES AUTHORITY. Not afraid to make assignments that will help others within the group to take on leadership roles.

GETTING LEADERSHIP EXPERIENCE

How can you get leadership experience? The Industrial Technology program or other areas in your school such as clubs offer leadership training to students willing to take advantage of the opportunities. Many programs provide leadership experience through student industrial education organizations such as the AMERICAN INDUSTRIAL ARTS STUDENT ASSOCIATION (AIASA) and the VOCATIONAL INDUSTRIAL CLUBS OF AMERICA (VICA). These develop leadership and personal abilities as they relate to the industrial-technical world, Fig. 2-12.

WHAT TO EXPECT AS A LEADER

Fig. 2-11. Good leadership brought the latest manufacturing processes like this master computer that continually monitors 58 welding robots into use and helps keep the company profitable. (Chrysler Corp.)

As a leader you will be expected to set a good example. The responsibility of leadership also

21

Fig. 2-12. Student organizations like the American Industrial Arts Student Association (AIASA) and Vocational Industrial Clubs of America, Inc. (VICA) offer many opportunities for students to develop leadership skills.

Fig. 2-13. Computer Aided Design (CAD) helped develop this advanced concept aircraft. Advanced technology such as this is creating jobs that did not previously exist. (Beechcraft)

means that you must get *all* members of the group involved in activities. This requires encouragement and tact. A good leader is not expected to have answers for every problem the group encounters. Members should be encouraged to recommend possible solutions. It is up to the leader to decide which to use.

Leadership is not easy or without difficulties. Decisions must be made that some group members may not like. People and even friends may have to be reprimanded or dismissed (fired) for being incompetent (doing poor work) or for trying to take advantage of the group.

If you think you have the traits of a strong leader (or believe you can develop them) and can handle the unpleasant tasks that are part of leadership, by all means look into joining groups where it will be possible for you to become a leader.

WHAT TO EXPECT WHEN YOU ENTER THE WORLD OF WORK

You will be very disappointed if you think that graduation means the end of training and study. Advancing technology, Fig. 2-13, means constantly developing new ideas, materials, processes, and manufacturing techniques that in turn are creating occupations and jobs that did not previously exist. It has been said that young graduates will be employed in the average of five different jobs during their lifetime, and three of them do not now exist!

To hold your job and advance in it, you will have to study to keep up-to-date with the knowledge and new skills needed by modern technology.

Industry expects a fair day's work for a fair day's pay. High manufacturing costs and competition from foreign made products have made this necessary. You will be expected to start work on time, and be on the job each work day. Sick leave was originated to help workers and, therefore, should not be abused.

Do your assigned work and never knowingly turn out a piece of substandard or faulty work, Fig. 2-14. Take pride in what you do. YOU must assume the responsibility for your actions. Industry is always on the lookout for bright young people who are not afraid to work and assume responsibility.

WHERE TO GET INFORMATION ABOUT CAREERS IN DRAFTING TECHNOLOGY

Information on careers can be found in many sources. The guidance office and industrial technology department in your school are the two

Fig. 2-14. Never knowingly turn out a piece of substandard work. Your work should be such that you can take pride in it. How would you feel if the work you did resulted in the disaster shown above?

good starting places. Almost all community colleges have CAREER INFORMATION CENTERS. They can furnish any needed occupational and career information.

Look in the OCCUPATIONAL OUTLOOK HANDBOOK published by the United States Department of Labor and available through the Government Printing Office. The Department of Labor also publishes the DICTIONARY OF OCCUPATIONAL TITLES which has job descriptions of most occupations. Both books can usually be found in the school library or guidance office.

The local STATE EMPLOYMENT SERVICE (it may have a different name in your state) is another excellent source of occupational and career information. They will also be able to tell you about local and statewide employment opportunities.

DRAFTING VOCABULARY

Calculations, Complex, Delegate, Employment, Extensive, Fixtures, Function, Interpret, Jigs, Manufacture, Metallurgy, Originates, Persistance, Properties, Raw materials, Revise, Simplification, Specifications, Specialization, Standards, Structural, Technical, Verbally.

TEST YOUR KNOWLEDGE — UNIT 2

Please do not write in the book. Place your answers on a sheet of notebook paper.

1. Drafters make _____.
2. The drafter usually starts as a _____ _____ and advances to a _____ _____ and then to a _____.
3. A good drafter should be able to draw _____ with accuracy and _____.
4. List the job classification relating to drafting and design occupations using the computer.
5. The industrial designer's main job is to __ _____.
6. _____ is a profession that offers a freedom not usually found in other professions.
7. _____ typically plan and design all kinds of structures and buildings. Most states

require them to be _____.

8. The engineer usually specializes in one of the many branches of the profession. List four types of engineering.

   a. _____

   b. _____

   c. _____

   d. _____

9. The modelmaker makes models, _____, or _____ to show the engineer's or designer's plans.

10. The _____ _____ must have a combination of technical and artistic abilities.

Match each word or phrase in the left column with the correct sentence in the right column.

11. ____ Leadership

12. ____ Leader

13. ____ Quality of leadership

14. ____ Unpopular decisions

   a. One of the difficulties of leadership.

   b. Determines whether an organization will be a success or be a failure.

   c. The ability to be a leader.

   d. A person who is in charge or in command.

OUTSIDE ACTIVITIES

1. Invite a representative from the local State Employment Service Office to visit your class, and discuss employment opportunities in the drafting occupations.

2. Make a study of the HELP WANTED COLUMNS in your daily newspaper for a period of two weeks. Prepare a list of drafting and related jobs available, salaries offered, and the minimum requirements for securing the jobs. How often are additional benefits such as insurance, hospitalization, etc., mentioned in the ads?

   Note the length of time (number of days) each ad is run in the column. Calculate the average number of days jobs are listed.

3. Summarize the information on the drafting occupations given in the OCCUPATIONAL OUTLOOK HANDBOOK (a Government publication) and make it available to the class.

4. *With your teacher's permission,* contact the AMERICAN INDUSTRIAL ARTS STUDENT ASSOCIATION* and VOCATIONAL INDUSTRIAL CLUBS OF AMERICA** for information on how to organize an Industrial Arts/Industrial Education club in your school. (Conduct a student survey to see if there is interest in forming such a club.)

*International Technology Education Association, 1914 Association Drive, Reston, VA 22091

**Vocational Industrial Clubs of America, Inc., P.O. Box 3000, Leesburg, VA 22075

Unit 3

SKETCHING

After studying this unit, you will develop and practice sketching skills and techniques. You will use the "alphabet of lines." You will sketch basic geometric shapes. You will demonstrate how the graph method can be used to enlarge or reduce the drawn size of an object.

SKETCHING is one of many drafting techniques. It is a quick way to show an idea that would be difficult to describe with words alone, Fig. 3-1.

A good sketch shows the shape of the object. It also provides dimensions and may include special instructions on how the object is to be made or finished.

Fig. 3-1. Most ideas begin as sketches. Sketching can be a quick way to visualize something that would be difficult to describe with words. Sketches can be simple or elaborate like this illustration of an "idea" automobile. While the car shown in this sketch will probably not reach production, many of the ideas incorporated in its design will. (General Motors Corp.)

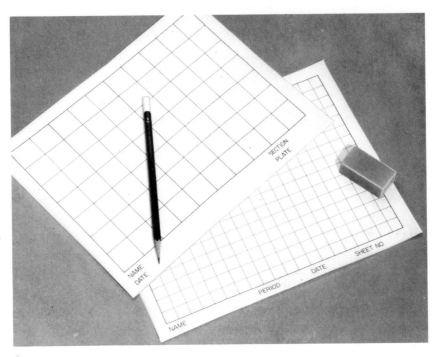

Fig. 3-2. Graph paper is useful in sketching.
A good eraser is also a necessity.

Sketching does not require a great deal of equipment. Properly sharpened F, 2H, or HB grade pencils and sheets of standard size paper (8 1/2 x 11 in.) are well suited for this purpose. See Fig. 3-2. The paper can be plain or cross sectioned (graph or squares). A good eraser is also needed.

In sketching, a line is drawn by making a series of short strokes, Fig. 3-3. It is recommended that light (thin) construction lines be drawn first. Corrections will then be easier to make. Horizontal lines are drawn from left to right. Vertical lines are drawn from the top down.

The instructions that follow tell and show you how to sketch basic geometric shapes. You will find that even the most complex drawings are made using a combination of these basic geometric shapes.

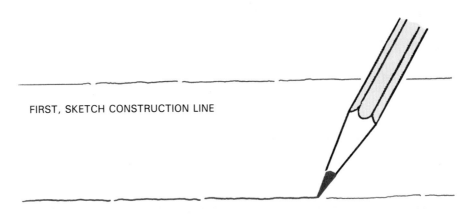

FIRST, SKETCH CONSTRUCTION LINE

SECOND, COMPLETE BY SKETCHING IN DESIRED WEIGHT LINE

Fig. 3-3. When sketching, a line is drawn by making a series of short pencil strokes.

ALPHABET OF LINES

A drafter uses lines of various weights (thicknesses) to make a drawing. Each line has a special meaning, Fig. 3-4. Contrast between the various line weights or thicknesses help to make a drawing easier to read. It is essential that you learn this ALPHABET OF LINES.

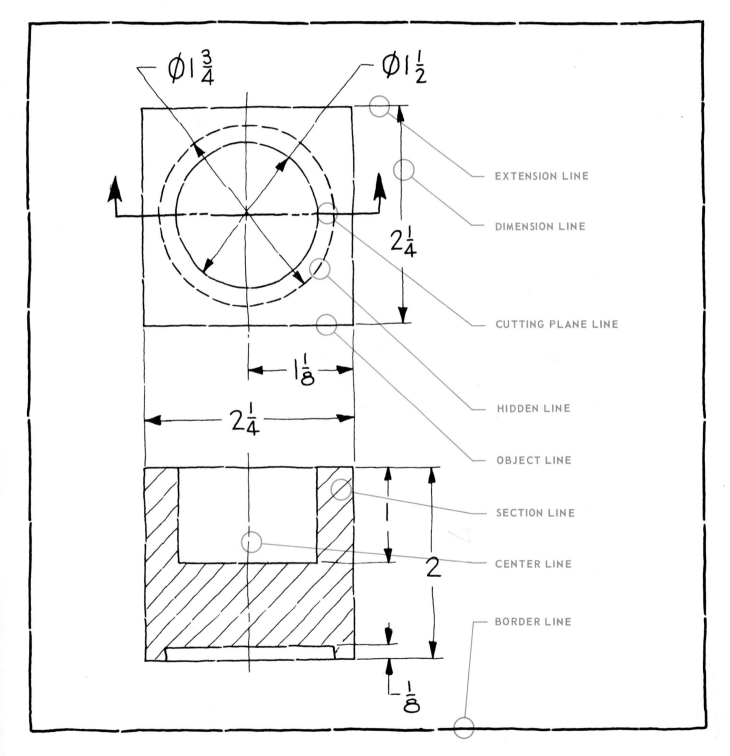

Fig. 3-4. The ALPHABET OF LINES as used in sketching.

CONSTRUCTION AND GUIDE LINE

CONSTRUCTION LINES are used to lay out drawings. GUIDE LINES are used when lettering to help you keep the lettering uniform in height. These lines are drawn lightly using a pencil with the lead sharpened to a long conical point.

BORDER LINE

The BORDER LINE is the heaviest (thickest) line used in sketching. First draw light construction lines as a guide, then go over them using a pencil with a heavy rounded point to provide the border lines.

OBJECT LINE

The OBJECT LINE is a heavy line, but slightly less in thickness than the border line. The object line indicates visible edges. In sketching object lines, use a pencil with a medium lead and a rounded point.

HIDDEN LINE

HIDDEN LINES are used to indicate or show the hidden features of a part. The hidden line is made up of a series of dashes (1/8 in.) with spaces (1/16 in.) between the dashes.

EXTENSION LINE

EXTENSION LINES are the same weight as dimension lines. These lines indicate points from which the dimensions are given. The extension line begins 1/16 in. away from the view and extends 1/8 in. past the last dimension line.

DIMENSION LINE

DIMENSION LINES generally terminate in arrowheads at the ends. They are usually placed between two extension lines. A break is made, usually in the center, to place the dimension. The dimension line is placed from 1/4 in. to 1/2 in. away from the drawing. It is a fine line and is drawn using a pencil sharpened to a long conical point.

CENTER LINE

CENTER LINES are made up of alternate long (3/4 in. to 1 1/2 in.) and short (1/8 in.) dashes with 1/16 in. spaces between. These are drawn about the same weight as dimension and extension lines, and are used to locate centers of symmetrical objects.

CUTTING PLANE LINE

A CUTTING PLANE LINE indicates where an object has been cut to show interior features. Two types are used: 1/4 in. dashes with 1/16 in. spacing; a long dash (3/4 to 1 1/2 in.), then two short dashes (1/8 in.) with 1/16 in. spacing. Draw the cutting plane line slightly heavier than an object line, using a pencil with a rounded point.

SECTION LINE

SECTION LINES are used when drawing inside features of an object to indicate the surfaces exposed by the cutting plane line. Section lines are also used to indicate general classifications of materials. These lines, light in weight, are drawn with a pencil sharpened to a long conical point.

HOW TO SKETCH A HORIZONTAL LINE

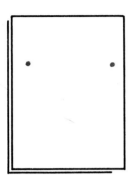

1. Mark off two points spaced a distance equal to the length of the line to be drawn. The points should be parallel to the top or bottom edge of your paper.

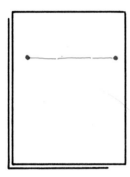

2. Move your pencil back and forth and connect these points with a construction line.

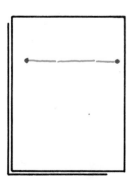

3. Start from the left point and sketch an object line to the right point. This line is sketched over the construction line.

HOW TO SKETCH A VERTICAL LINE

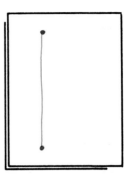

1. Mark off two points spaced a distance equal to the length of the line to be drawn. The two points should be parallel to the right or left edge of the sheet. Move your pencil back and forth and connect these points with a construction line.

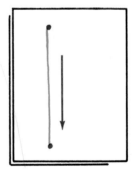

2. Start from the top point and sketch down and over the construction line to draw the desired line.

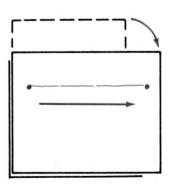

3. Vertical lines can also be sketched by rotating the paper into a horizontal position and proceeding as explained in HOW TO SKETCH A HORIZONTAL LINE.

HOW TO SKETCH AN INCLINED LINE

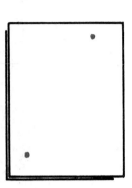

1. Mark off two points at the desired angle. Connect these points with a construction line.

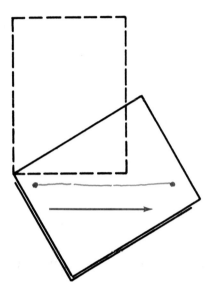

3. Inclined lines can also be sketched by rotating the sheet so the points are in a horizontal position. Sketch the line as previously described.

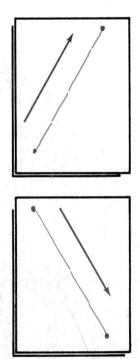

2. Sketch the desired weight line over the construction line. Sketch in the directions illustrated. Sketch up when the line inclines to the right. Sketch down when the line inclines to the left.

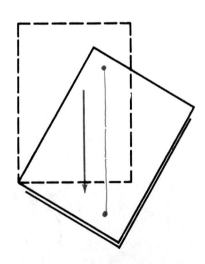

4. For some sketching problems, it may be easier to rotate the paper so the points are in a vertical position. Proceed as explained in HOW TO SKETCH A VERTICAL LINE.

HOW TO SKETCH SQUARES AND RECTANGLES

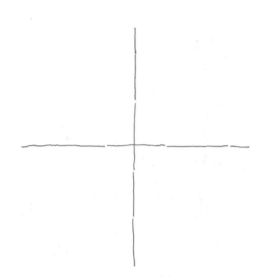

1. Sketch a horizontal line and a vertical line (axes).

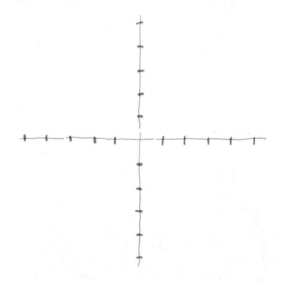

2. Begin at the intersection of these lines and lay out equal units on both lines in each direction. For example: If you want to draw a 2 in. square, you would estimate a unit of 1/4 in. and mark off four of these units on the vertical axis above and below the horizontal axis. Lay out the horizontal axis in the same manner.

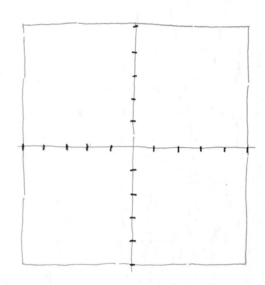

3. Sketch construction lines through the desired points.

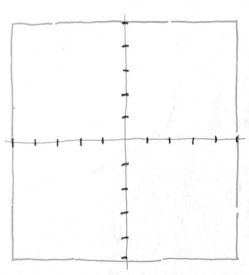

4. Go over the construction lines forming the square to produce the desired weight line.

5. Rectangles are sketched in the same way except that you will have more units on one axis (line) than the other axis (line).

HOW TO SKETCH ANGLES

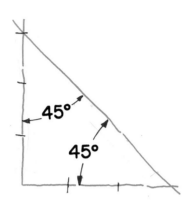

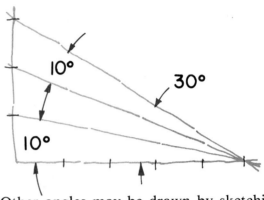

1. Sketch vertical and horizontal construction lines. These lines will form a 90 deg. or RIGHT ANGLE.

2. A 45 deg. angle is sketched by marking off an equal number of units on both lines. Connect the last unit of each line. This will form a 45 deg. angle with the vertical and the horizontal lines.

4. Other angles may be drawn by sketching an angle and subdividing this into the approximate number of degrees required. Example: Dividing a 30 deg. angle into thirds will give a 10 deg. angle.

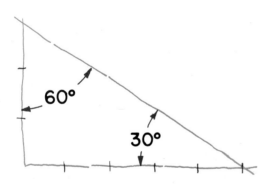

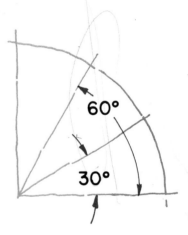

3. To sketch 30 and 60 deg. angles, mark off three units on one line and five units on the other line. Connecting the last unit on each line will give the required angles.

5. Another method used to develop angles in sketching, is to sketch a quarter circle and divide the resulting arc into the desired divisions. Example: Dividing the arc into three parts will give 30 and 60 deg. angles.

HOW TO SKETCH CIRCLES

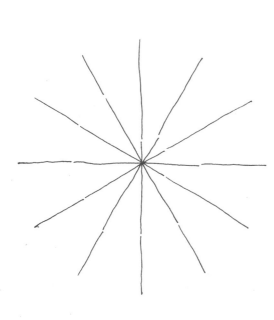

1. Sketch vertical, horizontal, and inclined axes.

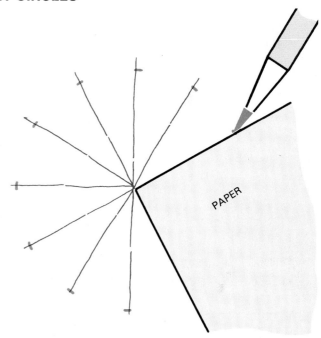

3. The radius units can be quickly and accurately located by marking off the desired radius on a piece of paper and using the paper as a measuring tool.

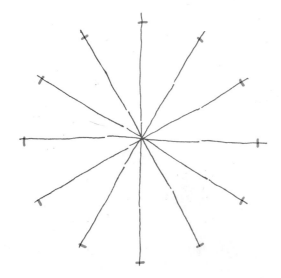

2. Mark off units equal to the radius of the required circle on each axis.

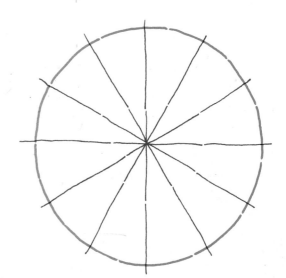

4. Sketch a construction line through the points. When satisfied with the construction line, fill it in with a line of the desired weight.

HOW TO SKETCH AN ARC

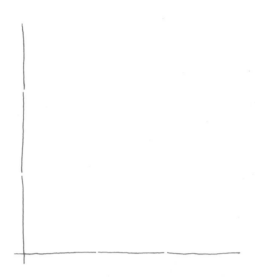

1. Sketch a right (90 deg.) angle. Use construction lines.

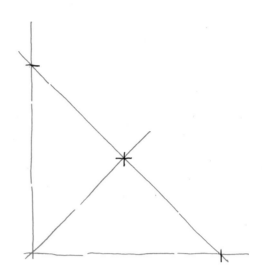

3. Divide this line into two equal parts. Starting from the point where the legs of the angle intersect, sketch a line through the dividing point of the diagonal line.

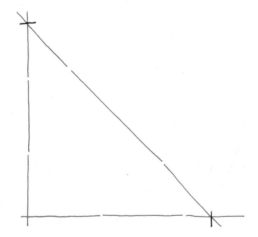

2. Units equal to the length of the desired radius are marked on each leg of the angle. Connect these points with a construction line.

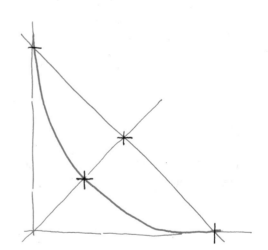

4. Mark off a point halfway between the diagonal line and the intersection of the legs of the angle. Sketch an arc through the three points as shown.

HOW TO SKETCH AN ELLIPSE

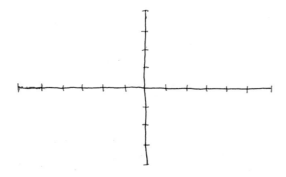

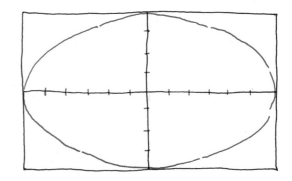

1. Sketch horizontal and vertical lines as shown. Mark off equal size units on the center lines to construct a rectangle with the dimensions equal to the major axis (the long axis) and the minor axis (the small axis) of the desired ellipse.

3. Lightly sketch arcs tangent to the lines that form the rectangle.

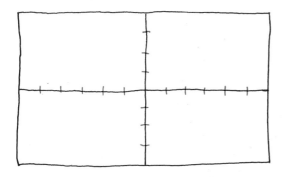

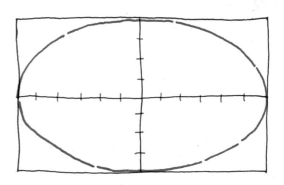

2. Construct the rectangle by sketching construction lines through the outer points.

4. When you are satisfied with the shape of the ellipse, complete it by going over the construction lines with lines of the desired weight.

HOW TO SKETCH A HEXAGON

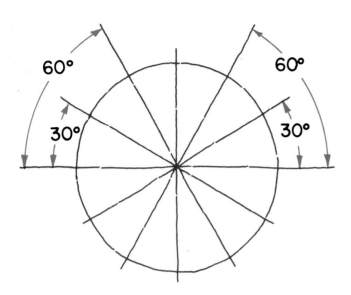

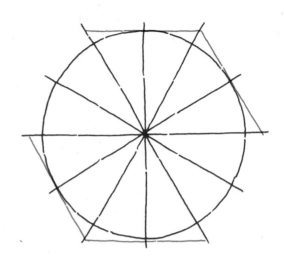

1. Sketch vertical and horizontal center lines, and inclined lines at 30 and 60 deg. Construct a circle with a diameter equal to the distance across the flats of the required hexagon. Use construction lines.

3. Sketch inclined parallel lines at 60 deg. and tangent to the circle at the point where the 30 deg. inclined line intersects the circle.

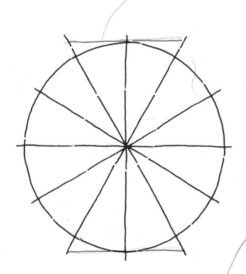

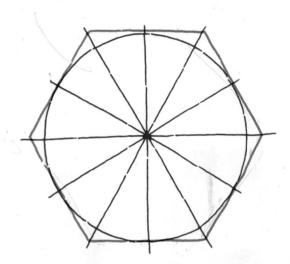

2. Sketch horizontal parallel lines at right angles (90 deg.) to the vertical center line. The lines are tangent to the circle at these points.

4. Complete the hexagon and go over the construction lines to produce the proper weight line.

HOW TO SKETCH AN OCTAGON

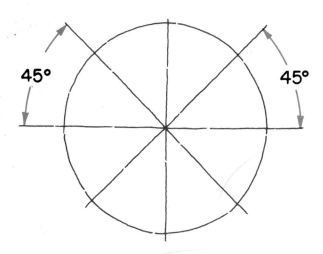

1. Sketch vertical and horizontal center lines and inclined lines at 45 deg. Construct a circle with a diameter equal to the distance across the flats of the required octagon. Use construction lines.

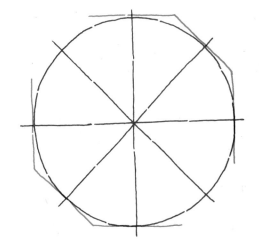

3. Sketch inclined parallel lines at 45 deg. and tangent to the circle at the point where the 45 deg. inclined lines intersect the circle.

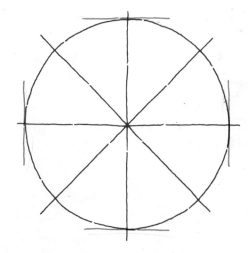

2. Sketch parallel lines tangent to the circle where the horizontal and vertical center lines intersect the circle.

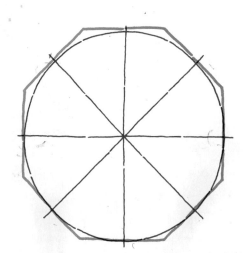

4. Complete the octagon and go over the construction lines to produce the desired weight line.

SHEET LAYOUT FOR SKETCHING

1. Sketch a 1/2 in. border around the edges of the paper. Use a construction line. The sheet should be 8 1/2 in. by 11 in. It may be plain or graph paper. Sketch in guide lines as shown in Fig. 3-5.

2. The edge of your drawing board or desk may be used as a guide in sketching the border and guide lines, Fig. 3-6. Place the pencil in a fixed position and move your fingers along the edge of the drawing board or desk.

3. Sketch a border line over the construction lines, letter in information as shown, Fig. 3-7, or as specified by your instructor.

4. Take your time and sketch in the border and information carefully and neatly.

Fig. 3-6. The edge of your darwing board can be used as a guide when sketching border lines. Note how the pencil is held.

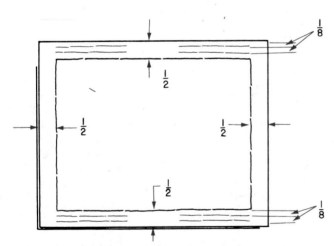

Fig. 3-5. Sketching border and lettering guide lines. Since no rule is used, these dimensions are only approximate.

SCHOOL OR CLASS   DATE

NAME   PLATE NO.

Fig. 3-7. Lettering in the required information.

ENLARGING OR REDUCING BY THE GRAPH METHOD

Drawings can be reduced or enlarged easily and quickly using this technique. The original drawing is blocked off into squares, Fig. 3-8. The squares are numbered as shown. The horizontal numbers are referred to as X-coordinates and the vertical numbers are Y-coordinates. After deciding how much larger or smaller the new drawing is to be made, draw squares of the new size on a blank sheet of paper. Add coordinate numbers to the new squares. Using the design in the original drawing as a guide, sketch the design into the larger or smaller blank squares.

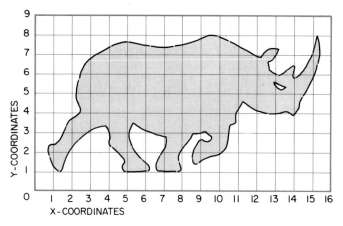

Fig. 3-8. Drawing to be enlarged has been blocked in.

EXAMPLE: A drawing of the design is to be enlarged to twice its original size. The original drawing is marked off into 1/4 in. (10.0 mm) squares. Square size will vary depending on the size of the original design. Number the squares as shown in Fig. 3-8. Another sheet is made up with squares that are twice the size of the 1/4 in. (10.0 mm) squares, or 1/2 in. (20.0 mm) squares. Iden-tify the large squares in the same manner as the smaller squares. Then, sketch in the details free-hand, Fig. 3-9.

Computer aided design and drafting (known as CADD) uses a similar system of coordinates in a grid pattern to enlarge, reduce, or duplicate original drawings. You will learn more about computer graphics in Unit 25.

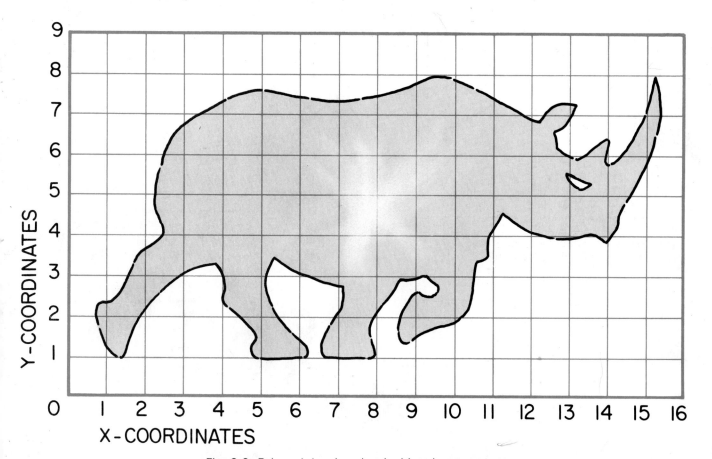

Fig. 3-9. Enlarged drawing sketched into larger squares.

DRAFTING VOCABULARY

Accuracy, Approximate, Arcs, Axis, Basic, Classification, Complex, Conical, Contrast, Coordinates, Diagonal, Diameter, Dimensions, Ellipse, Extends, Geometric, Hexagon, Horizontal, Inclined, Indicate, Intersection, Octagon, Parallel, Radius, Subdividing, Symmetrical, Tangent, Techniques, Terminate, Vertical.

TEST YOUR KNOWLEDGE—UNIT 3

1. In sketching, a line is drawn by making a series of _____ _____.
2. The heaviest line used in sketching is the _____ line.
3. Drawings can be _____ or _____ easily by using the graph method.
4. When sketching an inclined line, sketch up when line inclines to _____ and down when line inclines to _____.
5. Extension lines are the same weight as _____ lines.
6. In sketching, horizontal lines are drawn from _____ _____ _____; vertical lines from the _____ _____.
7. Dimension lines generally terminate in _____ at the ends.

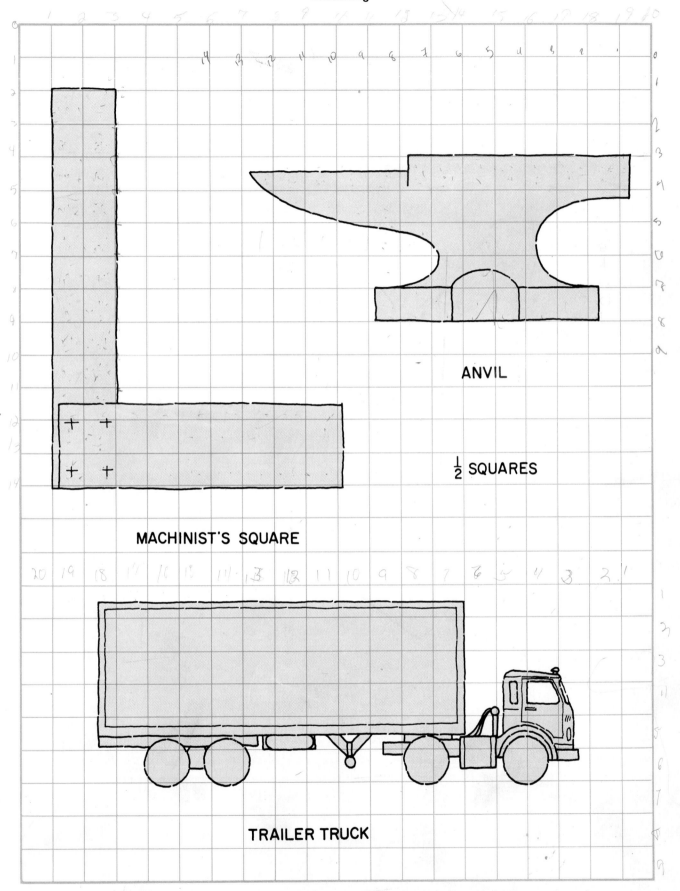

ANVIL

$\frac{1}{2}$ SQUARES

MACHINIST'S SQUARE

TRAILER TRUCK

PROBLEM SHEET 3-1. SKETCHING PROBLEMS. Practice your sketching techniques using these objects. You may enlarge or reduce by using the graph method.

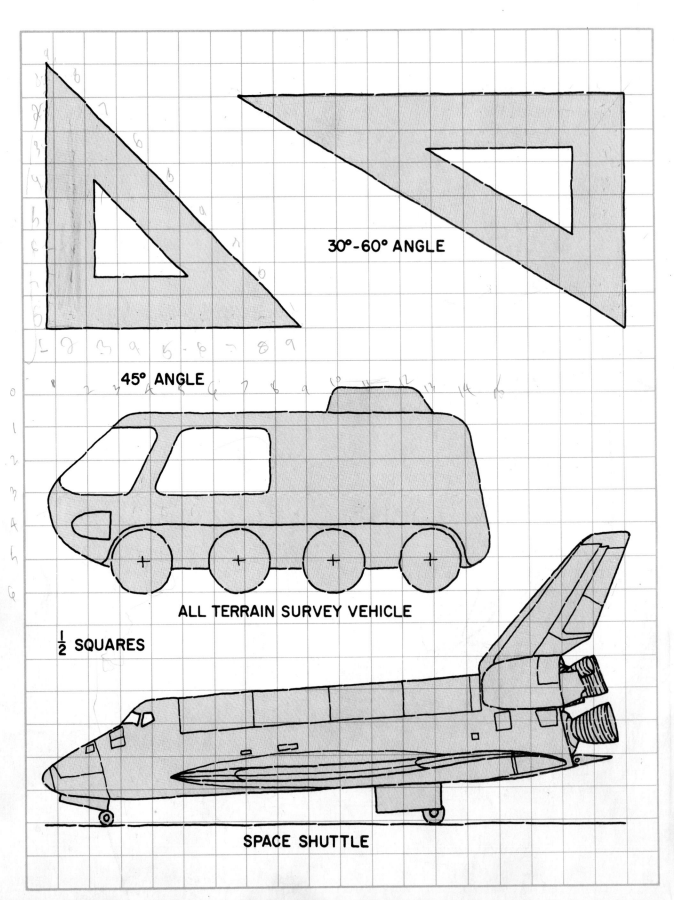

30°-60° ANGLE

45° ANGLE

ALL TERRAIN SURVEY VEHICLE

½ SQUARES

SPACE SHUTTLE

PROBLEM SHEET 3-2. SKETCHING PROBLEMS. Practice your sketching techniques using these objects. You may enlarge or reduce by using the graph method.

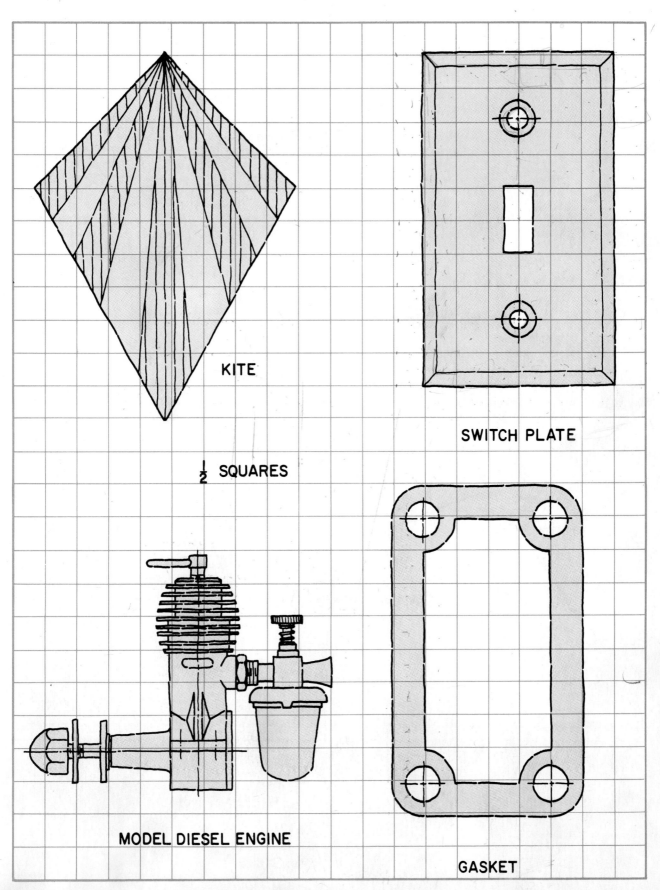

KITE

SWITCH PLATE

$\frac{1}{2}$ SQUARES

MODEL DIESEL ENGINE

GASKET

PROBLEM SHEET 3-3. SKETCHING PROBLEMS. Practice your sketching techniques using these objects. You may enlarge or reduce by using the graph method.

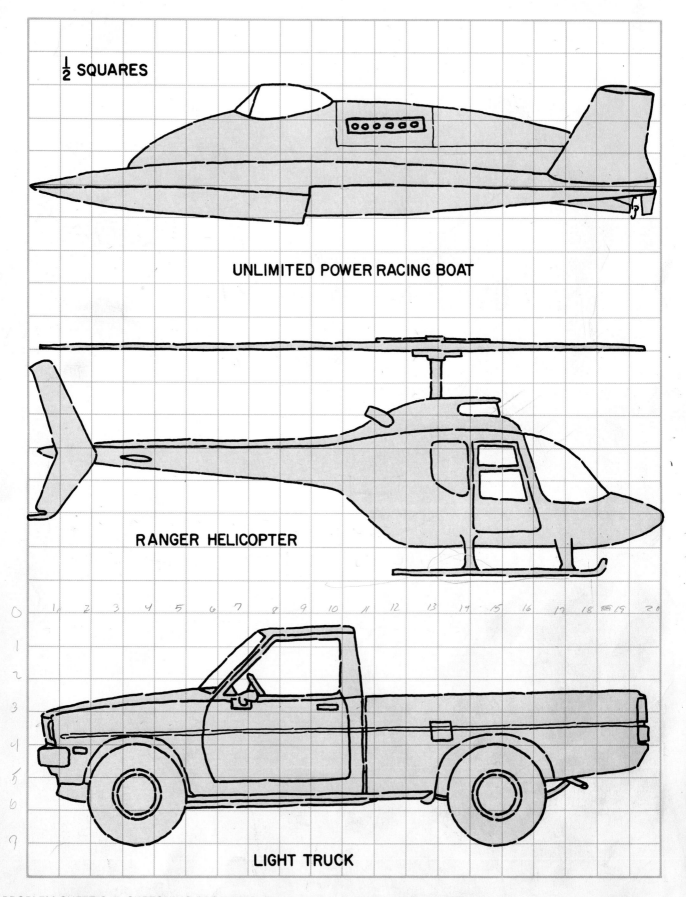

½ SQUARES

UNLIMITED POWER RACING BOAT

RANGER HELICOPTER

LIGHT TRUCK

PROBLEM SHEET 3-4. SKETCHING PROBLEMS. Practice your sketching techniques using these objects. You may enlarge or reduce by using the graph method.

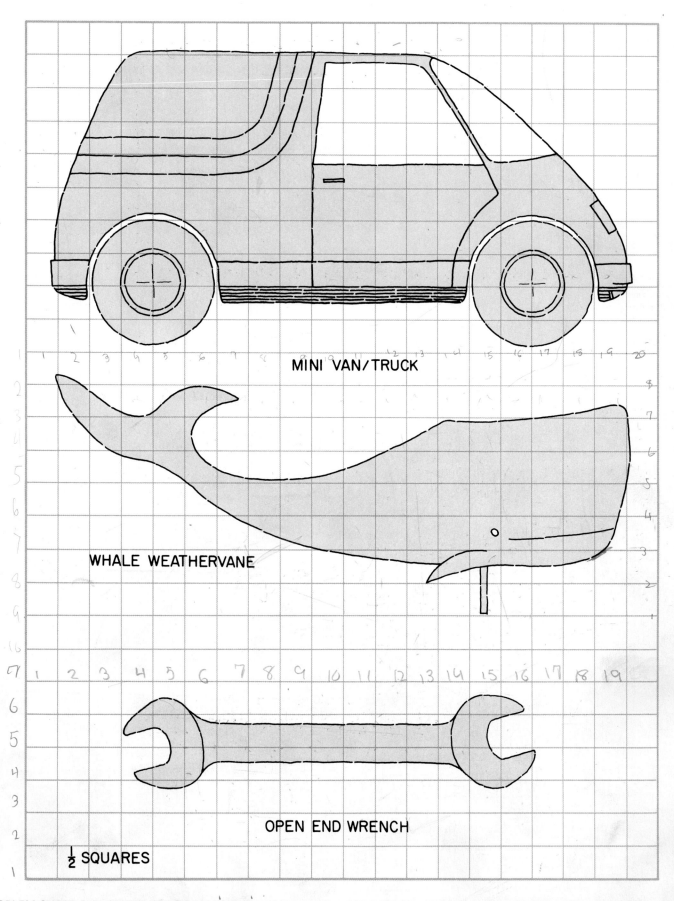

MINI VAN/TRUCK

WHALE WEATHERVANE

OPEN END WRENCH

½ SQUARES

PROBLEM SHEET 3-5. SKETCHING PROBLEMS. Practice your sketching techniques using these objects. You may enlarge or reduce by using the graph method.

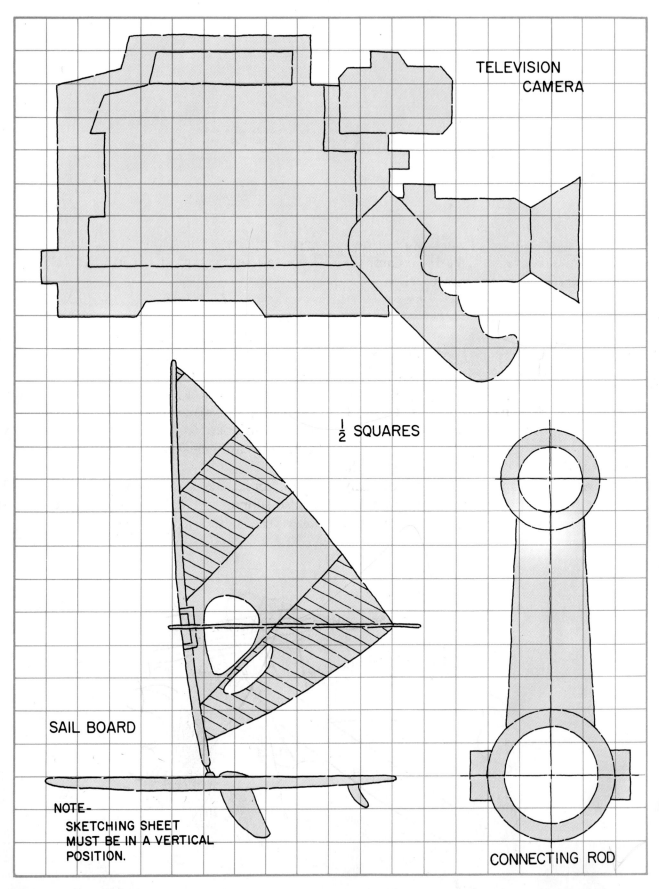

TELEVISION CAMERA

$\frac{1}{2}$ SQUARES

SAIL BOARD

NOTE-
SKETCHING SHEET
MUST BE IN A VERTICAL
POSITION.

CONNECTING ROD

PROBLEM SHEET 3-6. SKETCHING PROBLEMS. Practice your sketching techniques using these objects. You may enlarge or reduce by using the graph method.

Unit 4

DRAFTING EQUIPMENT

After studying this unit, you will be able to identify basic drafting equipment. You will name the tools correctly and identify them by sight.

Drafting tools must be in good condition to make first-rate drawings. Typical equipment found in many schools is shown in Fig. 4-1. It is always a good idea to check your drafting tools' condition before you use them.

The DRAWING BOARD provides the smooth, flat surface needed for drafting. The tops of many drafting tables are designed for this purpose. Individual drawing boards are manufactured in a variety of sizes. The majority of them are made from selected, seasoned basswood.

A right-handed drafter will use the left edge of the board as the working edge; a left-handed drafter uses the right edge. The working edge should be checked periodically for straightness.

Drafters often tape a piece of heavy paper or special vinyl board cover to the working face of the drawing board to protect its surface. The vinyl surface is easily cleaned.

Horizontal lines are drawn with the T-SQUARE. It also supports triangles when they are used to draw vertical and inclined lines.

Fig. 4-1. Drafting equipment typical of that found in many school drafting labs.

47

A T-square consists of two parts, the head and the blade or straight edge. The head is usually fixed solidly to the blade. However, a T-square with a protractor head and adjustable blade is available.

Clear plastic strips inserted in the blade edge of many T-squares make it easier to locate reference points and lines. The blade must never be used as a guide for a knife or other cutting tool.

If accurate line work is to be done, it is essential that the T-square head be held firmly against the working edge of the board.

It is recommended that the blade be left flat on the board or stored suspended from the hole in its end. This will keep warping or bowing of the blade to a minimum.

TRIANGLES

When supported on a T-square blade, the 30-60 degree and 45 degree TRIANGLES are used to draw vertical and inclined lines, Fig. 4-2. They are made of transparent plastic and are available in a number of different sizes.

To prevent warping, a triangle should be left flat on the drawing board when not being used.

Angles of 15 and 75 degrees can be drawn by combining the triangles as shown in Fig. 4-3.

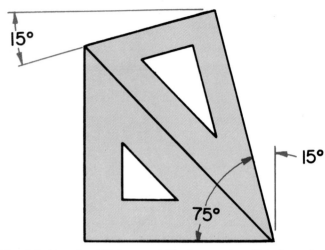

Fig. 4-3. Drawing 75 deg. and 15 deg. angles. Note how the two angles are combined to produce the required angles.

When drawing vertical lines, rest the triangle solidly on the T-square blade while holding the T-square head firmly against the working edge of the board.

DRAFTING MACHINES

Industry makes considerable use of DRAFTING MACHINES, Fig. 4-4. This device replaces both the T-square and triangles. The straight-edges can be adjusted to any angle using the built-in protractor blade. Drafting machines are often used in place of T-squares and triangles in school drafting rooms.

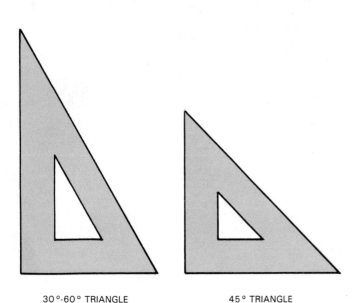

30°-60° TRIANGLE   45° TRIANGLE

Fig. 4-2. Triangles.

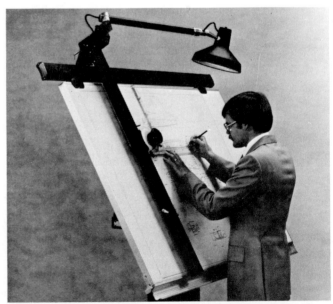

Fig. 4-4. Typical drafting machine. Note the adjustable protractor blade. (Vemco)

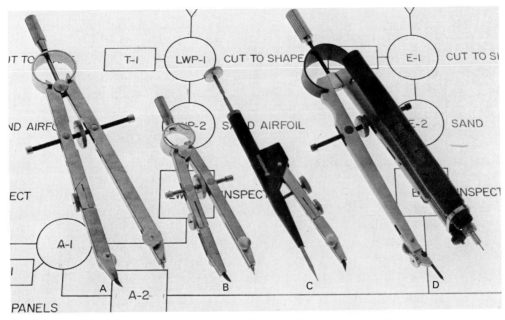

Fig. 4-5. Compasses. A—Big bow compass. B—Small bow compass. C—Drop bow compass used to draw very small arcs and circles. D—Big bow compass with inking attachment and pen.

COMPASS

In drafting, circles and arcs are drawn with a COMPASS, Fig. 4-5. The tool is held as shown in Fig. 4-6. For best results the lead should be adjusted so that it is about 1/32 in. (0.5 mm) shorter than the needle. Both legs will be the same length when the needle penetrates the paper. Fit the compass with lead that is one grade softer than the pencil used to make the drawing. The lead must be kept sharp.

Several compass attachments are available. A pen is substituted for inking, Fig. 4-7. The technical pen is shown in Fig. 4-5D. An extension is used to draw large circles. The BEAM COMPASS will do the job better, Fig. 4-8.

To set a compass to the size of the circle radius, first draw a straight line on a piece of scrap paper and then measure off the required distance equal to the radius. Set the compass on this line. Avoid setting a compass on a scale. "Sticking" the compass needle into the scale will eventually destroy the scale's accuracy.

DIVIDERS

Distances are subdivided and measurements are transferred with DIVIDERS, Fig. 4-9. Careful

Fig. 4-6. The correct way to hold a compass when drawing a circle.

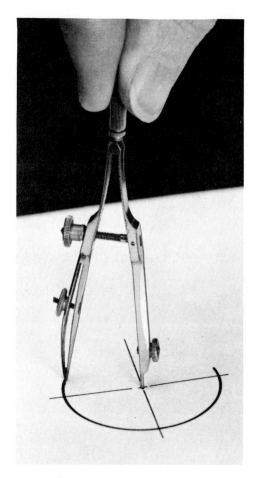

Fig. 4-7. Inking a circle using an ink compass.

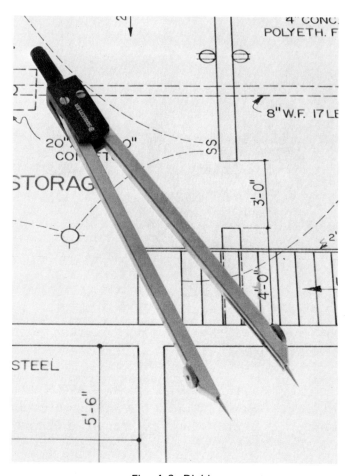

Fig. 4-9. Divider.

adjustment of the divider points is necessary. SAFETY NOTE: Be careful where and how you place dividers or compass after use. It is very painful to accidentally run the point into your hand when reaching for the tool.

PENCIL POINTER

It is not necessary to resharpen a drawing pencil every time it starts to dull. It can be repointed quickly with a PENCIL POINTER, Fig. 4-10. Use

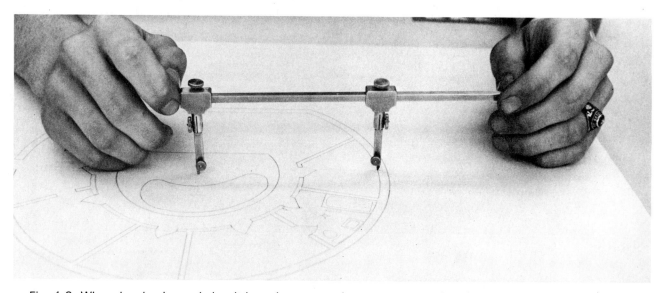

Fig. 4-8. When drawing large circles, it is easier to use a beam compass rather than a compass with extension.

Fig. 4-10. Mechanical pencil pointers.

Fig. 4-12. Examples of typical erasers used in the drafting room.

the wooden pencil sharpener only when the point becomes very blunt, or when it breaks.

Keep the pencil pointer clear of the drawing area when repointing a pencil. The graphite dust will smudge your paper when you attempt to remove eraser crumbs.

Many types of pencil pointers are available. The sandpaper pad, Fig. 4-11, may be found in the school drafting room. A piece of styrofoam cemented to the back of the pad can be used to remove graphite dust from the newly pointed pencil.

ERASERS

Many shapes and kinds of ERASERS are manufactured for use in the drafting room, Fig. 4-12. The type of material being drawn upon — paper, film, or vellum, will determine the type of eraser to be used.

NOTE! Always brush away eraser crumbs before starting to draw again.

ERASING SHIELD

Small errors or drawing changes can be erased without soiling a large section of the drawing if an ERASING SHIELD is employed, Fig. 4-13. Place a shield opening of the proper shape and size over the area to be changed and then erase. The

Fig. 4-11. Using a sandpaper pad to point a drafting pencil. Do not point your pencil over the drawing board.

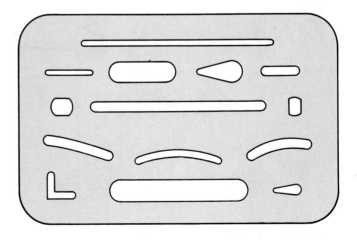

Fig. 4-13. Erasing shield.

erasure is made without touching other parts of the drawing.

One widely used shield is made from stainless steel. It is very thin, wear resistant, and does not stain or "smudge" drawings.

FASTENERS

Two preferred methods of attaching drafting media to a drawing board are shown in Fig. 4-14. Drafting tape is recommended as the most desirable. It does not puncture the paper or affect the surface of the drawing board.

Adhesive tape is sometimes used to attach large drawing sheets. Continued use may affect the working surface of the drawing board.

Thumb tacks and staples are to be avoided. These quickly destroy the smooth surface of a drawing board.

Fig. 4-15. Removing erasure crumbs from the drawing surface with a dusting brush.

DUSTING BRUSH

No matter how careful you are, some erasing crumbs and dirt particles will collect on the drawing area. They should be removed with a DUSTING BRUSH, Fig. 4-15. DO NOT REMOVE THEM WITH YOUR HAND. Using your hand will cause smudges and streaks.

Dusting brushes are available in a number of sizes and with natural or synthetic bristles.

PROTRACTOR

The PROTRACTOR shown in Fig. 4-16 is employed to measure and layout angles on drawings. Protractors are usually made from clear plastic and may be either circular or semicircular in shape. The degree graduations are scribed or engraved around the circumference of a protractor.

When measuring or laying out an angle, place the center lines of the protractor at the point (apex)

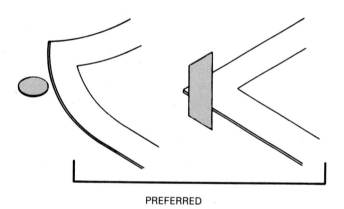

PREFERRED

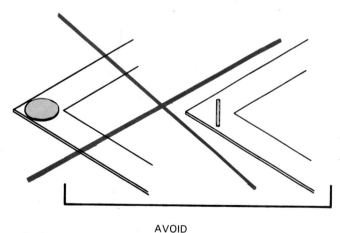

AVOID

Fig. 4-14. Methods of attaching paper to the drawing board. Staples and tacks are not recommended as they will eventually destroy the boards drawing surface.

Fig. 4-16. Protractor used in the drafting room.

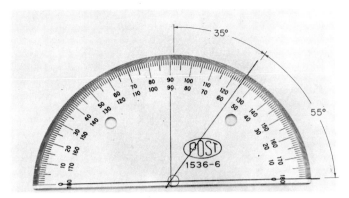

Fig. 4-17. Making a measurement with a protractor.

of the required angle as shown in Fig. 4-17. Read or mark the angle from the graduations on the circumference of the tool.

FRENCH CURVES

Curved lines that are not exactly circular in form are drawn with a FRENCH CURVE, Fig. 4-18. After the curved line is carefully plotted, it is drawn with a French curve as shown in Fig. 4-19.

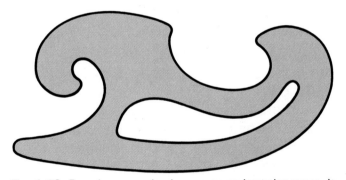

Fig. 4-18. French curve, also known as an irregular curve. It is available in many sizes and configurations.

The curves are made of transparent acrylic plastic. They range in size from a few inches to several feet in length and may be purchased individually or as a set. They are also known as IRREGULAR CURVES.

TEMPLATES

TEMPLATES are available in an almost unlimited range of shapes, Fig. 4-20. Made of thin, transparent plastic, they contain openings of different sizes and shapes. Most templates allow for the thickness of the pencil lead or pen tip.

Templates enable a drafter to do normally time consuming jobs with ease and accuracy, Fig. 4-21.

DRAFTING MEDIA

Drawings are made on many different materials —paper, tracing vellum, mylar film, etc. A heavyweight opaque paper that is white, buff, or pale green in color is used in many school drafting rooms.

While most papers take pencil lines well, they are often difficult to erase because the pencil point makes a depression in the paper when a line is drawn. Plan your work carefully to prevent mistakes.

Industry makes much use of tracing vellum and film because reproductions or prints must be made of all drawings.

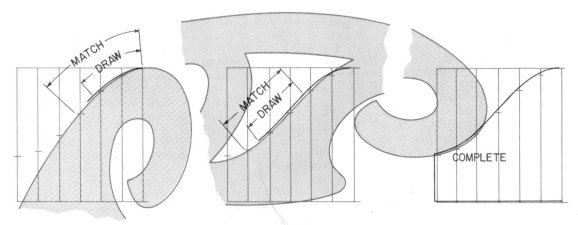

Fig. 4-19. Using a French Curve to draw an irregular curve.

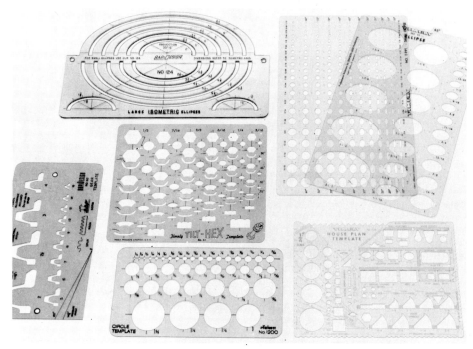

Fig. 4-20. A few of the many different types of templates used in a drafting room.

The bulk of the drawings used by industry are put on standard size drawing sheets. This makes them easier to file and identify. A listing of standard sheet sizes is shown in Figs. 4-22 and 4-23.

Designers, drafters, surveyors, engineers, and architects make considerable use of commercially prepared graph paper to make preliminary design studies, Fig. 4-24. Isometric grid paper makes it very easy to convert an orthographic drawing into an isometric drawing, Fig. 4-25.

Many types and sizes of grid patterns are available commercially in the form of prepunched loose leaf sheets (8 1/2 in. x 11 in.), other sheet sizes, and in roll form. See Fig. 4-26. The grid lines are printed in many colors — green, orange, black, and in a pale blue that will not reproduce when a drawing made on it is run through a print maker.

PENCILS

As most drawings are prepared with a pencil, it is important that the proper pencil be selected. The drawing media employed will determine the type pencil for best results. A conventional lead pencil is satisfactory with most papers and tracing vellums, while a pencil with plastic lead is necessary if the drawing is made on film.

The drafter can select from 17 grades of pencils that range in hardness from 9H (very hard) to 6B (very soft), Fig. 4-27. Many drafters use a 4H or 5H pencil for lay out work and a H or 2H pencil to darken the lines and to letter. In general, use a pencil that will produce a sharp, dense black line because this type line reproduces best on prints.

Avoid using a pencil that is too soft. It will wear rapidly, smear easily, and soil your drawing. Also,

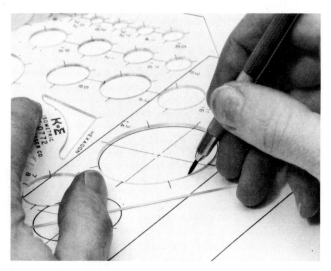

Fig. 4-21. Templates enable the drafter to do normally time consuming jobs with ease and accuracy.

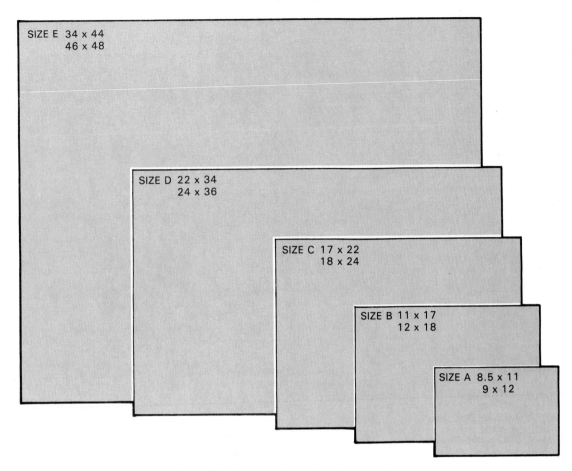

Fig. 4-22. Standard inch size drawing sheets.

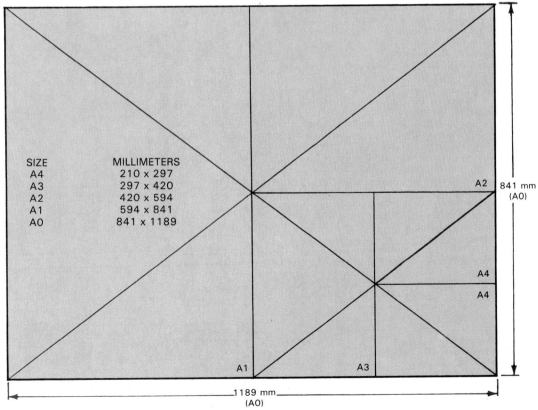

SIZE	MILLIMETERS
A4	210 x 297
A3	297 x 420
A2	420 x 594
A1	594 x 841
A0	841 x 1189

Fig. 4-23. Standard metric size drawing sheets.

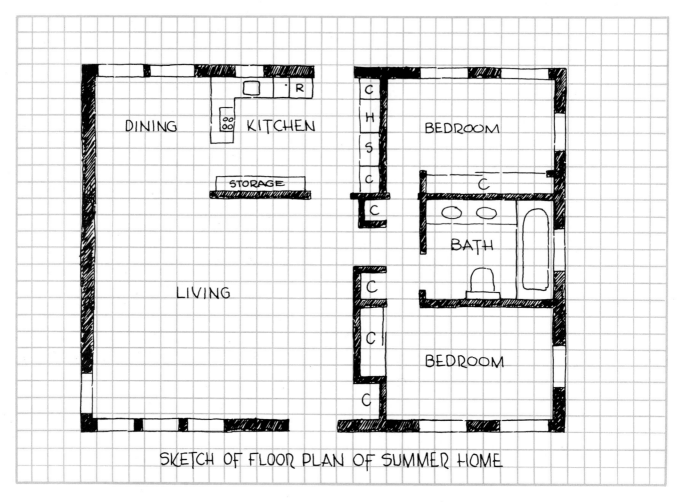

DINING KITCHEN R

STORAGE

C H S C BEDROOM

C

C C

LIVING

BATH

C

C

BEDROOM

C

SKETCH OF FLOOR PLAN OF SUMMER HOME

Fig. 4-24. Commercially prepared graph paper speeds up sketching. A floor plan sketched on graph paper. Each square equals one foot.

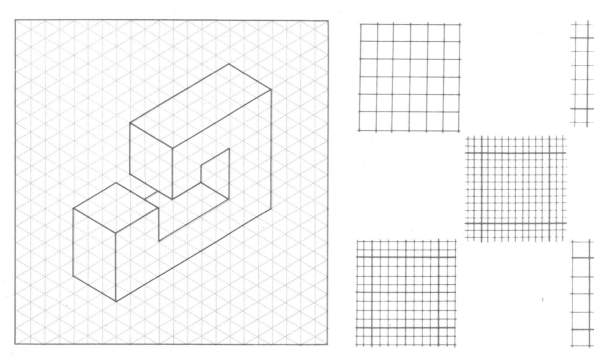

Fig. 4-25. An isometric drawing made on graph paper designed for that purpose.

Fig. 4-26. Graph paper is manufactured in many different grid sizes.

the line will be "fuzzy" and will not produce usable prints.

A conical shaped pencil point is preferred for most general purpose drafting. To sharpen the pencil, cut the wood away from the *unlettered* end. Sharpen with a knife or mechanical sharpener and point the lead with a pencil pointer, Fig. 4-28.

A semiautomatic pencil, Fig. 4-29, is usually preferred to a wood pencil. With this type pencil it is not necessary to cut away wood to expose the lead. The extended lead is shaped on a pencil pointer.

SCALES

A scale can be used to make measurements that are full size or larger or smaller than full size. The size of the object and drawing sheet size will determine the scale of a drawing.

SCALES are in constant use on the drawing board because almost every line on a mechanical drawing (a drawing made with instruments) must be a measured length, Fig. 4-30. Accurate drawings require accurate measurements.

Fig. 4-27. Pencils used in drafting are available in a wide range of hardness.

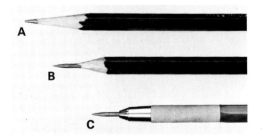

Fig. 4-28. Pencil points. A—Sharpened with a regular pencil sharpener. B—Sharpened with a knife and pointed on a sandpaper pad. C—Sharpened with a mechanical pencil pointer.

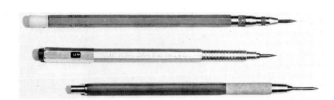

Fig. 4-29. A variety of semiautomatic drafting pencils. Only the lead needs to be replaced.

Because of the diversity of work that must be drawn, scales used by drafters are made in many shapes, lengths, and measurement graduations. They may be made of wood, plastic, or a combination of both materials. Graduations are printed on inexpensive scales, and are machine engraved (much more accurate) on the more costly ones.

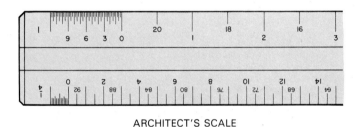

ARCHITECT'S SCALE

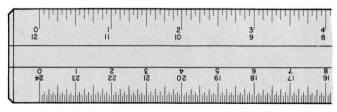

ENGINEER'S SCALE

MECHANICAL DRAFTER'S SCALE

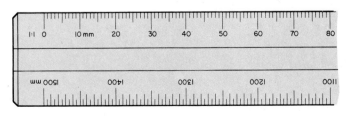

METRIC SCALE

Fig. 4-30. A few of the various types of the scales used in a drafting room.

ARCHITECTS' SCALE. All of the scales represent one foot. The scales are 1/8, 3/32, 3/16, 1/4, 3/8, 1/2, 3/4, 1, 1 1/2, and 3. The scale of a particular section is marked on the face at the end of each edge of the measuring device. Each division represents one foot (12 inches). The first division is used like a foot rule. For example, the first division of the 3/4 scale is divided into 24 parts. Each graduation equals 1/2 in.

One edge of this triangular scale is divided into 1/16 in. graduations.

ENGINEERS' SCALE. Used mostly where large reductions are required. It is divided into 10, 20, 30, 40, 50, and 60 units per inch.

MECHANICAL DRAFTER'S SCALE. Most commonly divided into the following graduations —full size, and 3/4, 1/2, 1/4, and 1/8 in. to one foot.

METRIC SCALE. This scale is now a required tool in the drafting room. Common divisions are in centimeters (cm) divided into millimeters (mm) and centimeters divided into half-millimeters (0.5 mm). Metric scales are also available in a number of enlargement ratios (2:1, 3 1/3:1, etc.) and reductions (1:2, 1:3, etc.).

COMPUTER DRAFTING EQUIPMENT

Much design and drafting work is done with computers, Fig. 4-31. Prints will be made on high speed plotters, Fig. 4-32, or drafting machines from information generated on a computer. A number of large firms and small companies utilize computer aided design and drafting (known as CAD or CADD).

The basic CADD system uses a KEYBOARD for inputting data and information, a CRT (cathode ray tube) to display the drawing, a DISK DRIVE or some other device to store and retrieve data, and the PROCESSOR which is the heart of the computer. To make the CADD system function, SOFTWARE is used to program the unit. The software allows the operator to design, draw, alter, erase, move, etc. any item on the screen or in memory. The selection of software is a major consideration in the operation of CADD.

Fig. 4-31. Many drafting rooms use a computer to develop the drawings needed to manufacture the many products we use. When needed, prints are made on high speed plotters of drafting machines from information generated on the computer.

Fig. 4-32. High speed plotter.
(Hewlett-Packard Marketing Communications)

The principles of drafting are common to both traditional drafting and CAD. Drafting technicians must have a working knowledge of basic drafting practices, standards, and procedures (the same ones you are now learning) before being able to use CAD to its fullest potential.

CAD and other computer graphics related to industry will be explained in more detail in Unit 23 — Computer Graphics.

DRAFTING VOCABULARY

Accurate, Adjustable, Bowing, Circular, Circumference, Constant, Conventional, Depression, Diversity, Extension, Graduations, Horizontal, Inclined, Opaque, Penetrate, Plotted, Protractor, Reference point, Reproduction, Ratio, Semiautomatic, Semicircular, Subdivided, Transparent, Vertical, Vinyl, Warping, Working edge.

TEST YOUR KNOWLEDGE — UNIT 4

Please do not write in the book. Place your answer on a sheet of notebook paper.
1. A drawing board provides the _____ _____ necessary for drafting.
2. A piece of heavy paper or a special vinyl board cover is sometimes attached to the working surface of the board to: (Check the correct answer.)
   a. ____ Provide a drawing surface.
   b. ____ Protect its surface.
   c. ____ Permit the pencil to draw better.
3. The T-square is used to draw _____.
4. Vertical and angular lines are drawn with _____.
5. A compass is used in drafting to draw _____ and _____.
6. A pencil is repointed with a _____.
7. The erasing shield is used to _____ _____
8. List the two preferred methods for attaching a drawing sheet to the board. Which methods cause damage to the board and are to be avoided?
9. Angles can be measured and laid out on drawings with a _____.
10. A ____H or ____H pencil is recommended for layout work.
11. An ____H or ____H pencil is recommended for darkening the lines and for lettering.
12. Why are scales important to drafters?
13. List the four types of scales described in the text.
14. What drafting concepts are common to both traditional board drafting and computer aided drafting?

OUTSIDE ACTIVITIES — UNIT 4

1. Examine the drafting equipment assigned to you and make a record of its condition. Notify your teacher if any of your tools need repair.
2. Prepare a bulletin board display on computer aided drafting (CAD).
3. Using catalogs of available drafting equipment, make a comparison of the cost of a fully equipped drafting room using drafting boards, drafting machines, triangles, etc. relative to the cost of computer aid drafting and design equipment.

Unit 5

DRAFTING TECHNIQUES

After studying this unit and drawing the problem sheets, you will be able to demonstrate important drafting skills. You will apply many of the drafting techniques a drafter uses every day. The measurement problems will help you become more accurate and proficient in making measurements. You will gain knowledge and comprehension of drafting techniques which may later be reproduced using CADD equipment.

You were introduced to the ALPHABET OF LINES in the Unit on Sketching. The lines you sketched can be drawn more uniformly and accurately with drafting instruments. See Fig. 5-1.

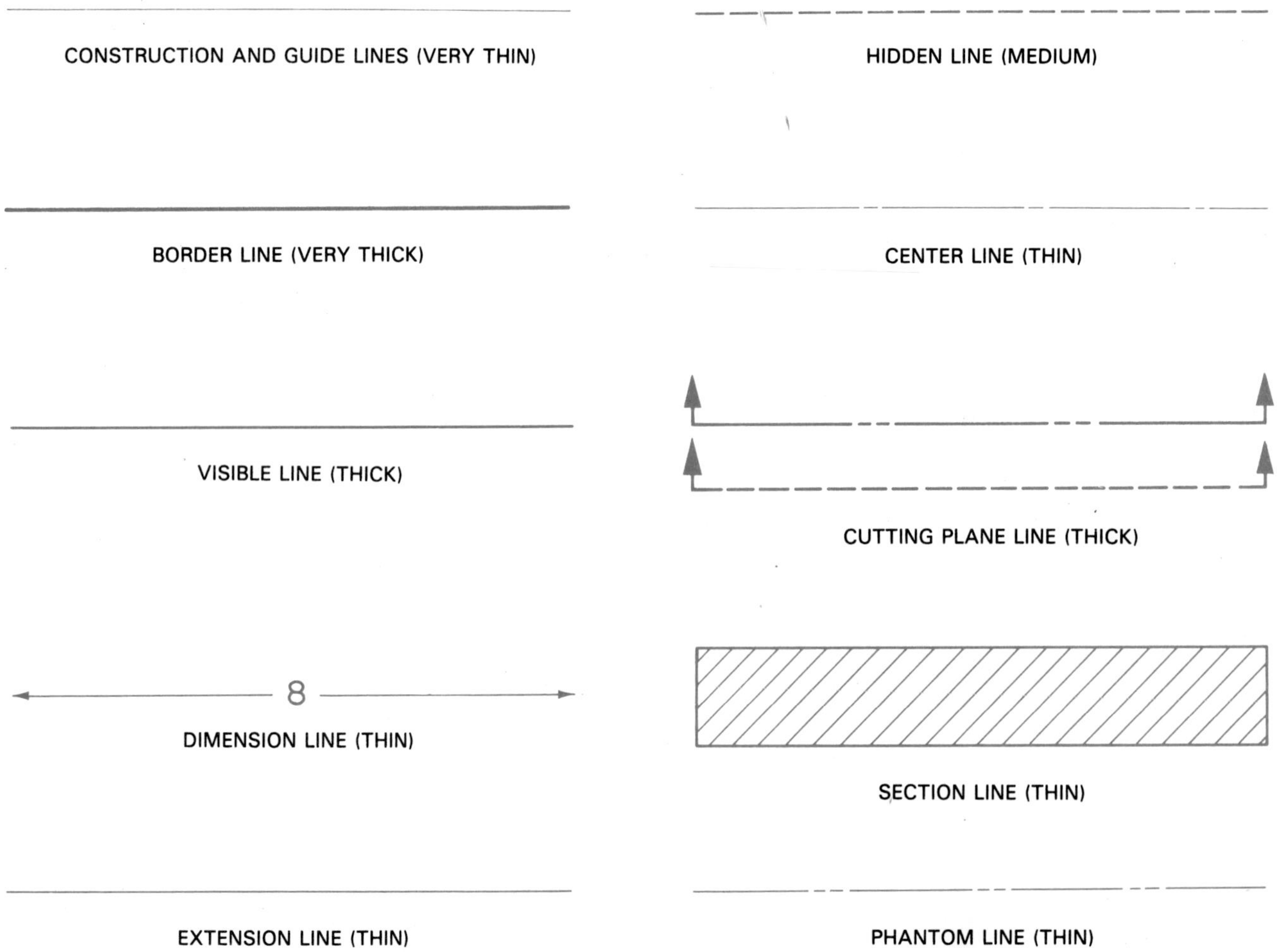

CONSTRUCTION AND GUIDE LINES (VERY THIN)

HIDDEN LINE (MEDIUM)

BORDER LINE (VERY THICK)

CENTER LINE (THIN)

VISIBLE LINE (THICK)

CUTTING PLANE LINE (THICK)

8

DIMENSION LINE (THIN)

SECTION LINE (THIN)

EXTENSION LINE (THIN)

PHANTOM LINE (THIN)

Fig. 5-1. The alphabet of lines.

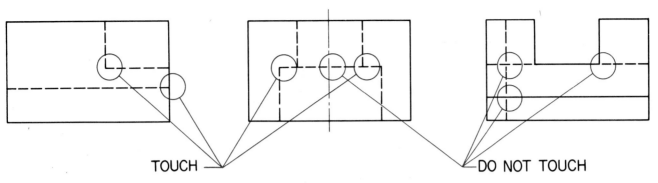

Fig. 5-2. Correct use of hidden lines.

To understand the LANGUAGE OF INDUSTRY, it is necessary that you know the characteristics of the various lines and the correct way to use them in a drawing. Fig. 5-2 shows an example of the correct use of hidden lines. In drafting room language, the characteristics and uses of lines are known as LINE CONVENTIONS.

Each type or kind of line has a specific meaning. It is most important that each line is drawn properly, is opaque, and is uniform in width throughout its entire length. See Fig. 5-3.

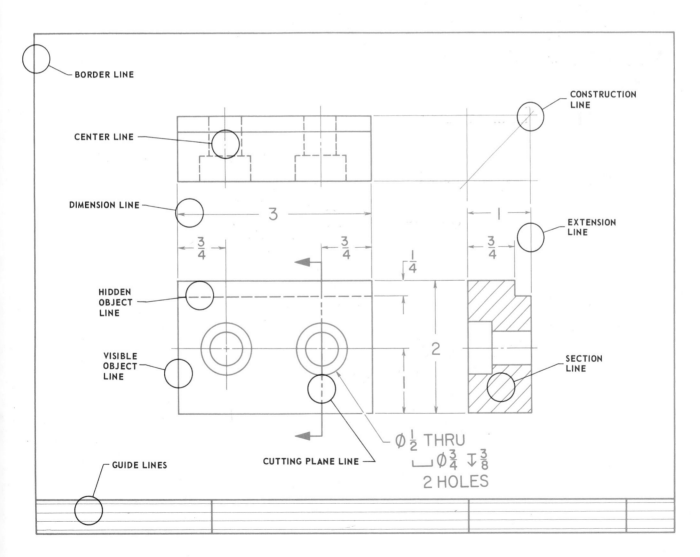

Fig. 5-3. How the various lines are used.

ALPHABET OF LINES

CONSTRUCTION AND GUIDE LINES (VERY THIN)

CONSTRUCTION LINES are drawn very lightly. They are used to block in drawings and as guide lines for lettering. They may be erased, if necessary, after they have served their purpose.

BORDER LINE (VERY THICK)

The BORDER LINE is the heaviest weight line used in drafting. It varies from 1/32 in. to 1/16 in. depending upon the size of the drawing sheet.

VISIBLE OBJECT LINE (THICK)

The VISIBLE OBJECT LINE (also VISIBLE LINE) is used to outline the visible edges of the object being drawn. They should be drawn so that the views stand out clearly on the drawing. All of the visible object lines on the drawing should be the same weight.

8

DIMENSION LINE (THIN)

The DIMENSION LINE is usually capped at each end with arrowheads and is placed between two extension lines. With few exceptions it is broken with the dimension placed at midpoint between the arrowheads. The dimension line is a light line a bit heavier than the construction line. It is placed 1/4 to 1/2 in. away from the drawing.

EXTENSION LINE (THIN)

The EXTENSION LINE is the same weight as the dimension line. It extends the dimension beyond the outline of the view so that the dimension can be read easily. The line starts about 1/16 in. beyond the object and extends about 1/8 in. past the last dimension line.

HIDDEN LINE (THICK)

The HIDDEN OBJECT LINE (also HIDDEN LINE) is used to show the hidden features of the object. It is drawn the same weight as the visible object line and is composed of short lines approximately 1/8 in. long separated by spaces approximately 1/16 in. They may vary slightly according to the size of the drawing. Hidden object lines should always start and end with a dash in contact with the visible object line. See Fig. 5-2 for the correct uses of the hidden object line.

CENTER LINE (THIN)

The CENTER LINE is used to indicate the center of symmetrical objects. It is a fine dark line composed of alternate long (3/4 in.) and short (1/8 in.) dashes with 1/16 in. spaces between the dashes. The center line should extend uniformly only a short distance beyond the circle or view. They start and end with long dashes and should not cross at the spaces between the dashes.

CUTTING PLANE LINE (THICK)

The CUTTING PLANE LINE is a heavy line. It is used to indicate where the sectional view will be taken. Two forms are recommended for general use. The first form is composed of a series of long (3/4 to 1 1/2 in.) and short (1/8 in. with 1/16 in. space) dashes. The second form is composed of equal dashes about 1/4 in. long with 1/16 in. spacing. The cutting plane line will be further explained in Unit 10 on Sectional Views.

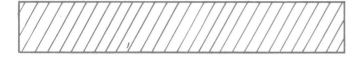

SECTION LINE (THIN)

SECTIONAL LINES are used when drawing the inside features of the object. They indicate material cut by the cutting plane line, and also indicate the general classification of the material. The lines are fine dark lines.

PHANTOM LINE (THIN)

The PHANTOM LINE is used to indicate alternate positions of moving parts and for repeated detail like threads and springs. It is a thin dark line made of long dashes (3/4 to 1 1/2 in.) long, alternated with pairs of short dashes 1/8 in. in length, with 1/16 in. spaces.

HOW TO MEASURE

Almost every line on a mechanical drawing must be a measured line. If your drawings are to be made accurately, YOU must be able to make accurate measurements.

Measurements are made in the drafting room with SCALES, Fig. 5-4. The term SCALE means

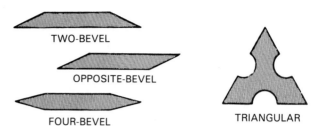

Fig. 5-5. Scales are available in a number of different shapes.

Fig. 5-6. The scale clip keeps the desired scale edge in an upright position. No time is wasted looking for the scale edge being used.

both the device or tool for making measurements, and the size to which the drawing is made.

Scales have graduations on the edges that show lengths used to indicate larger units of measure (as 1/4 in. equals one foot). While several different shapes of scales are available, Fig. 5-5, triangular shaped scales are most widely found in the school drafting room.

A SCALE CLIP provides a means to lift a triangular scale and keeps the desired scale edge in an upright position, Fig. 5-6.

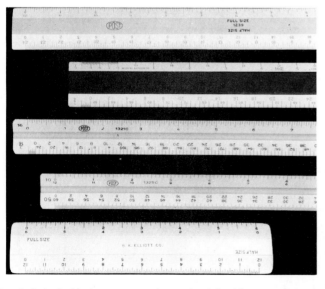
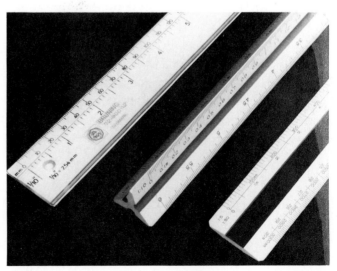

Fig. 5-4. Left. Various types of standard (inch) measurement scales found in a drafting room. Right. A few of the many types of metric drafting scales.

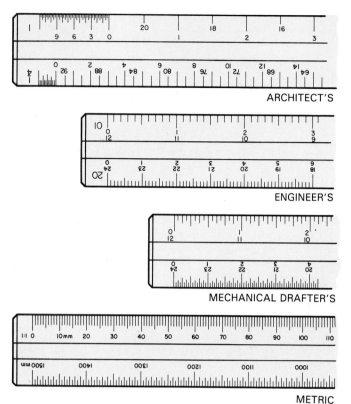

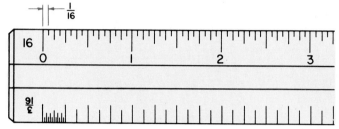

Fig. 5-9. One edge of the architect's scale is graduated in 1/16's of an inch.

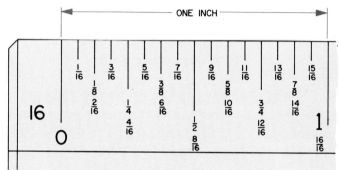

Fig. 5-10. The divisions of the 1/16 architect's scale are numbered for learning experiences.

Fig. 5-7. Scales in common usage in school drafting rooms.

There are four types of scales in common usage. They are classified as ARCHITECT'S, ENGINEER'S, MECHANICAL DRAFTER'S and METRIC scales, Fig. 5-7.

The triangular architect's scale (most often used in school drafting rooms) has six faces, (left) Fig. 5-8. Each is graduated differently.

You will find one face where the inch divisions are each divided into sixteen (16) parts. Each division is equal to one-sixteenth (1/16) of an inch, Fig. 5-9. To read the scale, imagine that the one-sixteenth (1/16) divisions are numbered as shown in Fig. 5-10. At first you may find it easier to count the spaces when you measure. However, after some practice this should not be necessary.

A section of the scale is shown in Fig. 5-11. How many of the measurements can YOU read correctly?

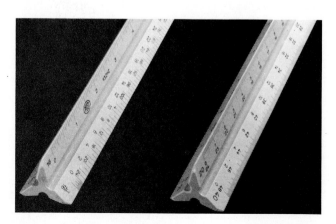

Fig. 5-8. The architect's scale is on the left and the engineer's scale is on the right.

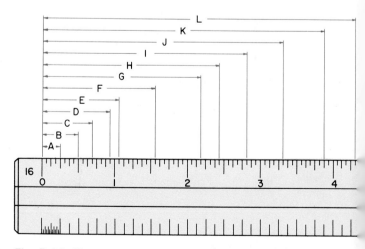

Fig. 5-11. How many can you answer correctly? On a piece of paper write the letters A to L. After each letter write the correct measurement. Reduce fractions to the lowest terms.

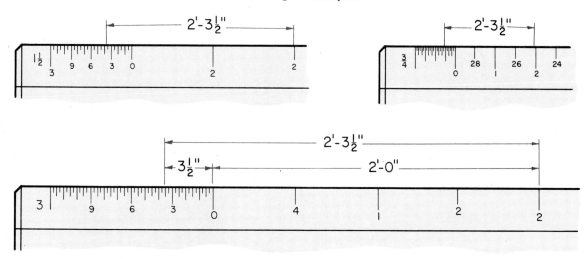

Fig. 5-12. Graduations on the scale edges indicate larger or smaller scale units of measure.

After you can read and make measurements accurately and quickly to one-sixteenth of an inch, note the remaining faces on the scale. Each is divided to represent one foot (12 in., 1'-0") of actual measurement reduced to a particular length.

There are two scales on each face and each is marked to show scale divisions of 3/32 and 3/16; 1/8 and 1/4; 3/8 and 3/4; and 1 1/2 and 3. For example, the face marked with a 3 means the foot (12 in.) has been reduced to 3 in. A drawing made using this scale would be one-quarter (1/4) actual size. See Fig. 5-12.

USING A METRIC SCALE

There are no architect's or engineer's scales in the metric system. Most drawings will be dimensioned in millimeters. Full size drawings will require a 1:1 ratio scale. Use a 1:2 scale when drawings are reduced to half size, Fig. 5-13. The preferred metric scales are shown in Fig. 5-14. For architectural drafting, reduction scales above 1:10 are most often used. See Unit 23 for more details.

Most metric drawings you will make are measured with a 1:1 metric scale. Each division is 1.0 mm, and the numbered lines are 10, 20, 30, etc., Fig. 5-15.

To make a measurement of 52.5 mm, for example, it is a simple matter to come out to the 50 division, then add 2.5 mm more, Fig. 5-16.

A section of a 1:1 scale is shown in Fig. 5-17. How many measurements can you read correctly?

COMMON DRAFTING SCALES

CUSTOMARY INCH		NEAREST ISOMETRIC EQUIVALENT (mm)
1:2500	(1 in. = 200 ft.)	1:2000
1:1250	(1 in. = 100 ft.)	1:1000
1:500	(1/32 in. = 1 ft.)	1:500
1:192	(1/16 in. = 1 ft.)	1:200
1:96	(1/8 in. = 1 ft.)	1:100
1:48	(1/4 in. = 1 ft.)	1:50
1:24	(1/2 in. = 1 ft.)	1:20
1:12	(1 in. = 1 ft.)	1:10
1:4	Quarter size (3 in. = 1 ft.)	1:5
1:2	Half-size (6 in. = 1 ft.)	1:2
1:1	Full-size (12 in. = 1 ft.)	1:1

Fig. 5-14. Preferred metric scales recommended by ISO (International Organization of Standards).

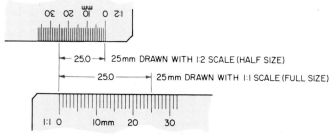

Fig. 5-13. Graduations on a metric scale also indicate larger or smaller scale units of measure.

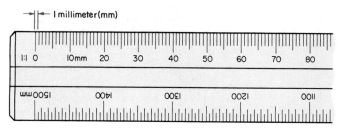

Fig. 5-15. Section of 1:1 metric scale. Each division is equal to 1.0 mm.

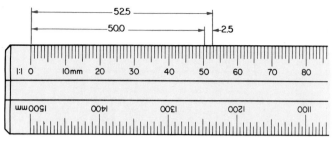

Fig. 5-16. Making a measurement of 52.5 mm.

Fig. 5-18. When making a measurement, observe the scale from directly above. Mark the desired measurement on the paper using a light perpendicular line made with a sharp pencil.

MAKING MEASUREMENTS

To make a measurement, observe the scale from directly above. Mark the desired measurement on the paper by using a light perpendicular line made with a sharp pencil, Fig. 5-18.

Keep the scale clean. Do not mark on it, or use it as a straight edge.

HOW TO DRAW LINES WITH INSTRUMENTS

Care must be taken to be sure lines are drawn to be the same weight. That is, lines must be uniform in width and darkness.

When lines are drawn using instruments, hold the pencil perpendicular to the paper, and inclined at an angle of about 60 degrees in the direction the line is being drawn. To keep the lines uniform in weight, especially if a long line is being drawn, rotate the pencil as you draw. Rotating the pencil will keep the point sharp.

Horizontal lines are drawn with a T-square, Fig. 5-19. The lines are drawn from left to right. Hold the T-square head firmly against the LEFT edge of the drawing board (left-handed drafters will use the RIGHT edge and draw lines from right to left). See Fig. 5-20.

Vertical lines may be drawn using a triangle and are drawn from bottom to the top of the sheet, Fig. 5-21. The base of the triangle must rest on the blade of the T-square. Use the left hand to hold the triangle in place.

Inclined lines are lines that are not vertical or horizontal. The drawing procedure depends on the direction of the slope. Lines that incline to the left are drawn more easily from the top downward, Fig. 5-22. Those that incline to the right should be drawn from the bottom upward, Fig. 5-23.

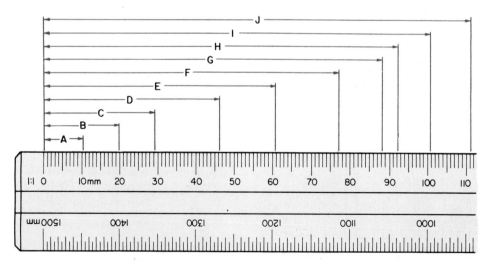

Fig. 5-17. How many can you answer correctly? On a piece of paper, write the letters A to J. After each letter, write the correct answer. Ask your instructor to check your answers.

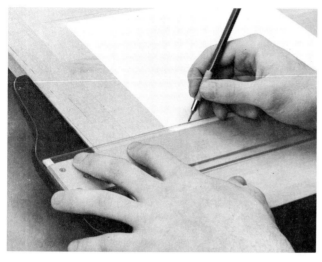

Fig. 5-19. Use the T-square to draw horizontal lines. Note how the T-square head is held against the edge of the drawing board. Incline the pencil about 60 degrees in the direction the line is being drawn.

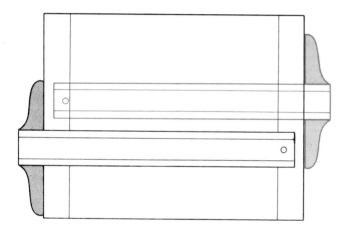

Fig. 5-20. Left-handed drafters use the right edge of the drawing board.

Fig. 5-21. Vertical lines are drawn using a triangle. The lines are drawn from the bottom to the top of the sheet. Incline the pencil in the direction of travel.

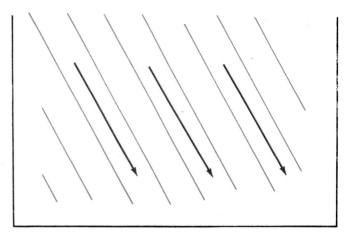

Fig. 5-22. Drawing lines that slant to the left, from the top down.

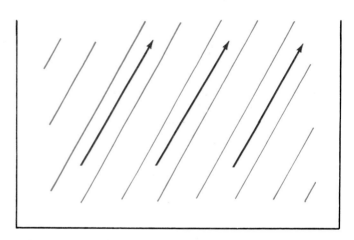

Fig. 5-23. Lines that incline to the right are drawn from the bottom up.

HOW TO DRAW A LINE PERPENDICULAR TO A GIVEN LINE USING INSTRUMENTS

To draw a line perpendicular to a given line using instruments, place the hypotenuse (long edge) of any triangle parallel to the given line, Fig. 5-24. Support the angle on the T-square or another triangle. Rotate the first triangle around the 90 degree corner to draw a line perpendicular to the given line.

Another technique used to draw a line perpendicular to a given line requires that you place either leg of the triangle parallel to the given line, Fig. 5-25. Support this triangle's hypotenuse side on the T-square or another triangle. Slide the first triangle on the supporting triangle or T-square, then draw the line. It will be perpendicular to the given line.

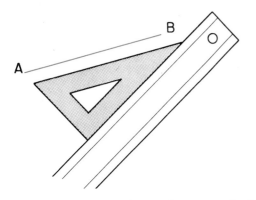

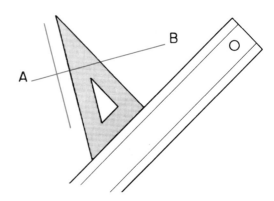

Fig. 5-24. How to draw a line perpendicular to a given line using a T-square and triangle. Be sure to hold the base (T-square) firmly in place when the triangle is moved.

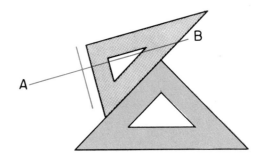

Fig. 5-25. How to draw a line perpendicular to a given line using two triangles.

HOW TO ERASE

When drawing, every effort should be made to prevent mistakes. However, even the best drafter must occasionally make changes on a drawing. This will require erasing.

Some suggestions to follow when erasing:
1. Keep your hands and instruments clean. This will help to keep "smudges" to a minimum.
2. Use an ERASING SHIELD whenever possible, Fig. 5-26. Select an opening that will expose only the area to be erased. The shield will protect the rest of the drawing while the erasure is made.
3. Clean all eraser crumbs from the board immediately after making an erasure. Remove them with a brush (preferred) or clean cloth, NOT YOUR HANDS.
4. Hold the paper with your free hand when erasing. This will keep the paper from wrinkling.
5. Erasing will remove the lead but will not take away the pencil grooves from the paper. Avoid deep, wide grooves by first blocking in all views with light construction lines.

HOW TO USE A COMPASS

In general drafting work, circles and arcs are drawn with a compass. Care must be taken so that the line drawn with the compass is the same weight as the line produced with the pencil. To accomplish this, it is recommended that the compass lead

Fig. 5-26. An erasing shield protects the area around where the correction or erasure is being made.

should be one or two grades softer than that of the pencil.

Sharpen the lead and adjust the point as shown in Fig. 5-27. Do not forget to adjust the point after each sharpening.

To SET A COMPASS, draw a line on scrap paper that is equal in length to the desired radius. Adjust the compass on this line, Fig. 5-28. Avoid setting a compass on the scale. The point will eventually ruin the division lines on the scale.

To DRAW A CIRCLE, rotate the compass in a clockwise direction, Fig. 5-29. Incline the tool in the direction of rotation. Start and complete the circle on a centerline. When drawing a series of concentric circles (circles with the same center), draw the smallest circle first.

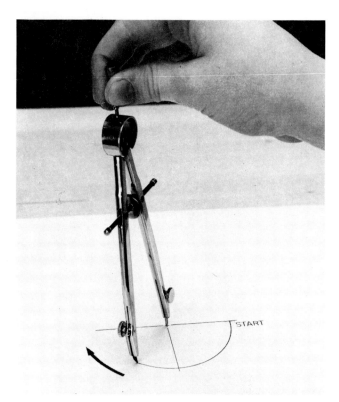

Fig. 5-29. Drawing a circle. Note that the compass is inclined in the direction of rotation. Start drawing the circle on a centerline if one is available.

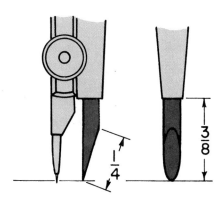

Fig. 5-27. Sharpening and adjusting the compass lead. Readjust point each time the lead is sharpened.

ATTACHING A DRAWING SHEET TO THE BOARD

The drawing sheet should be attached to the board with drafting tape. Tape is preferred because it does not damage the board. Before attempting to attach the paper, remove all eraser crumbs.

To attach the paper, place the sheet on the board as shown in Fig. 5-30. Left-handed drafters should use the right-hand portion of the board.

Place the T-square on the board with the head firmly against the left edge. Slide it up until the top of the blade is in line with the top edge of the drawing sheet, Fig. 5-31. Position the sheet so the top edge is parallel with the T-square blade. Fasten the sheet to the board. Larger sheets may also require fasteners on the bottom corners.

The following procedure is recommended when positioning and attaching A-size (8 1/2 x 11 in.) and A4 (210 x 297 mm) size drawing sheets. Position the sheet on the board as shown in Fig. 5-30.

Fig. 5-28. Adjust the compass to the proper radius size on a measured line. Never set it to size directly on the scale.

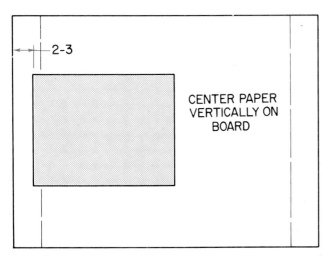

Fig. 5-30. Locating the drawing sheet on the board.

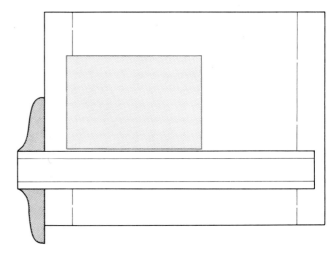

Fig. 5-32. The drawing sheet can also be aligned on the board by placing the bottom edge of the sheet on the edge of the T-square.

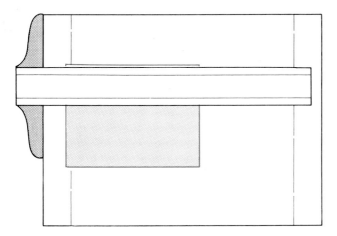

Fig. 5-31. Aligning the top of the drawing sheet with a T-square.

Place the T-square head firmly against the left edge of the board. Align the drawing sheet by sliding the T-square blade until it contacts the bottom edge of the paper, Fig. 5-32. Align the sheet with this edge. Fasten the sheet to the board.

Lightweight paper, like tracing vellum, is slightly more difficult to attach to the board. It has a tendency to wrinkle. This paper is aligned using the sequence shown in Fig. 5-33.

DRAFTING SHEET FORMAT

Most drafting rooms use a standard format in the layout of their drawing sheets. In general, the format consists of the border, title block, and standards notes, Fig. 5-34. A border is included to define the drawing area of the sheet. The title

block and standard notes provide information that is necessary for the manufacture and/or assembly of the object described on the drawing sheet. A preprinted title block is shown in Fig. 5-35.

Most industrial firms employ standard drawing sheet sizes. Drawings made on standard sheet sizes are easier to file. They also present less difficulty when prints are made from them.

With few exceptions, the drawings in this text should be drawn on 8 1/2 x 11 in., or 11 x 17 in. size sheets. The small sheet is known as an A-size sheet, and the large sheet as a B-size sheet. Corresponding metric sheet sizes would be A4 (210 x 297 mm) and A3 (297 x 420 mm) size sheets.

Plan your work carefully. Avoid using a large sheet when a smaller size sheet will do.

RECOMMENDED DRAWING SHEET FORMATS

The following drawing sheet formats are suggested. The first is similar to the sketching sheet format, Fig. 5-36. It should be used when class time is limited.

A more formal sheet format is prepared by the following: Put a 1/2 in. (12.5 mm) border on the sheet, Fig. 5-37. Use a short light pencil stroke, not a dot, as the guide for locating the border line. The short light guide mark should be covered when the border line is drawn. Allow another 1/2 in.

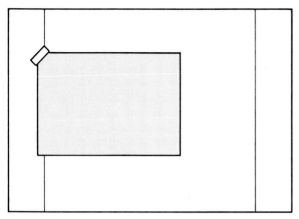

1. USE T-SQUARE TO ALIGN THE SHEET ON BOARD. FASTEN UPPER LEFT CORNER OF SHEET WITH TAPE.

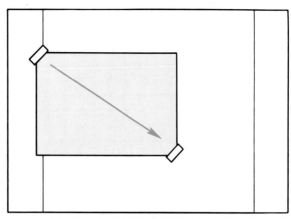

2. SMOOTH TO THE LOWER RIGHT CORNER. ATTACH SHEET.

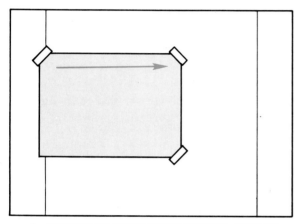

3. SMOOTH TO UPPER RIGHT CORNER. ATTACH SHEET.

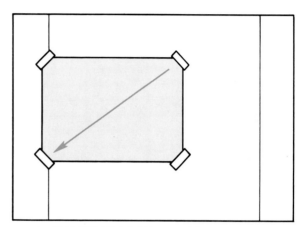

4. SMOOTH TO LOWER LEFT CORNER. FINISH ATTACHING SHEET TO BOARD.

Fig. 5-33. Sequence recommended for attaching light weight papers (such as tracing vellum) to the board.

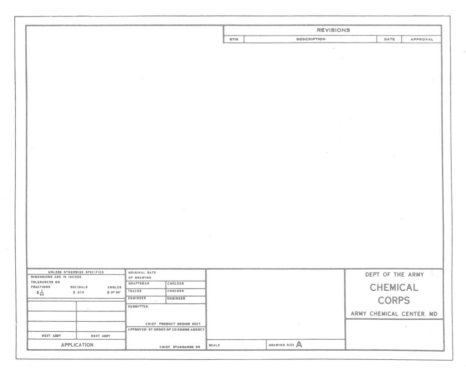

Fig. 5-34. Preprinted drawing sheets save a great deal of time for the drafter.

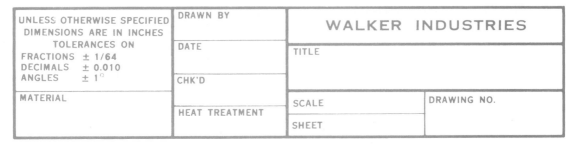

UNLESS OTHERWISE SPECIFIED DIMENSIONS ARE IN INCHES TOLERANCES ON FRACTIONS ± 1/64 DECIMALS ± 0.010 ANGLES ± 1°	DRAWN BY	WALKER INDUSTRIES	
	DATE	TITLE	
	CHK'D		
MATERIAL		SCALE	DRAWING NO.
	HEAT TREATMENT	SHEET	

Fig. 5-35. Preprinted title block. When it is used, only the border has to be drawn.

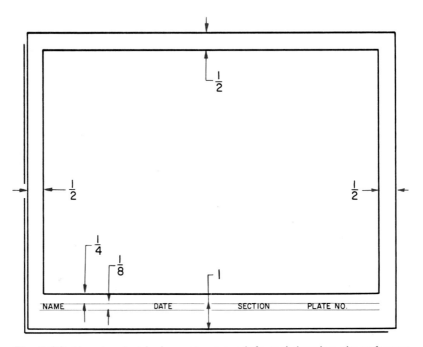

Fig. 5-36. How to start laying out a more informal drawing sheet format.

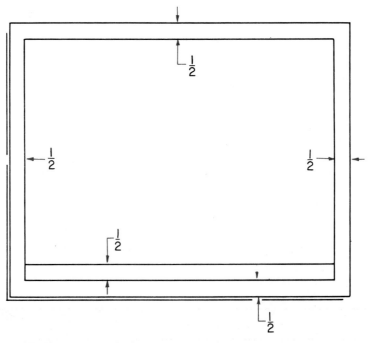

Fig. 5-37. How to start laying out a more formal drawing sheet format.

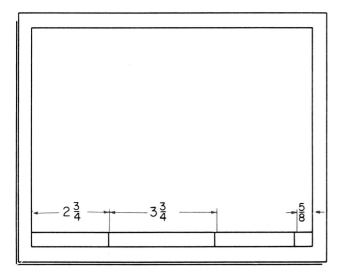

Fig. 5-38. Horizontal drawing sheet format.

(12.5 mm) for the title block. Divide the title block as shown in Figs. 5-38 and 5-39. Guide lines for the title block are drawn, Fig. 5-40. They should be drawn very lightly. Letter in the necessary information, Fig. 5-41.

It is recommended that modern duplicating equipment be used to prepare preprinted drawing sheets. The sheet format and the title block should be designed to meet the specific requirements of the drafting department and the industry. Prescribed tolerances can be printed on the sheets. Preprinted drafting sheets will save a great deal of time by eliminating repetitive and time consuming drawing operations.

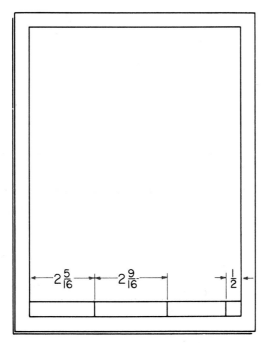

Fig. 5-39. Vertical drawing sheet format.

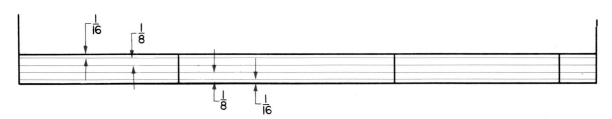

Fig. 5-40. Layout of guide lines needed for lettering.

Fig. 5-41. Suggested information to be lettered on drawing sheets for problems in this text.

DRAFTING VOCABULARY

Accomplish, Accurate, Alphabet of lines, Alternate, Characteristic, Clockwise, Composed, Concentric, Consists, Contacts, Dimension, Divisions, Features, Format, Graduations, Hypotenuse, Inclined, Line convention, Opaque, Perpendicular, Preprinted, Positioning, Procedure, Recommend, Scale, Sequence, Series, Standard, Suggested, Technique, Tendency, Uniform, Wrinkle.

TEST YOUR KNOWLEDGE—UNIT 5

1. Identify these lines:

   a. _____

   b. _____

   c. _____

   d. ◄————— 8 —————►

   e. – – – – – – – – – – – –

   f. ——— – ——— – ———

   g.

   h.

   i. ——— – – ——— – – ———

2. In drafting room language, the characteristics of the above lines and their correct use are known as _____ _____.
3. Why is it important for a drafter to be able to measure accurately?
4. In drafting, the term SCALE has two meanings. What are they?
5. Prepare sketches which show four different shapes of scales available.
6. List three types of scales in common use.
7. In drafting, horizontal lines are drawn using the _____.
8. Vertical lines are drawn using _____.
9. When erasing, the _____ is often used to protect surrounding areas.
10. Circles and arcs are drawn with a _____.
11. The recommended method of attaching drawing sheets is _____ _____.

OUTSIDE ACTIVITIES

1. Secure samples of drawing sheet formats used by industry. After studying these industrial examples, design a sheet format for the school's drafting department.
2. Obtain a compass and demonstrate to the class the proper way to sharpen the lead, adjust the length of the point, set the compass to the proper radius, and draw a circle.

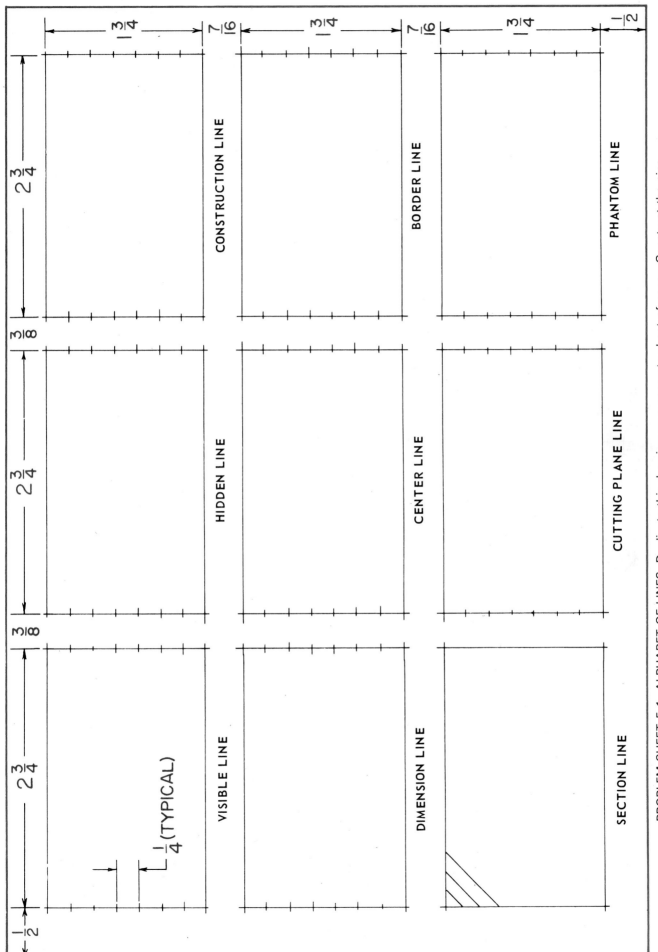

VISIBLE LINE

HIDDEN LINE

CONSTRUCTION LINE

¼ (TYPICAL)

DIMENSION LINE

CENTER LINE

BORDER LINE

SECTION LINE

CUTTING PLANE LINE

PHANTOM LINE

PROBLEM SHEET 5-1. ALPHABET OF LINES. Duplicate this drawing on a separate sheet of paper. Construct the nine different lines called for.

$\frac{1}{2}$ (TYPICAL)

$\frac{3}{4}$

$1\frac{1}{2}$ $2\frac{1}{8}$ $2\frac{5}{8}$ $3\frac{1}{16}$ $3\frac{5}{16}$ $3\frac{9}{16}$ $4\frac{3}{16}$ $5\frac{7}{16}$ $6\frac{1}{4}$ $7\frac{3}{8}$ $7\frac{11}{16}$ $6\frac{7}{8}$ $7\frac{15}{16}$

PROBLEM SHEET 5-2. MEASURING PRACTICE. Duplicate this drawing on a separate sheet of paper.

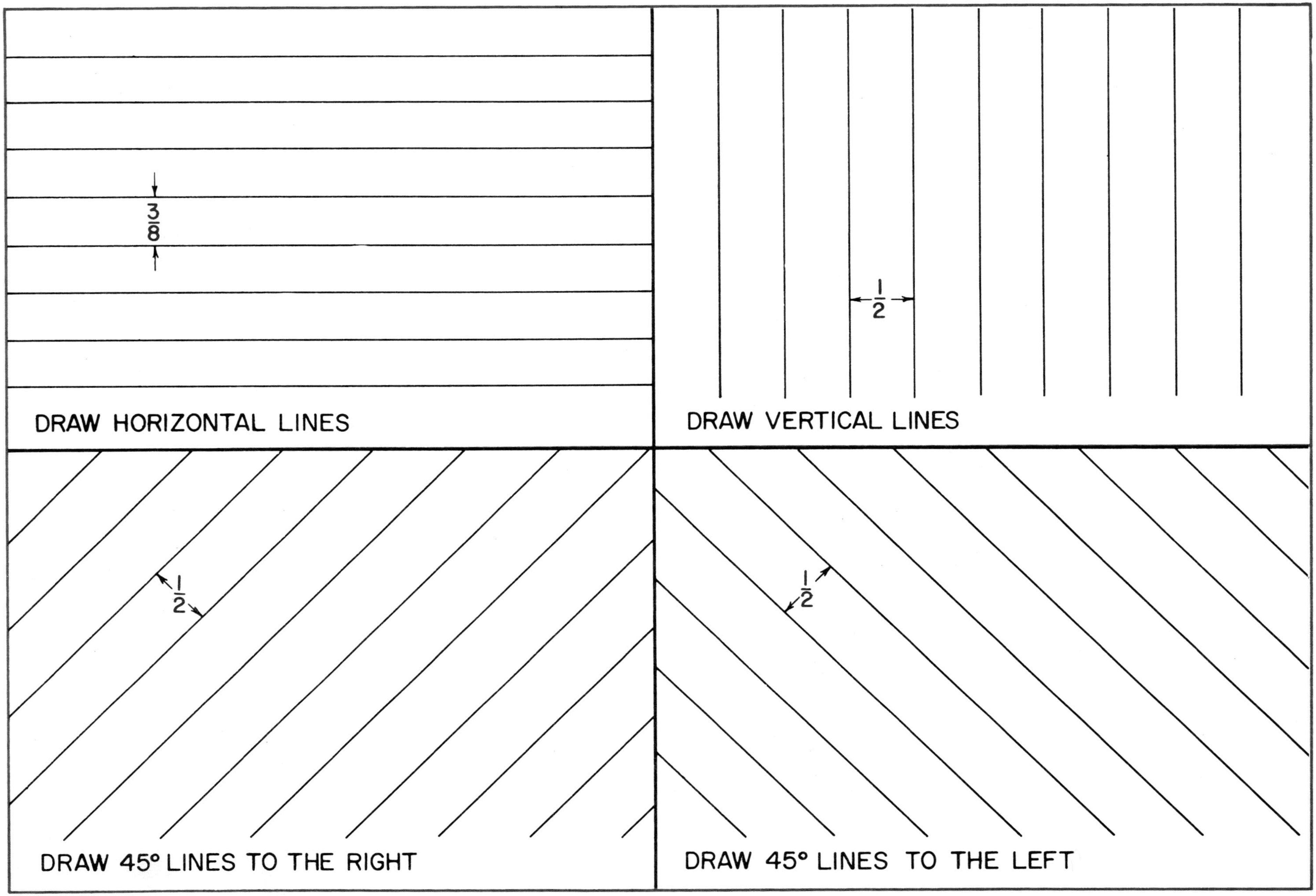

DRAW VERTICAL LINES

$\frac{1}{2}$

DRAW 45° LINES TO THE LEFT

DRAW HORIZONTAL LINES

$\frac{3}{8}$

DRAW 45° LINES TO THE RIGHT

$\frac{1}{2}$

$\frac{1}{2}$

PROBLEM SHEET 5-3. INSTRUMENT PRACTICE.

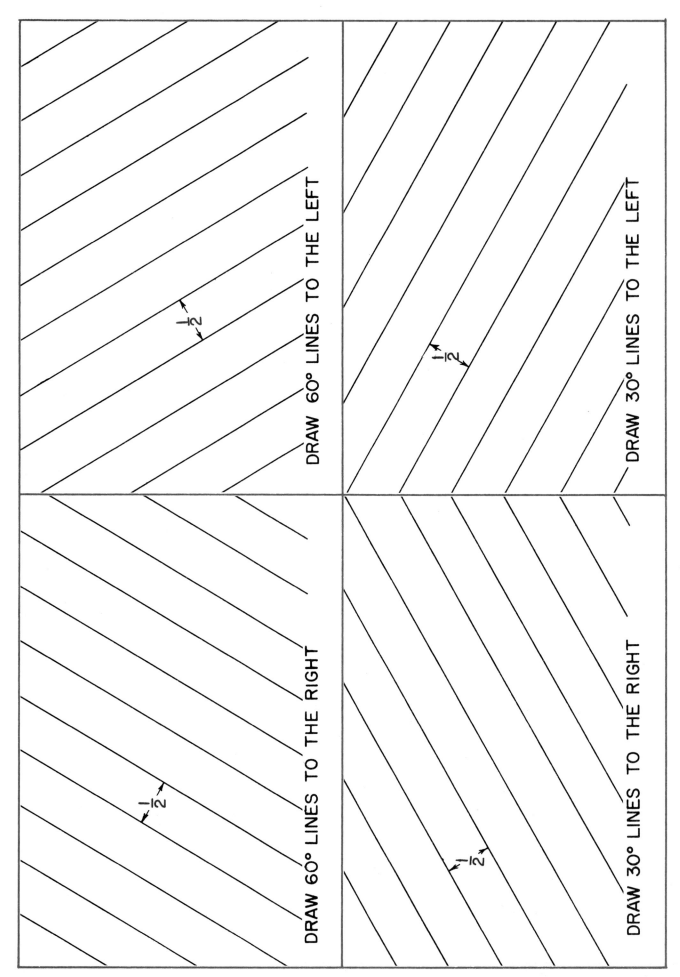

DRAW 60° LINES TO THE LEFT

DRAW 30° LINES TO THE LEFT

DRAW 60° LINES TO THE RIGHT

DRAW 30° LINES TO THE RIGHT

$\frac{1}{2}$

$\frac{1}{2}$

$\frac{1}{2}$

$\frac{1}{2}$

PROBLEM SHEET 5-4. INSTRUMENT PRACTICE.

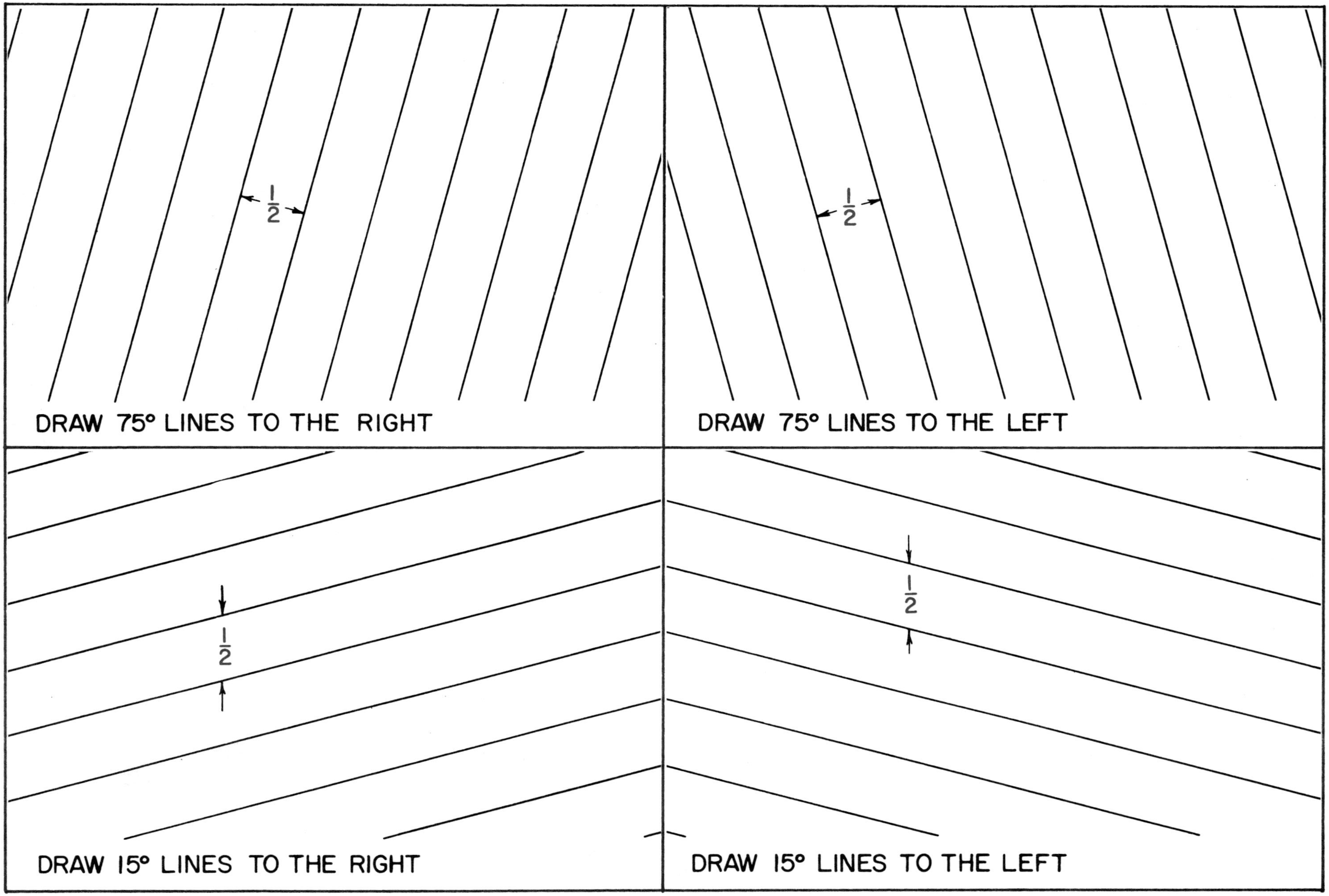

DRAW 75° LINES TO THE LEFT

$\frac{1}{2}$

DRAW 15° LINES TO THE LEFT

DRAW 75° LINES TO THE RIGHT

$\frac{1}{2}$

DRAW 15° LINES TO THE RIGHT

$\frac{1}{2}$

$\frac{1}{2}$

PROBLEM SHEET 5-5. INSTRUMENT PRACTICE.

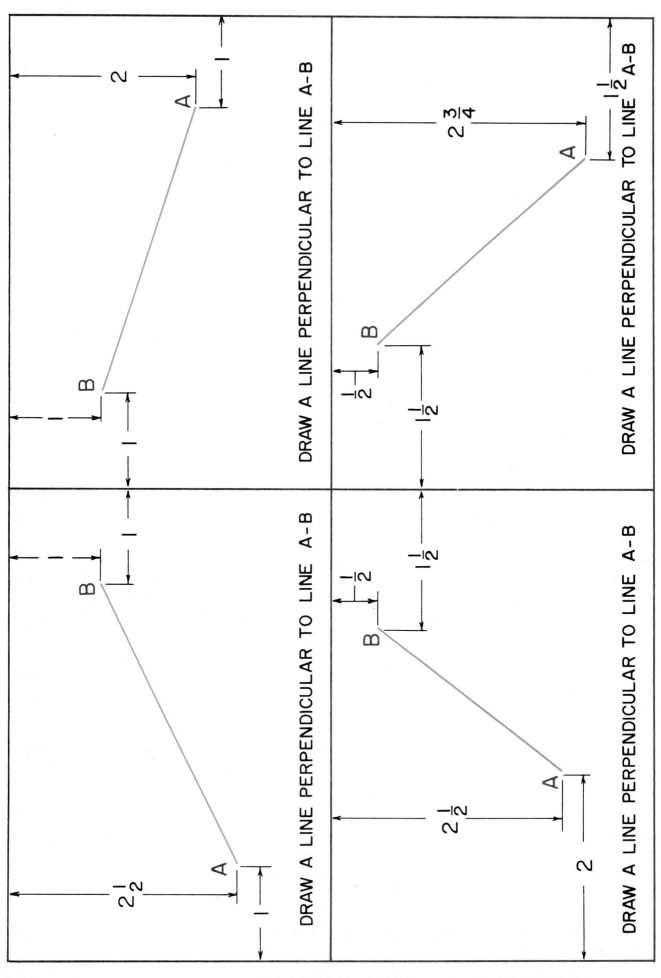

DRAW A LINE PERPENDICULAR TO LINE A-B

DRAW A LINE PERPENDICULAR TO LINE A-B

DRAW A LINE PERPENDICULAR TO LINE A-B

DRAW A LINE PERPENDICULAR TO LINE A-B

PROBLEM SHEET 5-6. INSTRUMENT PRACTICE.

Unit 6

BASIC GEOMETRIC CONSTRUCTION

After studying this unit, you will recognize basic geometric shapes. You will accurately construct basic geometric shapes using drafting equipment. You will be able to apply an understanding of geometric shapes to the design of everyday items. By practicing geometric construction techniques, you will become better prepared to solve drafting problems.

Whether we realize it or not, we see geometry in use every day, Fig. 6-1. The design of the aircraft that flies overhead and the automobile that passes on the street is based on geometrics. Buildings and bridges utilize squares, rectangles, triangles, circles, and arcs in their construction. Every drawing employed to manufacture a structure or product is composed of one or more geometric shapes.

Fig. 6-1. Geometry is used in the design of most products and structures. How many geometric figures can you identify in the above photos? (Aircraft—Grumman Aerospace Corp., Train—French National Railroad, Race Car—Champion Spark Plug Company, Camera—Minolta Corp.)

Fig. 6-2. Geometry is the basis of all computer generated drawings. This is a "wireframe" model of a proposed aircraft. (Evans & Sutherland)

Geometry is the basis of all computer generated drawings, Fig. 6-2. The first stage in developing a design is to generate a mesh or wire frame model of the design using basic geometric shapes.

It is important that you acquire the ability to visualize and draw the basic geometric shapes presented in this Unit, Fig. 6-3. This will aid you in solving drafting problems. It will also provide you an opportunity to improve your skill with drafting instruments.

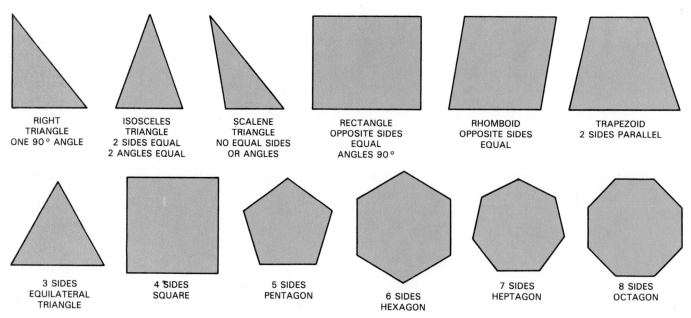

Fig. 6-3. Basic geometric shapes on which drawings are based.

DRAWING BASIC GEOMETRIC PROBLEMS

Size DOES NOT enter into the solution of most geometric problems. For that reason, no dimensions are given for the solution of the basic geometric problems. Most of them require little space in their solution; therefore, several problems may be included on a single drawing sheet. A suggested sheet layout is shown in Fig. 6-4.

Drawing sheets can be made more interesting and attractive if the geometric figures are constructed with colored pencils.

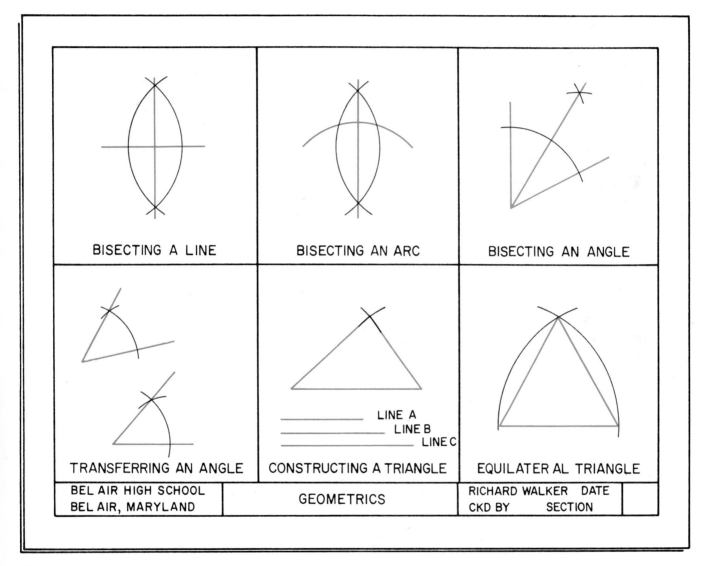

Fig. 6-4. Suggested sheet layout for drawing basic geometric problems. Several figures may be placed on a single sheet.

HOW TO BISECT OR FIND THE MIDDLE OF A LINE

(The bisecting line will be at right angles [90 deg.] to the given line.)
1. Let line A-B be the line to be bisected.
2. Set your compass to a distance larger than one half the length of the line to be bisected. Using this setting as the radius and the end of the line at A as the center point, draw arc C-D. Using the same compass setting but the end of the line at B as the center point, draw arc E-F.
3. Draw a line through the points where the arcs intersect. This line will be at right angles (90 deg. or perpendicular) to and bisect the original line A-B.

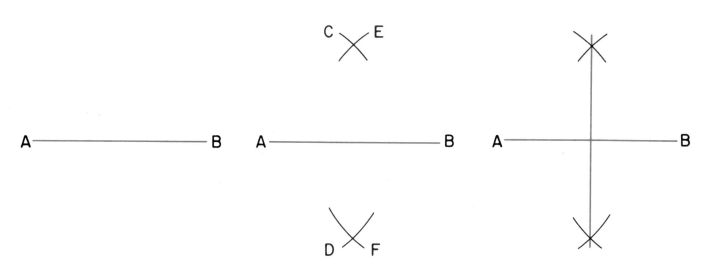

HOW TO BISECT AN ARC

1. An arc or part of a circle is bisected by the same method described in HOW TO BISECT OR FIND THE MIDDLE OF A LINE.

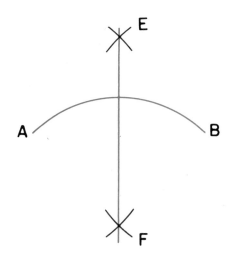

HOW TO BISECT AN ANGLE

1. Let lines A-B and B-C be the angle to be bisected.
2. With B as the center, draw an arc intersecting the angle at D and E.
3. Using a compass setting greater than one half D-E as centers, draw intersecting arcs. A line through this intersection and B will bisect the angle.

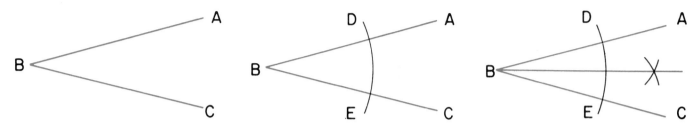

HOW TO TRANSFER OR COPY AN ANGLE

1. Let lines A-B and B-C be the angle to be transferred or copied.
2. Locate the new position of the angle and draw line A'-B'.
3. With B as the center point, draw an arc of any convenient radius on the given angle. This arc intersects the given angle at D and E.
4. Using the same radius and B' as the center draw arc D'-E-'.
5. Set your compass equal to D-E. With point D' as a center and D-E as the radius, strike an arc which intersects the first arc at point E'.
6. Draw a line through the intersecting arcs to complete the transfer of the given angle.

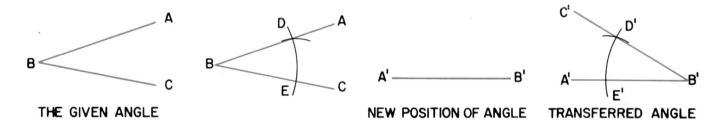

THE GIVEN ANGLE   NEW POSITION OF ANGLE   TRANSFERRED ANGLE

HOW TO CONSTRUCT A TRIANGLE FROM GIVEN LINE LENGTHS

1. Let lines A, B, and C be the sides of the required triangle.
2. Draw a line that is equal in length to line A.
3. Set your compass to a length equal to line B. Use one end of line A as a center and strike an arc. Reset your compass to a length equal to line C and with the other end of line A as a center strike another arc.
4. Connect the ends of line A to the points where the two arcs intersect.

LINE C

LINE B

LINE A

HOW TO CONSTRUCT AN EQUILATERAL TRIANGLE

(A triangle having all sides equal in length.)
1. Let line A-B be the length of the sides of the triangle.
2. With A as the center and with the compass setting equal to the length of line A-B, strike the arc B-C. Using the same compass setting but with B as the center strike the arc A-D. These arcs intersect at E.
3. Complete the triangle by connecting A to E and B to E.

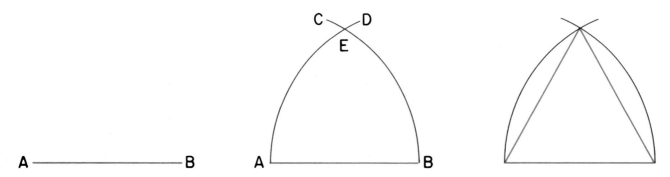

HOW TO DRAW A SQUARE WITH THE DIAGONAL GIVEN

1. Draw a circle with a diameter equal to the length of the diagonal.
2. Connect the points where the center lines intersect the circle.

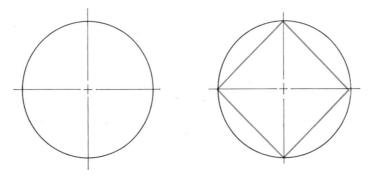

HOW TO DRAW A SQUARE WITH THE SIDE GIVEN

1. Draw a circle with a diameter equal to the length of the side.
2. Draw tangents at a 45 deg. to the center lines.

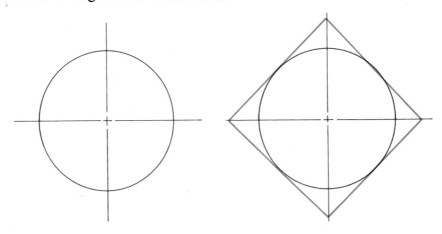

HOW TO CONSTRUCT A PENTAGON OR FIVE POINT STAR

1. Draw a circle. Let A-C and B-D be the center lines and O be the point where the center lines intersect.
2. Bisect the line (radius) O-B. This will locate point E.
3. With E as the center and with the compass set to the radius E-A strike the arc A-F.
4. The distance A-F is one-fifth circumference of the circle. Set your compass or dividers to this distance and circumscribe the circle starting at A. Connect these points as a pentagon or as a five point star.

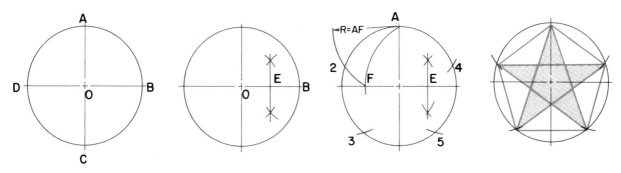

HOW TO CONSTRUCT A HEXAGON (FIRST METHOD)

1. Draw a circle.
2. With a compass setting equal to the radius of the drawn circle:
   a. With point A as the center, swing an arc to locate points B and F. With point D as the center, swing an arc and locate points C and E.
   b. Start at point A and move around the circle locating points B, C, D, E, and F in sequence.
3. Connect the points with a straightedge to complete the hexagon.

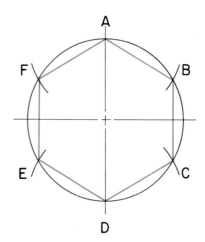

HOW TO CONSTRUCT A HEXAGON (SECOND METHOD)

1. Draw a circle.
2. Use the 30-60 deg. triangle and draw construction lines at 60 deg. to the vertical center line and tangent to the circle.
3. Draw two vertical construction lines tangent to the circle. Fill in the construction lines with visible object lines to complete the hexagon.
4. The same basic technique can be used to draw (inscribe) a hexagon inside the circle.

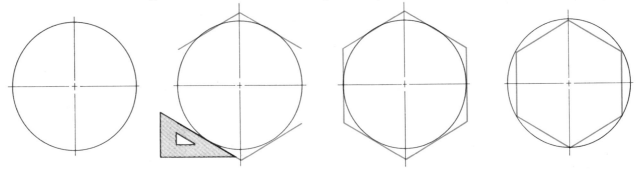

HOW TO DRAW AN OCTAGON USING A CIRCLE

1. Draw a circle with a diameter equal to the distance across the flats of the desired octagon.
2. Draw the vertical and horizontal lines tangent to the circle. Use construction lines.
3. Complete the octagon by drawing the 45 deg. angle lines tangent to the circle. Fill in the construction lines with visible object lines.

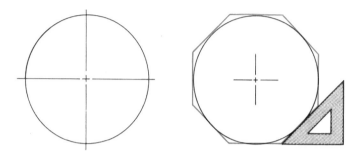

HOW TO DRAW AN OCTAGON USING A SQUARE

1. Draw a square with the sides equal in length to the distance across the flats of the required octagon. Draw the diagonals A-C and B-D. The diagonals intersect at O.
2. Set your compass to radius A-O and with corners A, B, C, and D as centers, draw arcs that intersect the square at points 1, 2, 3, 4, 5, 6, 7, and 8.
3. Complete the octagon by connecting point 1 to 2, 2 to 3, 3 to 4, 4 to 5, 5 to 6, 6 to 7, 7 to 8, and 8 to 1 with object lines.

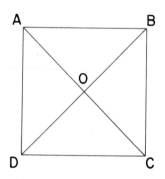
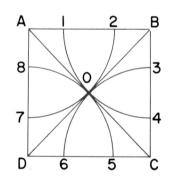
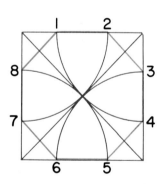

HOW TO DRAW AN ARC TANGENT TO TWO LINES AT A RIGHT ANGLE

1. Let A-B and B-C be the lines that form the right angle.
2. Set your compass to the radius of the required arc and using B as the center strike arc D-E. With D and E as centers and with the same compass setting draw the arcs that intersect at O.
3. With O as the center and with the compass at the same setting draw the required arc. It will be tangent to lines A-B and B-C at points D and E.

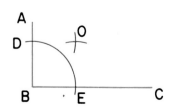

HOW TO DRAW AN ARC TANGENT TO TWO STRAIGHT LINES

1. Let lines A-B and C-D be the two straight lines.
2. Set your compass to the radius of the arc to be drawn tangent to the two straight lines and with points near the ends of the lines A-B and C-D as centers strike two arcs on each line.
3. Draw straight construction lines tangent to the arcs.
4. The point where the two lines intersect (O) is the center for drawing the required arcs.

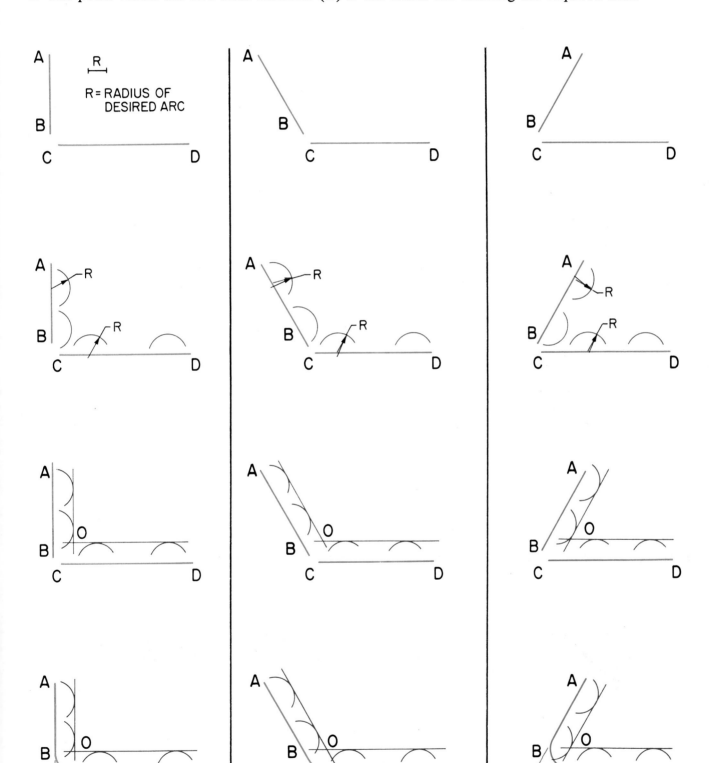

HOW TO DRAW AN ARC TANGENT WITH A STRAIGHT LINE AND A GIVEN ARC

1. Draw the given arc r and straight line a-b in proper relation to one another.
2. Draw line A-B a distance equal to the radius (R) of the desired arc from and parallel to the straight line a-b.
3. Draw arc C-D by setting your compass to a radius equal to R + r. This arc intersects with line A-B at point O.
4. Using point O as the center and with the compass set to the desired radius R draw the required arc. It will be tangent to the given arc and to the straight line a-b. Fill in construction lines with object lines.

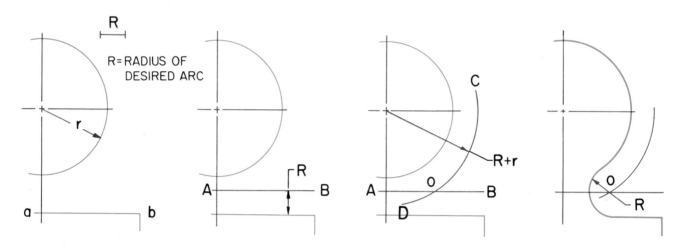

HOW TO DRAW TANGENT ARCS

1. Draw the two arcs ra and rb in proper relation to each other that require the tangent arc to join them. Let O and X be the centers of these arcs.
2. Set the compass to a distance equal to the radius R + ra. Using X as the center draw arc A-B. Reset the compass to a distance equal to the radius R + rb. Using O as the center draw arc C-D. These arcs intersect at Y.
3. Set the compass to the radius of the required arc R. With Y as the center draw the desired arc. This arc will be tangent to the given arcs.

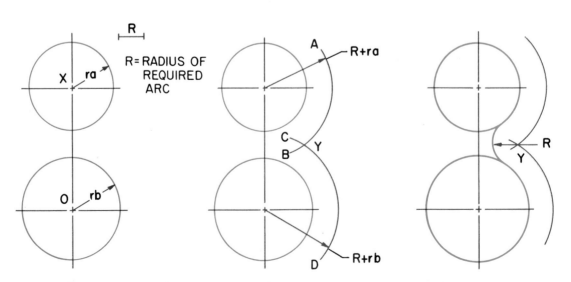

HOW TO DIVIDE A LINE INTO A GIVEN NUMBER OF EQUAL DIVISIONS

1. PROBLEM: Divide line A-B into five (5) equal parts.
2. With the T-square and angle draw vertical line B-C. Use construction lines.
3. Locate the scale with one point at A. Adjust the scale until a multiple of the divisions required (in this case five 1 in. divisions) lies between A and vertical line B-C.
4. Make vertical points at each of the five 1 in. divisions. Project vertical lines from these points parallel to line B-C. These vertical lines divide line A-B into five equal parts.

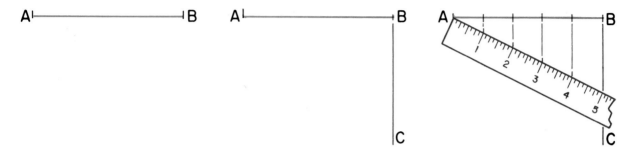

HOW TO DRAW AN ELLIPSE USING CONCENTRIC CIRCLES

1. Draw two concentric circles. The diameter of the large circle is equal to the length of the large axis of the desired ellipse. The diameter of the small circle is equal to the length of the small axis of the desired ellipse.
2. Divide the two circles into twelve equal parts. Use your 30-60 deg. triangle.
3. Draw horizontal lines from the points where the dividing lines intersect the small circle. Vertical lines are drawn from the points where the dividing lines intersect the large circle.
4. Connect the points where the vertical and horizontal lines intersect with a French curve.

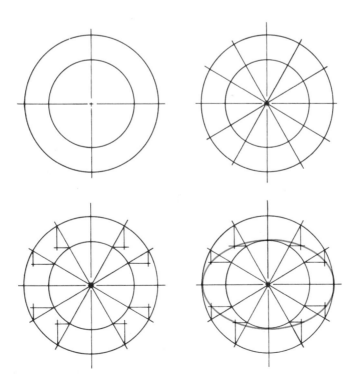

HOW TO DRAW AN ELLIPSE USING THE PARALLELOGRAM METHOD

This method is satisfactory for drawing large ellipses.

1. Let line A-B be the major axis and line C-D be the minor axis of the required ellipse.
2. Construct a rectangle with sides equal in length and parallel to the axes.
3. Divide A-O and A-E into the same number of equal parts.
4. From C draw a line to pont 1 on line A-E. Draw a line from D through point 1 on line A-C. The point of intersection of these two lines will establish the first point of the ellipse. The remaining points in this section are completed as are similar points in the other three sections (quadrants).
5. Connect the points with a French curve to complete the ellipse.

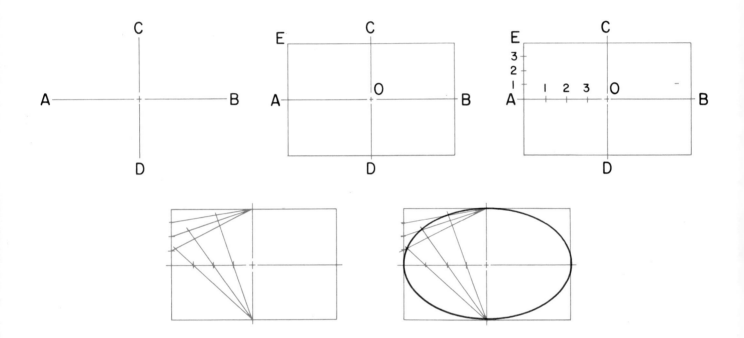

HOW TO CONSTRUCT AN ELLIPSE USING THE FOUR-CENTER APPROXIMATION METHOD

1. Given major axis A-B and minor axis C-D intersecting at point O.
2. Draw diagonal C-B. With point O as the center and O-C as the radius, strike an arc. The arc will intersect line O-B at E.
3. With the radius E-B, and using C as the center, strike an arc that intersects line C-B at F.
4. Construct a perpendicular bisector of F-B and extend it to intersect the major and minor axes at G and H.
5. Points G and H are the centers for two of the arcs needed to construct the ellipse. With O as the center, use a compass to locate points J and K. They are symmetrical with points G and H.
6. Draw a line from H extending through J and from K through G and J.
7. Using G and J as centers, strike arcs G-B and J-A.
8. With H and K as centers strike arcs H-C and K-D.
   These four arcs will be tangent to each other and form a four-center approximate ellipse.

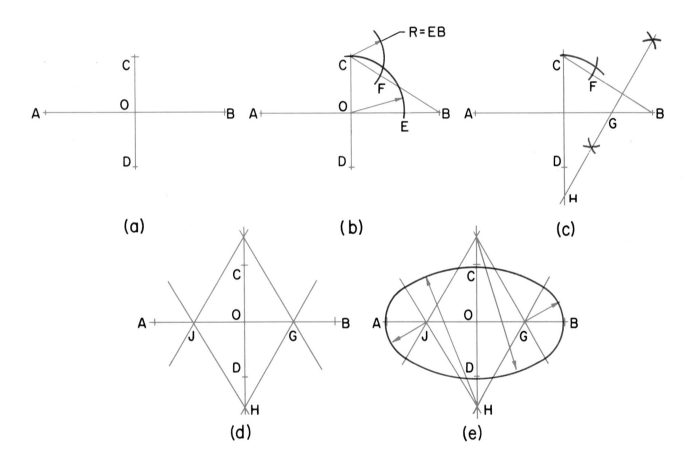

(a)   (b)   (c)

(d)   (e)

DRAFTING VOCABULARY

Arc, Axes, Axis, Basic, Bisect, Circle, Circumscribe, Concentric, Construct, Diagonal, Diameter, Division, Ellipse, Equilateral triangle, Geometric, Hexagon, Horizontal, Intersect, Length, Major axis, Minor axis, Octagon, Parallel, Parallelogram, Pentagon, Quadrant, Radius, Rectangle, Solution, Square, Tangent, Technique, Transfer, Triangle, Vertical, Visualize.

OUTSIDE ACTIVITIES

1. Prepare a bulletin board display that illustrates the use of geometric shapes in buildings and bridges.
2. Develop a list of everyday items that make use of geometric shapes. For example: Nut and bolt heads are round, square, and hexagonal.
3. Secure a photo or a drawing of a modern airplane. Place a sheet of tracing vellum over it and sketch in the various geometric shapes used in its design.
4. Do the same using a photo or drawing of a late model automobile.
5. Carefully examine a bicycle. Make a list of all the geometric shapes used in the design and construction of the bicycle. Be sure to look at all of the component parts.

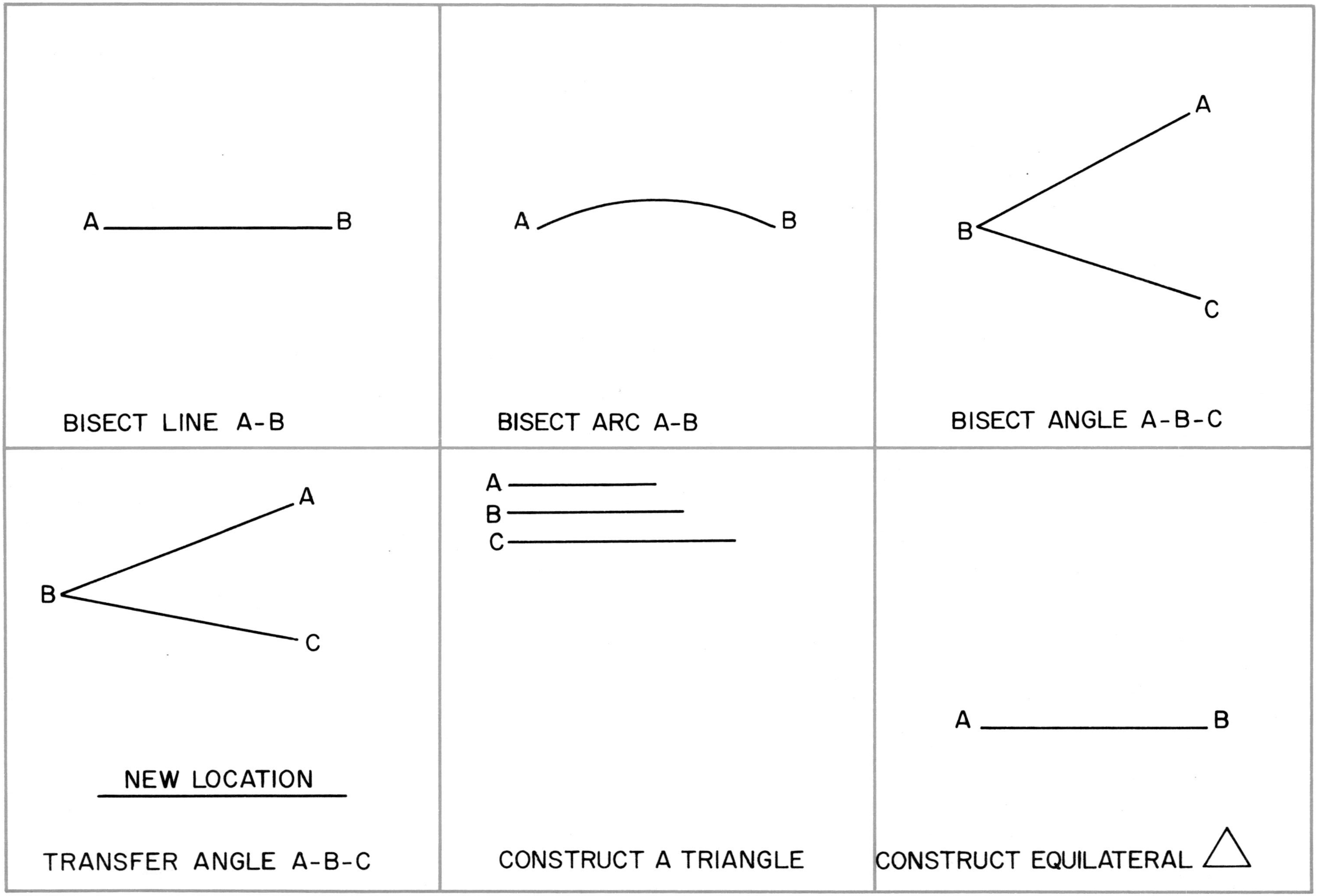

BISECT ANGLE A-B-C

A

B

C

CONSTRUCT EQUILATERAL △

A ——— B

BISECT ARC A-B

A

B

CONSTRUCT A TRIANGLE

A ————
B ———
C —————

PROBLEM SHEET 6-1. GEOMETRICS.

BISECT LINE A-B

A ——— B

TRANSFER ANGLE A-B-C

NEW LOCATION ———

A

B

C

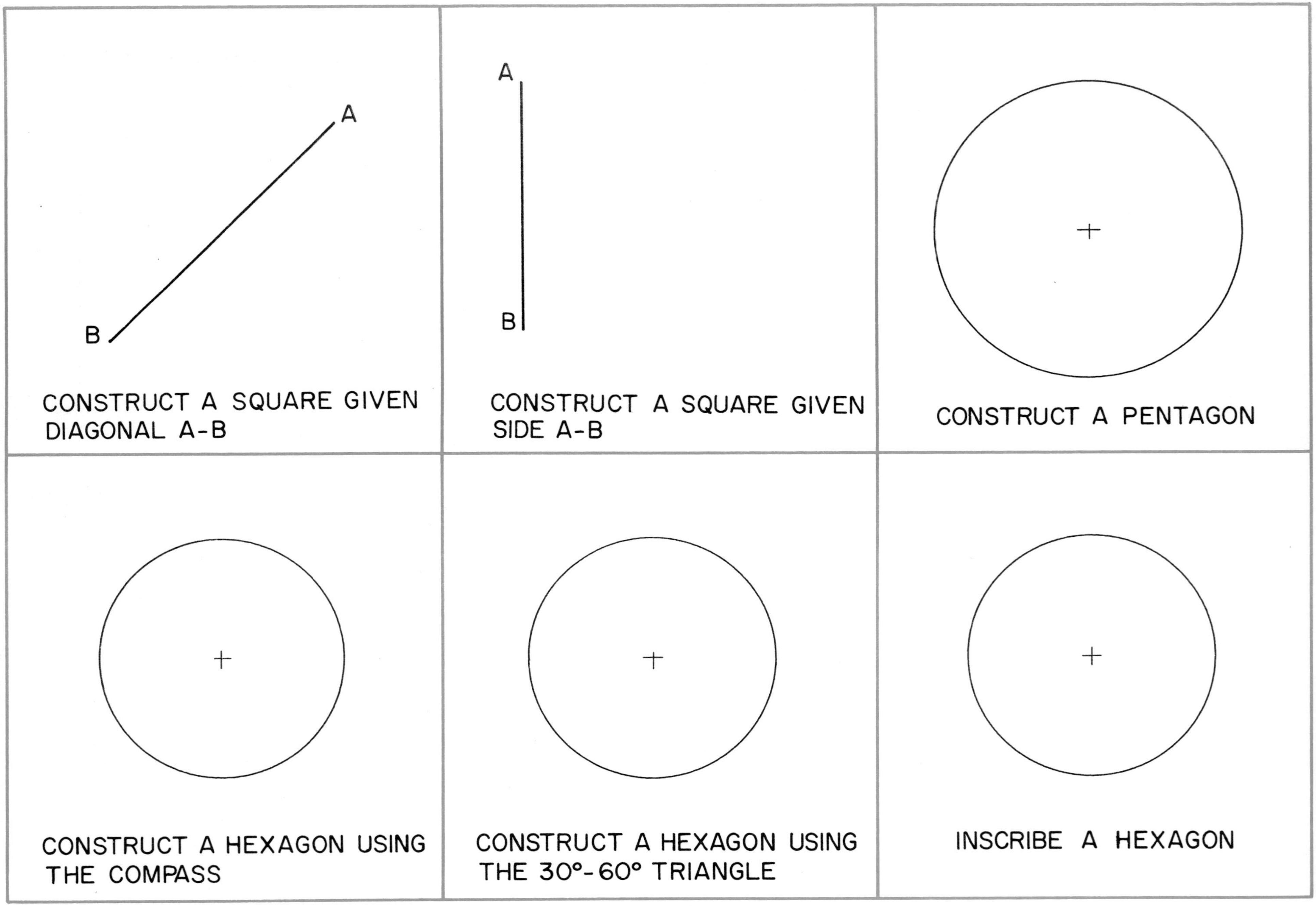

CONSTRUCT A PENTAGON

INSCRIBE A HEXAGON

CONSTRUCT A SQUARE GIVEN
SIDE A-B

A

B

CONSTRUCT A HEXAGON USING
THE 30°-60° TRIANGLE

CONSTRUCT A SQUARE GIVEN
DIAGONAL A-B

A

B

CONSTRUCT A HEXAGON USING
THE COMPASS

PROBLEM SHEET 6-2. GEOMETRICS.

96

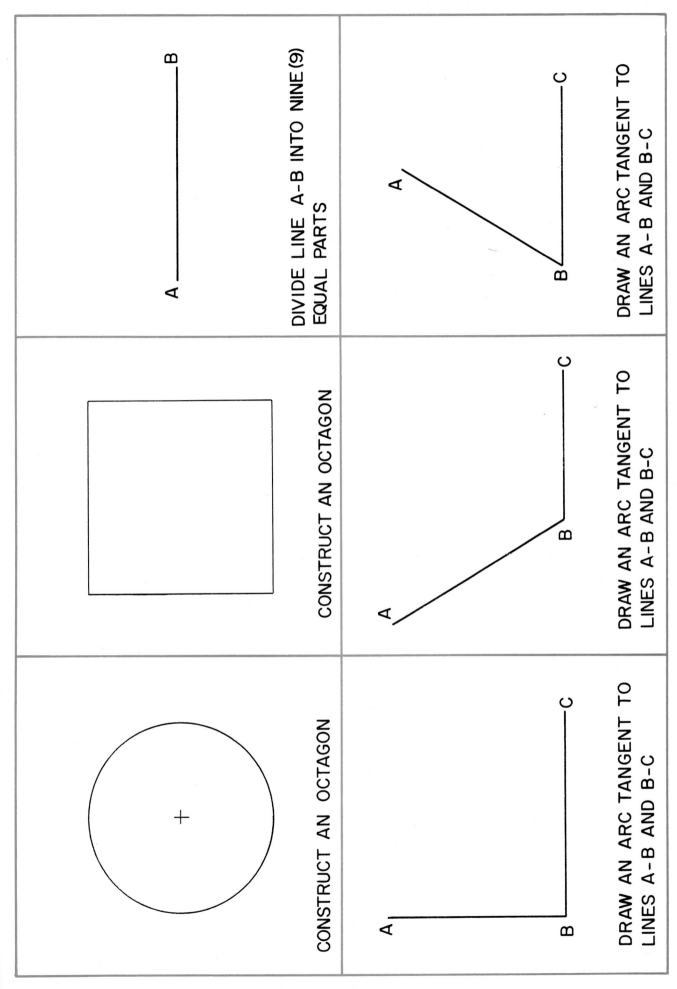

DIVIDE LINE A-B INTO NINE (9) EQUAL PARTS

DRAW AN ARC TANGENT TO LINES A-B AND B-C

CONSTRUCT AN OCTAGON

DRAW AN ARC TANGENT TO LINES A-B AND B-C

CONSTRUCT AN OCTAGON

DRAW AN ARC TANGENT TO LINES A-B AND B-C

PROBLEM SHEET 6-3. GEOMETRICS.

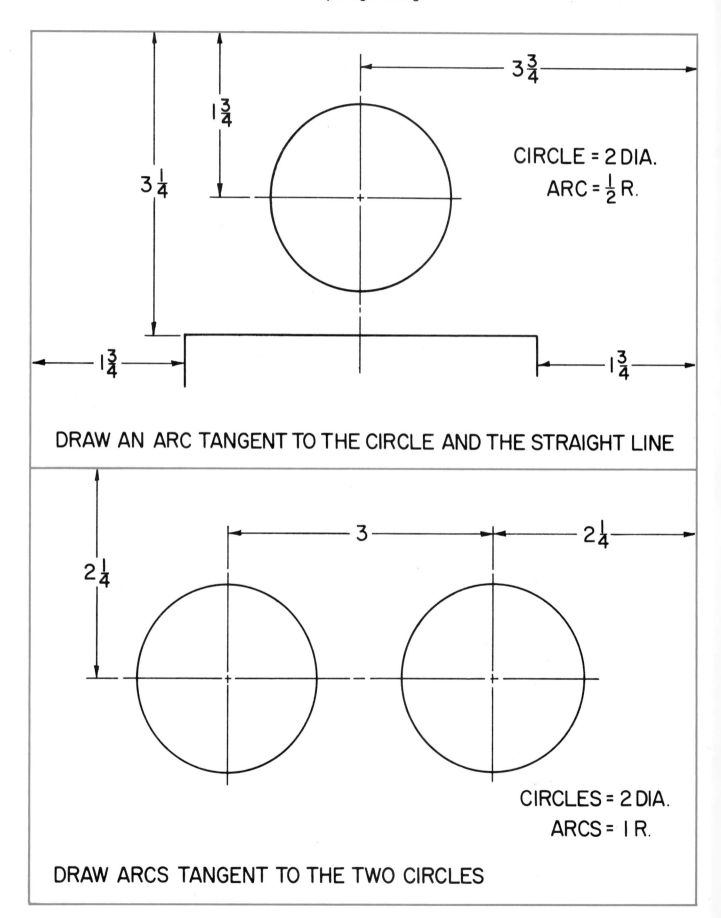

CIRCLE = 2 DIA.

ARC = $\frac{1}{2}$ R.

DRAW AN ARC TANGENT TO THE CIRCLE AND THE STRAIGHT LINE

CIRCLES = 2 DIA.

ARCS = 1 R.

DRAW ARCS TANGENT TO THE TWO CIRCLES

PROBLEM SHEET 6-4. Use a vertical sheet layout. Divide the work area as shown. Dimensions do not have to be put on your finished drawing.

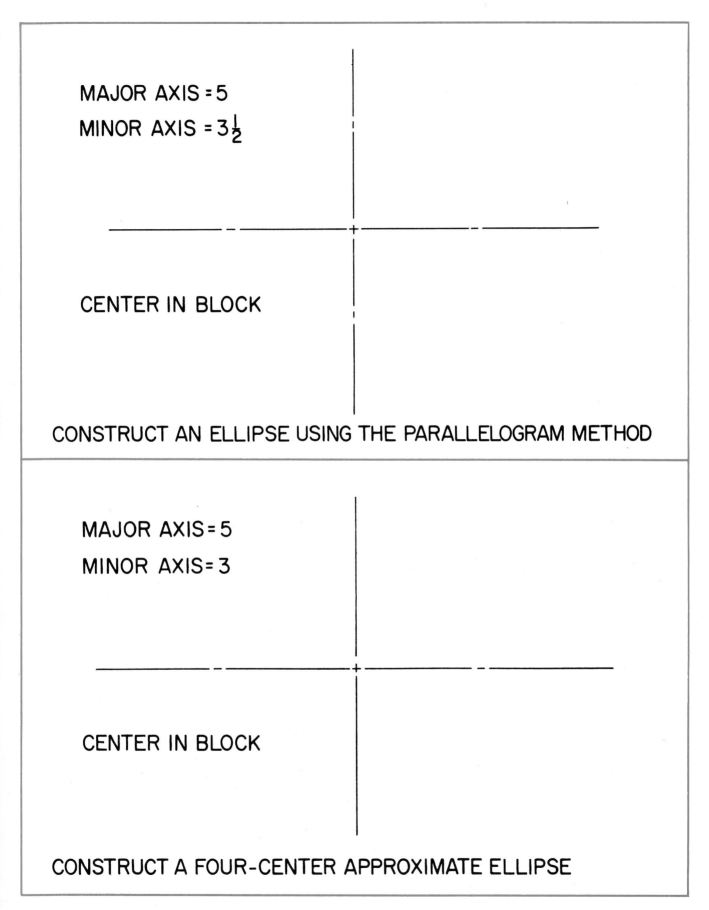

MAJOR AXIS = 5

MINOR AXIS = $3\frac{1}{2}$

CENTER IN BLOCK

CONSTRUCT AN ELLIPSE USING THE PARALLELOGRAM METHOD

MAJOR AXIS = 5

MINOR AXIS = 3

CENTER IN BLOCK

CONSTRUCT A FOUR-CENTER APPROXIMATE ELLIPSE

PROBLEM SHEET 6-5. Use a vertical sheet layout. Divide the work area as shown. Dimensions do not have to be put on your finished drawing.

AIRCRAFT INSIGNIA OF THE WORLD

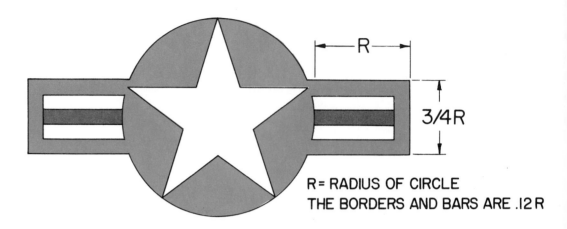

R = RADIUS OF CIRCLE
THE BORDERS AND BARS ARE .12 R

UNITED STATES

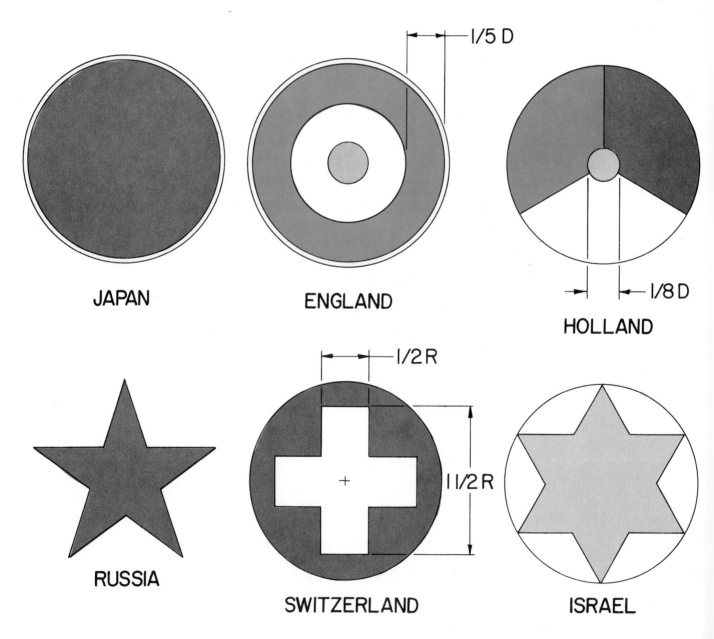

JAPAN

ENGLAND

HOLLAND

RUSSIA

SWITZERLAND

ISRAEL

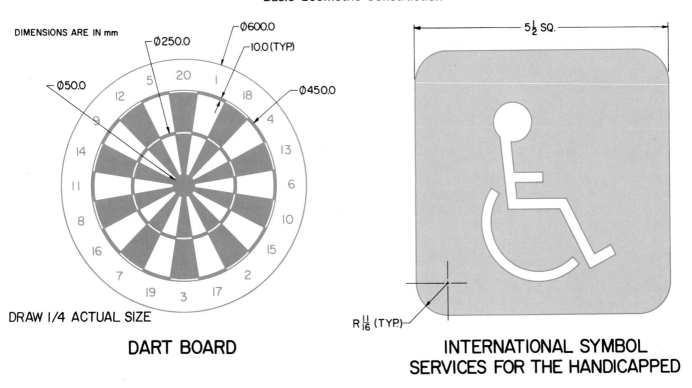

DIMENSIONS ARE IN mm

Ø600.0
Ø250.0
10.0 (TYP.)
Ø50.0
Ø450.0

DRAW 1/4 ACTUAL SIZE

DART BOARD

5½ SQ.

R 11/16 (TYP.)

INTERNATIONAL SYMBOL
SERVICES FOR THE HANDICAPPED

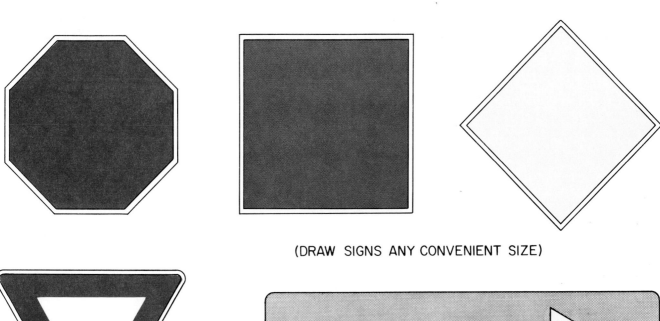

(DRAW SIGNS ANY CONVENIENT SIZE)

TRAFFIC WARNING SIGNS
HOW IS EACH SIGN USED?

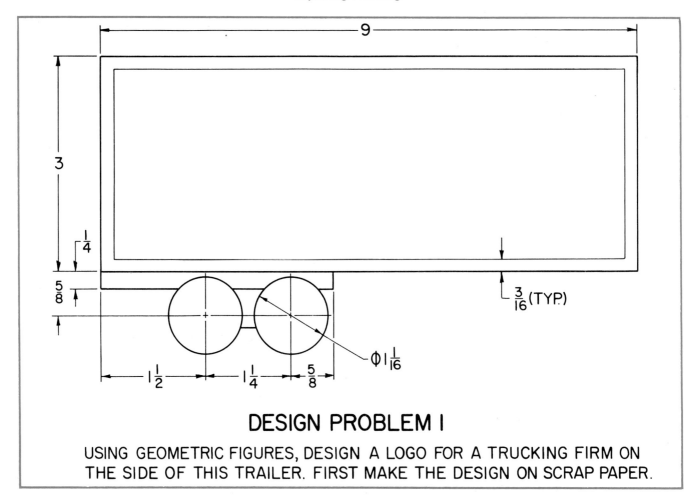

DESIGN PROBLEM I

USING GEOMETRIC FIGURES, DESIGN A LOGO FOR A TRUCKING FIRM ON THE SIDE OF THIS TRAILER. FIRST MAKE THE DESIGN ON SCRAP PAPER.

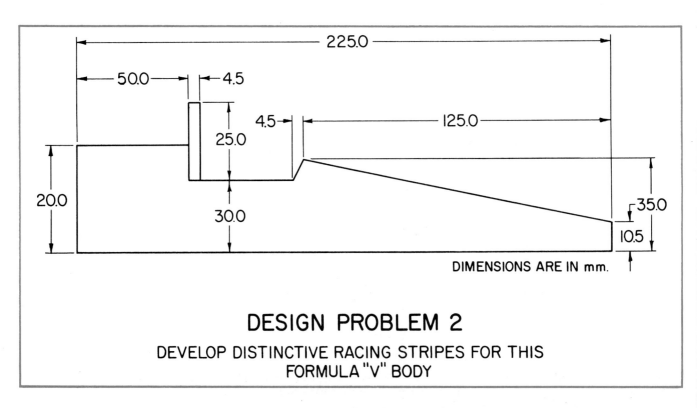

DIMENSIONS ARE IN mm.

DESIGN PROBLEM 2

DEVELOP DISTINCTIVE RACING STRIPES FOR THIS
FORMULA "V" BODY

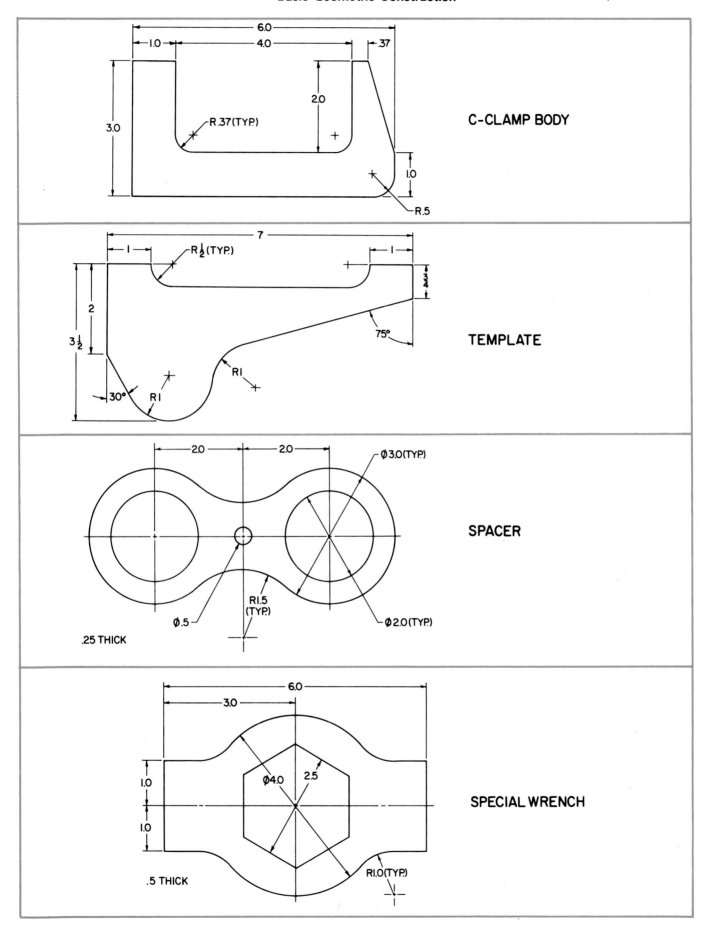

C-CLAMP BODY

TEMPLATE

SPACER

SPECIAL WRENCH

GEOMETRIC PROBLEMS. Draw the C-CLAMP BODY, TEMPLATE, SPACER and SPECIAL WRENCH full size. Center each problem on the drawing sheet. It will not be necessary to draw top and/or side views.

Unit 7

LETTERING

After studying this unit, you will produce neat lettering that will be easily read and understood. You should be able to select and demonstrate various mechanical lettering devices. You will be able to use a selection of preprinted lettering. You will be able to describe the application of CAD lettering.

LETTERING is used on drawings to give dimensions and other pertinent information needed to fully describe the item. The lettering must be neat and legible if it is to be easily read and understood.

A drawing will be improved by good lettering. A good drawing will look sloppy and unprofessional if the lettering is poorly done.

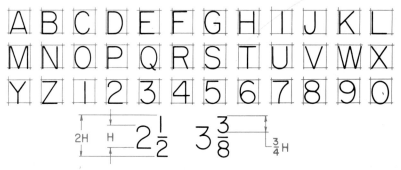

VERTICAL SINGLE STROKE GOTHIC ALPHABET

INCLINED SINGLE STROKE GOTHIC ALPHABET

RECOMMENDED SEQUENCE FOR MAKING SINGLE STROKE GOTHIC ALPHABET

Fig. 7-1. The Single Stroke Gothic Alphabet.

104

ONLY ONE FORM OF LETTERING SHOULD
APPEAR ON A DRAWING.

*AVOID COMbINING SEVERAL fORMS Of
LETTERING.*

Fig. 7-2. Avoid mixing several forms of lettering on a drawing.

SINGLE STROKE GOTHIC ALPHABET

The American National Standards Institute (ANSI) recommends that the SINGLE STROKE GOTHIC ALPHABET be the accepted lettering standard. It can be drawn rapidly and is highly legible, Fig. 7-1. It is called single stroke lettering not because each letter is made with a single stroke of the pencil (most letters require several strokes to complete), but because each line is only as wide as the point of the pencil or pen.

Single stroke lettering may be vertical or inclined. There is no definite rule stating that it should be one way or the other. However, mixing styles on a drawing should be avoided, Fig. 7-2.

YOU can do first class lettering if you learn the basic shapes of the letters, the proper stroke sequence for making them, and the recommended spacing between letters and words. You must also practice regularly.

LETTERING WITH A PENCIL

A H or 2H pencil is used by most drafters for lettering. Sharpen the pencil to a sharp conical point. Rotate it as you letter, Fig. 7-3, to keep the point sharp and the letters uniform in weight and line width. Resharpen the pencil when the lines become wide and "fuzzy."

GUIDE LINES

Good lettering requires the use of GUIDE LINES, Fig. 7-4. Guide lines are very fine lines made with a "needle sharp" 4H or 6H pencil. Guide lines should be drawn so lightly they will not show up on a print made from the drawing. VERTICAL GUIDE LINES, Fig. 7-5, may be used to assure that the letters will be vertical. Use INCLINED GUIDE LINES, drawn at 67 1/2 deg. to the horizontal line, where inclined lettering is to be used, Fig. 7-6.

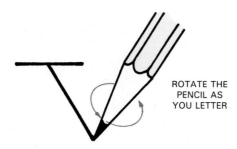

ROTATE THE
PENCIL AS
YOU LETTER

Fig. 7-3. Rotate the pencil point as you letter to keep the point
sharp and lettering uniform in weight.

ALWAYS USE GUIDE LINES WHEN
LETTERING. THEY ARE NEEDED.

Fig. 7-4. Guide lines must be used when lettering to keep the letters uniform in height.

VERTICAL GUIDE LINES HELP IN KEEPING
LETTERS UNIFORMLY VERTICAL.

Fig. 7-5. Vertical guide lines.

INCLINED GUIDE LINES HELP KEEP
INCLINED LETTERING UNIFORM.

Fig. 7-6. Inclined guide lines.

SPACING

In lettering, proper spacing of the letters is important. There is no hard and fast rule which indicates how far apart the letters should be spaced. The letters should be placed so spaces between the letters appear to be about the same. Adjacent letters with straight lines require more space than curved letters. Letter spacing is judged by eye rather than by measuring, Fig. 7-7.

Spacing between words and between sentences is another matter. The spacing between words and sentences should be equal to the height of the letters, Fig. 7-8.

DEVALUATION   DEVALUATION

SPACED BY MEASURING   SPACED VISUALLY

Fig. 7-7. Letter spacing is judged by eye rather than by measuring.

WORDS AND LETTERS MUST BE CLEARLY
SEPARATED. SPACING BETWEEN WORDS
AND SENTENCES IS EQUAL TO THE HEIGHT
OF THE LETTER USED.

Fig. 7-8. Spacing between words and sentences.

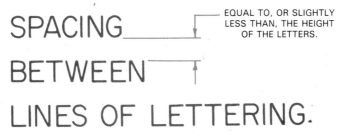

SPACING ⎯⎯⎯⎯ — EQUAL TO, OR SLIGHTLY
LESS THAN, THE HEIGHT
OF THE LETTERS.

BETWEEN

LINES OF LETTERING.

Fig. 7-9. Spacing between lines of lettering.

Spacing between lines of lettering should be equal to, or slightly less than, the height of the letters, Fig. 7-9.

LETTER HEIGHT

On most drawings, letters 1/8 inch high will be satisfactory. Titles are usually 3/16 to 1/4 inch high. To make the information easier to read, these sizes may be increased on large drawings.

LETTERING AIDS AND DEVICES

The BRADDOCK LETTERING TRIANGLE, shown in Fig. 7-10, and the AMES LETTERING INSTRUMENT, Fig. 7-11, are devices that may be used as aids for drawing guide lines. The numbers engraved below each series of holes in the triangle indicate the height of the letters in thirty-seconds. The series marked 3 indicates that the guide lines will be 3/32 inch apart; 4 indicates that they will be 4/32 or 1/8 inch apart. The numbers on the disk of the Ames lettering instrument are rotated until they are even with a line engraved on the base of the tool. The numbers also indicate the spacing of the guide lines in thirty-seconds.

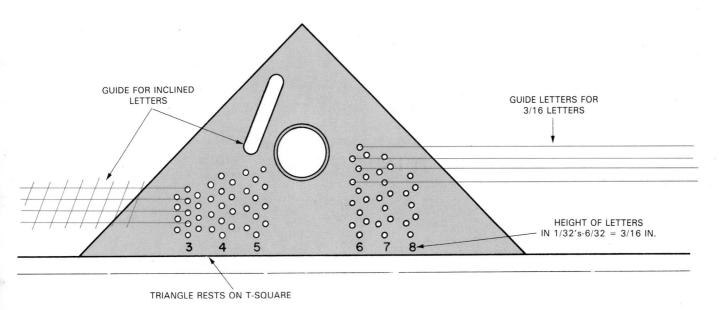

Fig. 7-10. The Braddock lettering triangle.

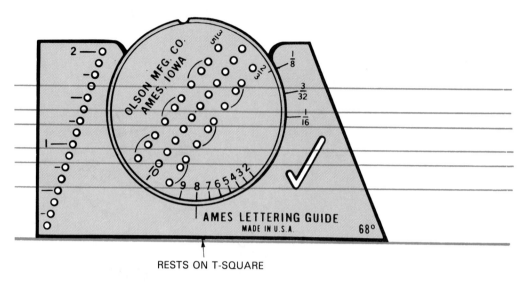

RESTS ON T-SQUARE

Fig. 7-11. The Ames lettering instrument.

MECHANICAL LETTERING DEVICES

Hand lettering is expensive because it requires so much time. For this reason, industry utilizes many mechanical devices to save time and improve the legibility of the letters and figures.

Many drafting and engineering offices utilize special typewriters to put information on drawings, Fig. 7-12. This releases highly skilled drafters for other work.

Fig. 7-12. A special typewriter may be used to put information on drawings.
(Mechanical Enterprises, Inc.)

Fig. 7-13. Mechanical type lettering device. Many styles of lettering are available.

With the use of microfilm, where the drawings are reduced and enlarged photographically, it is necessary for lettering to be highly legible. Good lettering can be done easily and rapidly with mechanical lettering devices, such as shown in Fig. 7-13. A stylus (pin) following a design cut into a metal or plastic matrix (pattern), guides a pen to draw individual letters. India ink is used in the pen. Many different letter and symbol patterns are available.

The Varigraph® is another type of mechanical lettering instrument, Fig. 7-14. It can be adjusted to produce many types and sizes of letters.

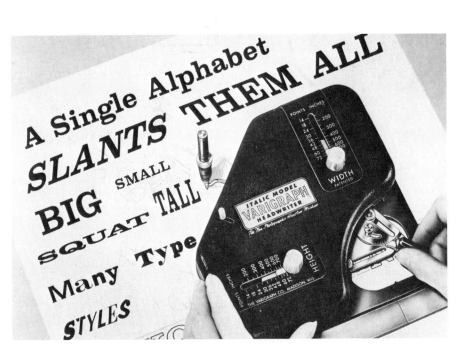

Fig. 7-14. Type of mechanical lettering instrument that can be adjusted to produce many sizes and styles of type faces. (Varigraph, Inc.)

LEADERSHIP IN THE
18 leadership in the creat
LEADERSHIP IN
24 leadership in the
LEADERSHIP
24 leadership in
LEADERS
30 leadership

Leaders
Leadershi
leadership in
Leadership In T

LEADERSHIP IN THE
24 leadership in the cre
LEADERSHIP IN
36 leadership in
LEADERSHIP IN THE
24 leadership in the
LEADERSHIP

Fig. 7-15. Preprinted lettering. Many styles and sizes are available. (Formatt)

PREPRINTED LETTERING

A wide selection of alphabets and symbols are available in preprinted lettering, Fig. 7-15.

Rub-on materials are also known as DRY TRANSFER MATERIALS, Fig. 7-16. The lettering is transferred from the backing sheet by lightly rubbing over the design with a burnisher. This is done after the letter has been positioned on the drawing sheet.

PRESSURE SENSITIVE MATERIALS offer lettering attached to a plastic support sheet, Fig. 7-17. The drafter cuts the letter or figure away from the backing sheet. After it is positioned, a slight pressure will adhere it to the drawing sheet.

CAD Lettering

This unit is concerned with helping you to develop the skills and dexterity to letter clearly and accurately and, eventually, rapidly on your drawings.

Fig. 7-16. Left. Transfer letters are applied by placing the letter in position and burnishing the back of the sheet with a smooth object. Right. Removing the lettering sheet after the letter has been applied. The guide line (not used on all transfer lettering) is erased after the entire line of lettering has been applied.

Fig. 7-17. Left. Removing a CUT-OUT LETTER from the lettering sheet. Be careful not to cut through the backing sheet. Right. Applying the letter to the sheet. The guide line is cut away after the entire line of lettering has been set into position.

The lettering on drawings produced by computer aided drafting (CAD) is similar to the single stroke gothic lettering you are now learning. However, they are first generated on the screen by keyboarding in the appropriate text and numerals. The shape and size of the letters and numerals produced on the drawing printout is determined by the CAD program used.

More information on CAD lettering can be found in Unit 24, Computer Graphics.

DRAFTING VOCABULARY

Adjacent, Burnisher, Definite, Device, Engraved, Gothic, Indicate, Legible, Matrix, Microfilm, Needle sharp, Pertinent, Sequence, Stencil, Stylus, Uniform, Utilize.

TEST YOUR KNOWLEDGE—UNIT 7

Please do not write in the text. Place your answers on a sheet of notebook paper.
1. Lettering is used on drawings to give _____ and other pertinent _____ needed to fully describe the item.
2. Why must lettering on a drawing be neat and legible?
3. The single stroke gothic letter is recommended because it can be drawn _____ and is highly _____.
4. The single stroke lettering on a drawing sheet may be _____ or _____ because there is no definite rule stating that it has to be one or the other.
5. A _____ or _____ pencil is usually used for lettering.
6. Why should guide lines be used when lettering?
7. Name three types of lettering aids and mechanical lettering devices. Briefly describe how each works.
8. Why does industry use mechanical lettering devices?

OUTSIDE ACTIVITIES

1. Using signs and posters from your school's bulletin board, make a display of different lettering examples. Point out good spacing and poor spacing. Check to see if lettering styles are mixed in each example.
2. Research the variety of lettering styles, alphabets, and symbols available with mechanical lettering devices.
3. Letter the sentence "Good lettering technique requires practice and concentration." by hand, using Vertical Single Stroke Gothic. Letter the same sentence using a mechanical lettering device such as the LEROY lettering equipment. Time yourself and report to the class which is faster and which is easier to read.
4. Demonstrate the use of transfer lettering by producing a safety poster for another classroom in your school.

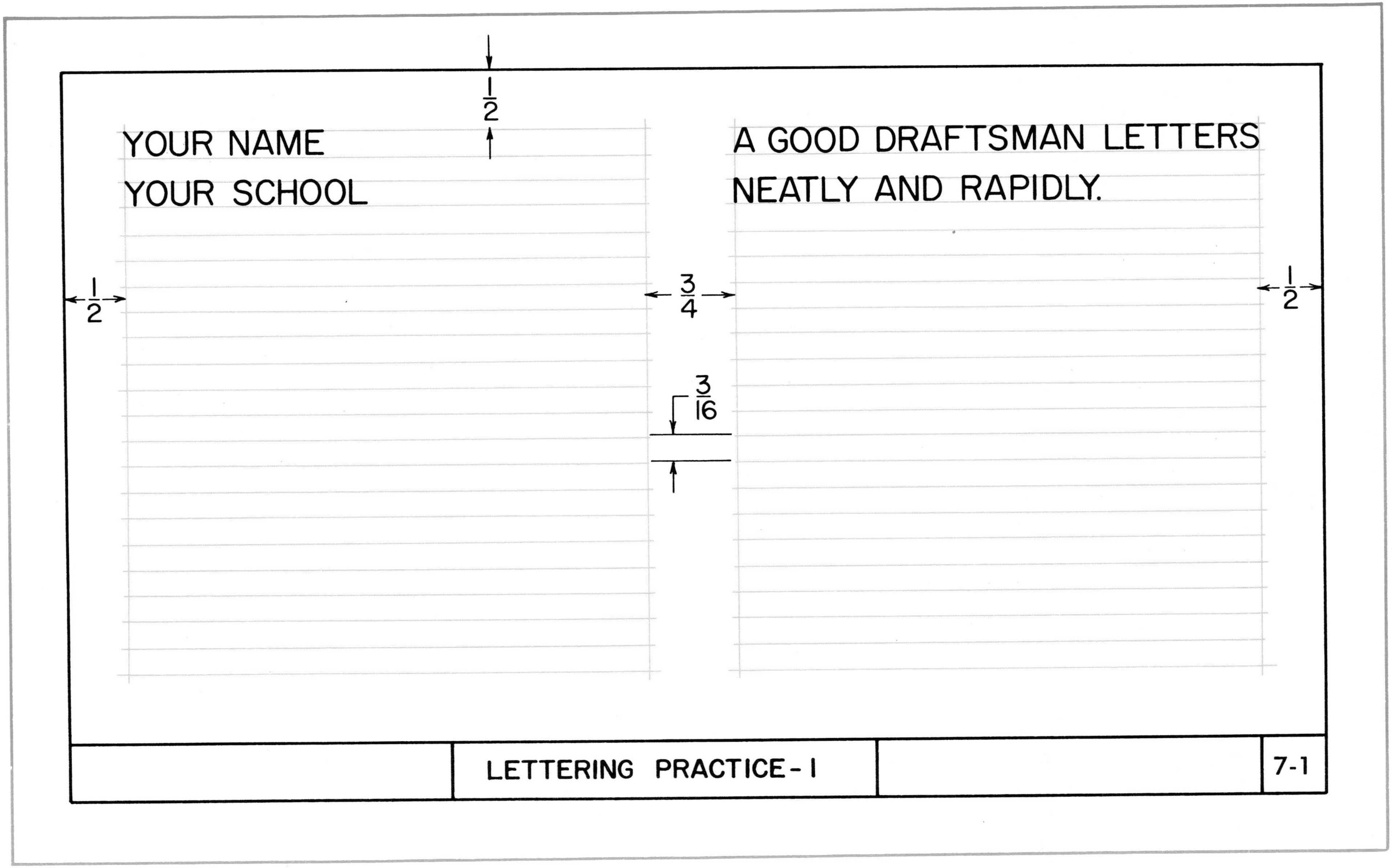

A GOOD DRAFTSMAN LETTERS
NEATLY AND RAPIDLY.

$\frac{1}{2}$

$\frac{3}{4}$

$\frac{3}{16}$

$\frac{1}{2}$

YOUR NAME
YOUR SCHOOL

$\frac{1}{2}$

7-1

LETTERING PRACTICE – 1

PROBLEM SHEET 7-1. LETTERING PRACTICE.

THE QUICK RED FOX JUMPED

OVER THE LAZY BROWN DOG.

$1 2 3 4 5 6 7 8 9 0 \quad \frac{1}{2} \quad \frac{1}{16} \quad \frac{3}{8} \quad \frac{5}{32}$

LETTERING PRACTICE-2

7-2

PROBLEM SHEET 7-2. LETTERING PRACTICE.

"ONE SMALL STEP FOR A MAN,
ONE GIANT LEAP FOR MANKIND."

"HERE MEN FROM THE PLANET
EARTH FIRST SET FOOT UPON
THE MOON JULY 1969,A.D.   WE
CAME IN PEACE FOR ALL MAN-
KIND."

LETTERING PRACTICE-3

7-3

PROBLEM SHEET 7-3. LETTERING PRACTICE.

1 2 3 4 5 6 7 8 9 0 $\frac{1}{2}$ $\frac{3}{4}$ $\frac{5}{8}$ $\frac{7}{9}$

PACK EACH BOX WITH SEVEN
DOZEN GIANT JUGS.

LETTERING PRACTICE-4

7-4

PROBLEM SHEET 7-4. LETTERING PRACTICE.

SELECT A FAVORITE SAYING
OR QUOTATION AND LETTER
IT IN $1\frac{1}{8}$, $\frac{3}{16}$ AND $\frac{1}{4}$ VERTICAL
OR INCLINED LETTERS.

| LETTERING PRACTICE-5 | 7-5 |

PROBLEM SHEET 7-5. LETTERING PRACTICE.

Unit 8

MULTIVIEW DRAWINGS

After studying this unit, you will recognize that most industrial drawings are of the multiview type. You will comprehend the use of orthographic projection to develop multiview drawings. You will select the views needed and locate them on the drawing sheet. You will be able to transfer points between views.

When a drawing is made with the aid of instruments, it is called a MECHANICAL DRAWING. Straight lines are made with a T-square and triangle or a drafting machine. Circles, arcs, and curved lines are drawn with a compass, French curve, or suitable template.

Drawings can also be generated on a computer by Computer Aided Design and Drafting (CAD or CADD), Fig. 8-1. They are converted into hard copy (paper drawings) on a high speed plotter. Many large firms use CAD almost exclusively.

Regardless of the technique, whether traditional or computer aided, the principles of drafting remain the same. The drafter must be familiar with the standards and procedures necessary to develop the graphics that will accurately describe the part. It is possible to use advanced computer technology to transfer the CAD generated information directly to a computer controlled machine tool that will manufacture the part. No drawings are necessary.

Fig. 8-1. Computer aided design or CAD permits design changes to be made rapidly. The engineer, designer, or drafter can see the modifications immediately by calling them onto the computer screen. Some programs will also produce performance changes, if any, that will result from the design changes. (Beech Aircraft Corp.)

This process is called CAD/CAM—computer aided design/computer aided manufacturing.

A great many of the drawings used by industry are in the form of MULTIVIEW DRAWINGS. That is, more than one view is required to give an accurate shape and size description of the object being drawn or generated. In developing the needed views, the object is normally viewed from six directions, as shown in Fig. 8-2.

The various directions of sight will give the FRONT, TOP, RIGHT SIDE, LEFT SIDE, REAR, and BOTTOM VIEWS, Fig. 8-3. To obtain the views, think of the object as being enclosed in a hinged glass box, Fig. 8-4, with the views projected into the side of the box.

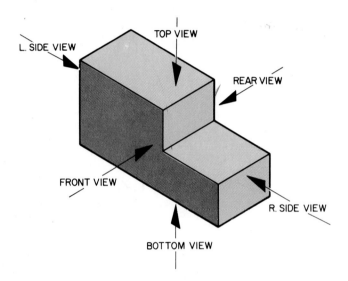

Fig. 8-2. An object is normally viewed from six different directions.

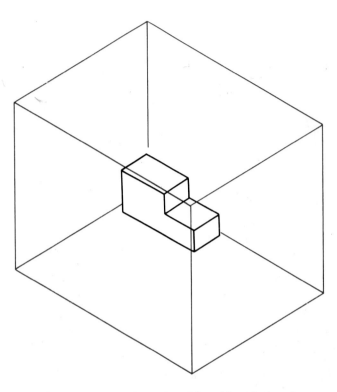

Fig. 8-4. The object is enclosed in a hinged glass box.

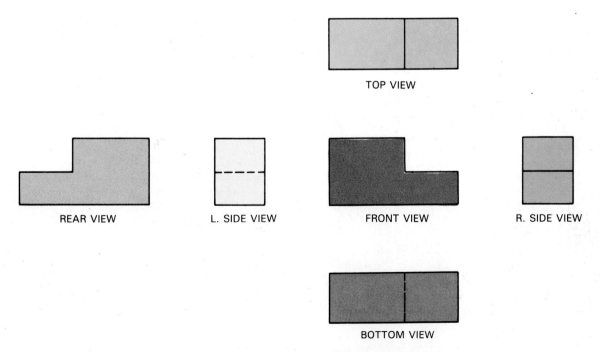

Fig. 8-3. The six directions of sight give these views.

The method for developing multiview drawings is called ORTHOGRAPHIC PROJECTION. It permits three dimensional objects to be shown on a flat surface having only two dimensions. It reveals the width, depth, and height of the object.

Orthographic projection forms the basis for engineering drawing. Two methods of projection are used, Fig. 8-5. THIRD ANGLE PROJECTION is preferred in the United States and several other nations. FIRST ANGLE PROJECTION is typically used in Europe.

With third angle projection, the object is drawn AS VIEWED IN THE GLASS BOX, Fig. 8-6. That is, the views are projected to the six sides of the box. The projected views are drawn as shown when the box is opened out.

With first angle projection, the object is drawn AS IF THE OBJECT WERE PLACED ON EACH SIDE OF THE GLASS BOX. That is, as

if the object is viewed as projected onto the drawing surface. See Fig. 8-7.

The ISO (International Organization for Standardization) symbols shown in Fig. 8-8 are employed to indicate the projection system of a given drawing.

SELECTING VIEWS TO BE USED

As can be seen on Figs. 8-6 and 8-7, at least six views of an object can be drawn. This does not mean that all six of the views must be used, or are needed. Only those views required to give a shape description of the object should be drawn. Any view that repeats the same shape description as another view can be eliminated, Fig. 8-9.

In many instances, two or three views are sufficient to show the shape of an object.

Those views of a drawing showing a large

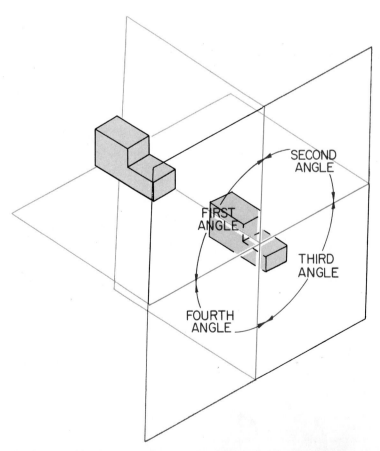

Fig. 8-5. Two methods of projection are generally employed: first angle projection in Europe; third angle projection in the United States and many other countries.

120

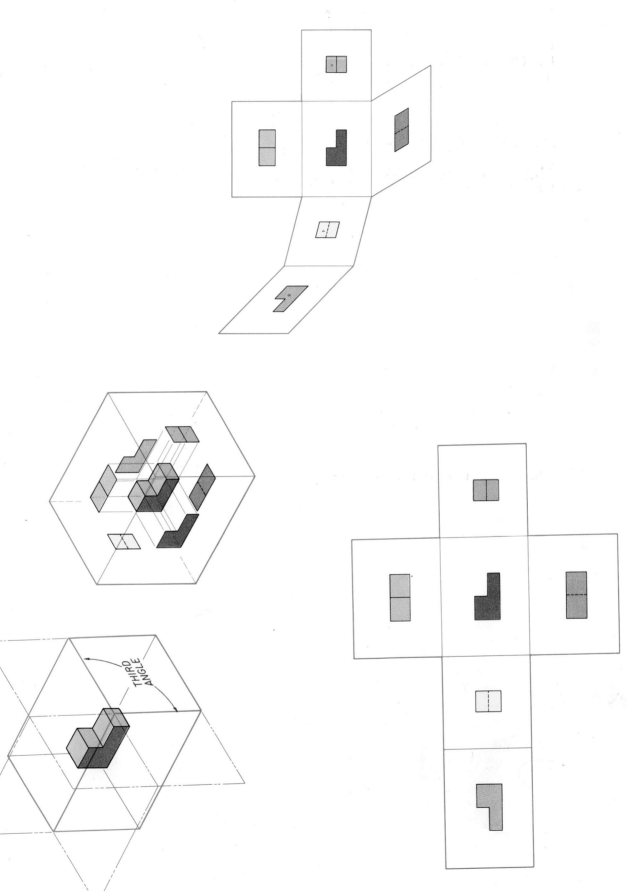

Fig. 8-6. A graphic explanation of third angle projection.

Fig. 8-7. A graphic explanation of first angle projection.

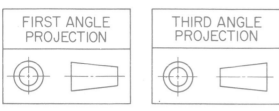

Fig. 8-8. The appropriate ISO symbol is placed on a drawing to show which method of projection was used on that sheet.

number of hidden lines are used only when absolutely necessary. The use of too many hidden lines on a drawing tends to make the drawing confusing to the person reading the drawing or fabricating the part. Use another view, without as many hidden lines, as in Fig. 8-10. Remember, the goal is to communicate clearly.

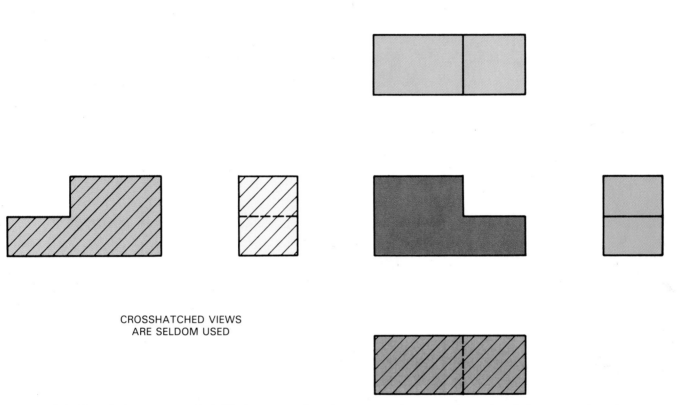

CROSSHATCHED VIEWS
ARE SELDOM USED

Fig. 8-9. Not all views are needed. Eliminate any view that repeats the same shape description as another view.

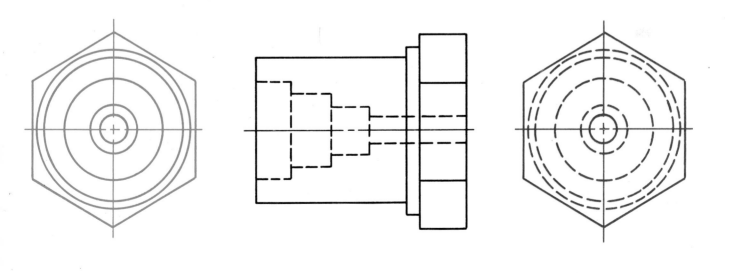

PREFERRED

AVOID

Fig. 8-10. Views showing a large number of hidden lines are used only if absolutely necessary. Too many hidden lines tend to make the drawing confusing.

TRANSFERRING POINTS

Each view will show a minimum of two dimensions. Any two views of an object will have at least one dimension in common. Time can be saved if a dimension from one view is projected to the other view instead of measuring the dimension a second time, Fig. 8-11. Transfer the points with construction lines.

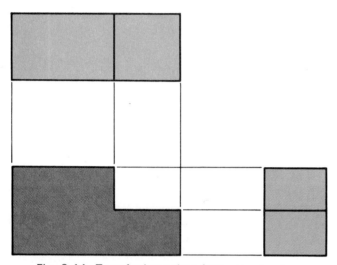

Fig. 8-11. Transferring points from view to view.

Additional time can be saved in transferring the depth of the top view to the side view. Two methods of projection are shown in Fig. 8-12. Projection provides for greater accuracy in the alignment of the views. It is faster than measuring each view separately with a scale or dividers.

HOW TO CENTER A DRAWING ON THE SHEET

A drawing looks more professional if the views are evenly spaced and centered on the drawing sheet. Centering the views on a sheet is not difficult if the following procedure is used:

1. Examine the object to be drawn. Observe its dimensions—width, depth, and height, Fig. 8-13. Determine the position in which the object will be drawn.
2. Measure the working area of the sheet AFTER the border and title block have been drawn. It should measure 7 in. by 10 in., if you used an 8 1/2 in. by 11 in. drawing sheet. Refer to how to prepare a drawing sheet on page 70.
3. Allow one inch between views.
4. To locate the front view, add the width of the front view, one inch spacing between views,

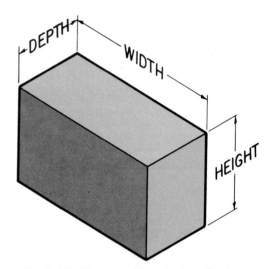

Fig. 8-13. How an object is described.

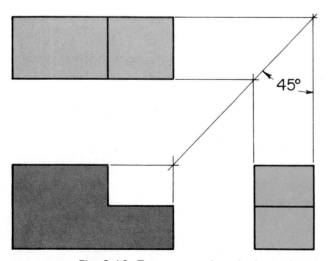

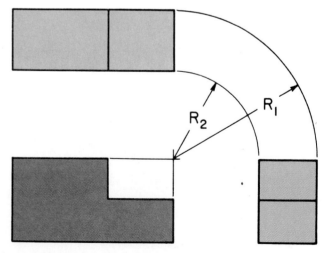

Fig. 8-12. Two accepted methods employed to transfer the depth of the top view to the side view.

and the depth of the right side view.

Subtract this total from the horizontal width of the working surface (10 in.). Divide this answer by 2. This will be the starting point for laying out the sheet horizontally.

Using the object shown in Fig. 8-14, for example, it would be as follows:

Width of front view = 5 in.
Space between views = 1 in.
Depth of right side view = 1 1/2 in.
Total = 7 1/2 in.

Width of working
surface = 10 in.
Total dimensions of
views and spacing = −7 1/2 in.
Result = 2 1/2 in.

Divide 2 1/2 inches by 2 = 1 1/4 in.
This is the distance in from left border line to locate starting point for drawing.

5. Measure in 1 1/4 in. from the left border line. Draw a vertical construction line through this point.

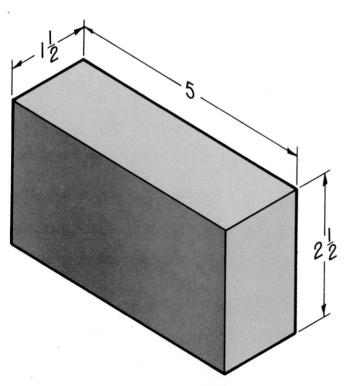

Fig. 8-14. Object used as an example for centering views on a drawing sheet.

6. From the above line, measure over a distance equal to the width of the front view. Draw another vertical construction line. See Fig. 8-15.

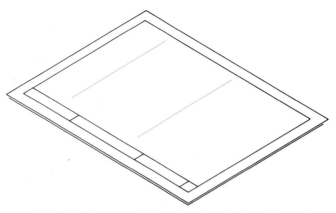

Fig. 8-15. The first step in locating the front and top views on the drawing sheet.

7. The same procedure is followed to center the views vertically. The height of the front view and the depth of the top view are used. A one inch space will separate the views. Add these distances together. Subtract the sum from the vertical working space (7 in.). Divide the answer by 2.

For example:

Height of front view = 2 1/2 in.
Space between views = 1 in.
Depth of top view = 1 1/2 in.
Total = 5 in.

Height of working
surface = 7 in.
Total dimensions of
views and spacing = −5 in.
Result = 2 in.

Divide 2 in. by 2 = 1 inch
This is the distance up from lower border line to locate starting point for drawing.

8. Measure up 1 in. from the lower border line. Draw a horizontal construction line through this point.

9. From the above line, measure up the height of the front view and mark a point. Mark a point at the one inch spacing that separates

the views. Mark one more point for the depth of the top view. Draw construction lines through these points. See Fig. 8-16.

10. Use either the 45 deg. angle method or the radius method to transfer the depth of the top view to the right side of the object, Fig. 8-17.

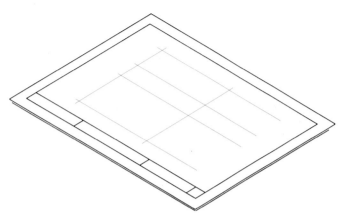

Fig. 8-16. The front and top views blocked in with construction lines.

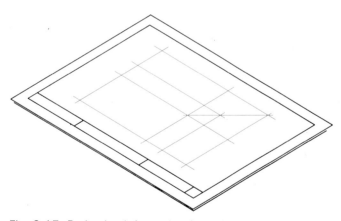

Fig. 8-17. Projecting information from the top view to block in the right side view.

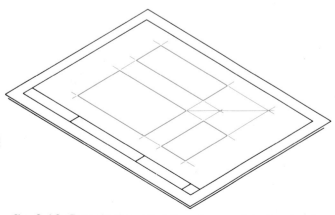

Fig. 8-18. Draw in the object lines to complete the drawing. Construction lines may be erased.

11. Draw in the right side view. Use construction lines.
12. Complete the drawing by going over the construction lines. Use the correct weight for the type of line (object line, hidden object line, center line, etc.), Fig. 8-18.

DRAFTING VOCABULARY

Direction of sight, First angle projection, Horizontal, ISO, Mechanical drawing, Multiview drawing, Orthographic projection, Shape description, Third angle projection, Three dimensional, Transferring points, Vertical.

TEST YOUR KNOWLEDGE—UNIT 8

Please do not write in the text. Place your answers on a sheet of notebook paper.

1. A drawing is said to be a MECHANICAL DRAWING when it is drawn using _____.
2. What does the term CADD mean?
3. A drawing that uses two or more views to describe an object is known as a _____ _____.
4. In developing the needed views, the object being drawn is normally viewed from six directions. What are the six views that will be seen?
5. The method employed to develop these six views is called _____ _____.
6. Each of the views will show a minimum of _____ dimensions. Any two of the views will have at least _____ dimension in common.
7. Orthographic projection permits _____ dimensional objects to be shown on a flat surface having only _____ dimensions.

OUTSIDE ACTIVITIES

1. Collect props for the class to draw using instruments. One prop should require only a two-view drawing; another prop should require a three-view drawing. Find other props which require more than three views to give a complete shape description.
2. Build a hinged box out of clear plastic which can be used to demonstrate the unfolding of an object into its multiview parts; the front, top, bottom, and sides. Place a prop inside the

box, trace the profile of the object on the side of the plastic box with chalk, then unfold the box to show the multiview projections.

3. Make a large poster for your drafting room showing the step-by-step procedure to follow in centering a drawing on a sheet.

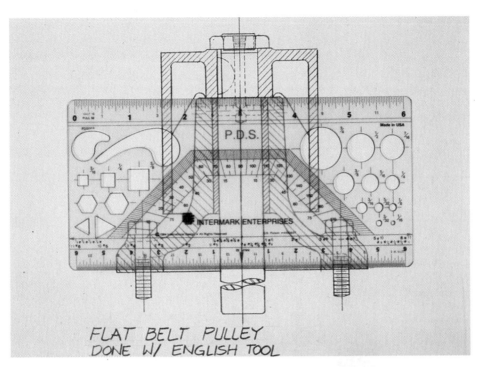

FLAT BELT PULLEY
DONE W/ ENGLISH TOOL

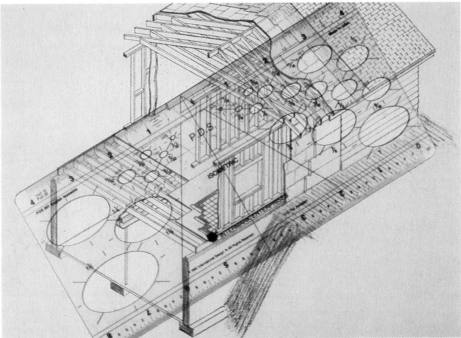

Typical examples of a compact, portable drawing system which allows a person to make accurate drawings. The clear plastic lets you look through the tool to line up marks and lines on the paper with marks and lines on the tool. With this visual referencing, you can construct parallel and perpendicular lines, 45 deg. and 90 deg. angles. Used in conjunction with the scales on the tool's edges, the protractor, and the beam compass, all angles, line lengths, and circles can be constructed. (Intermark Enterprises)

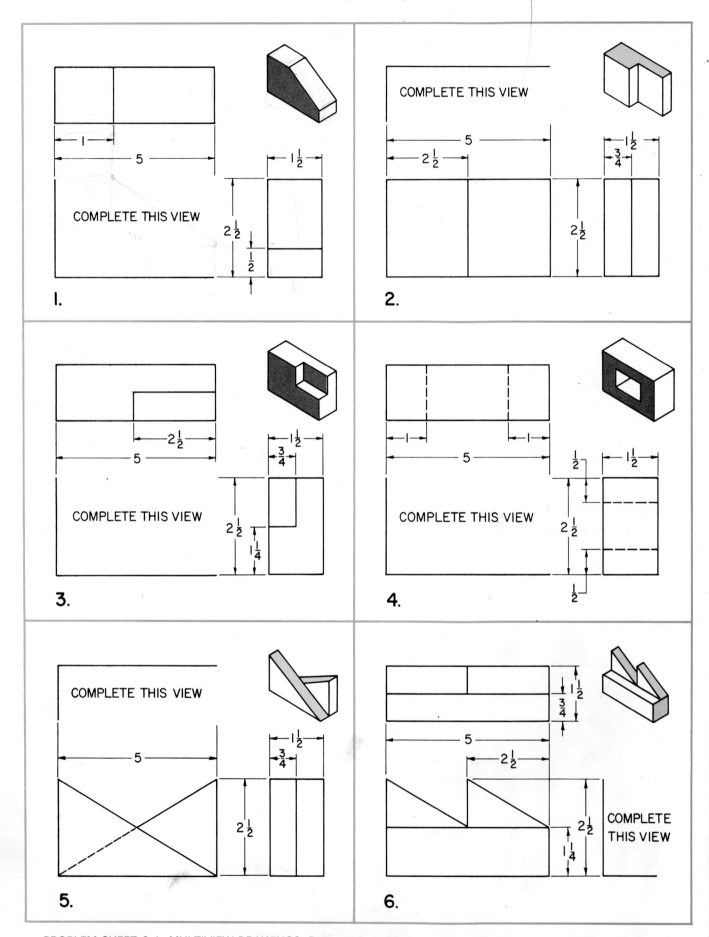

1.

2. COMPLETE THIS VIEW

3.

4.

5. COMPLETE THIS VIEW

6.

PROBLEM SHEET 8-1. MULTIVIEW DRAWINGS. Draw each problem on a separate sheet and complete as indicated.

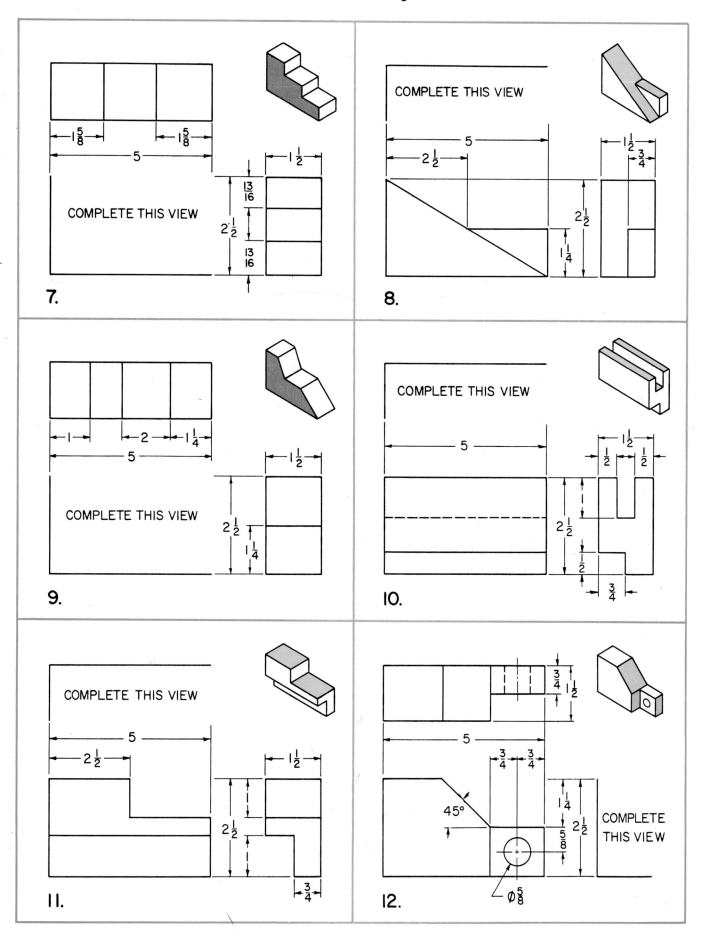

7.

$1\frac{5}{8}$ $1\frac{5}{8}$ 5 $1\frac{1}{2}$

COMPLETE THIS VIEW

$\frac{13}{16}$ $2\frac{1}{2}$ $\frac{13}{16}$

8.

COMPLETE THIS VIEW

5 $2\frac{1}{2}$ $1\frac{1}{2}$ $\frac{3}{4}$

$2\frac{1}{2}$ $1\frac{1}{4}$

9.

1 2 $1\frac{1}{4}$ 5 $1\frac{1}{2}$

COMPLETE THIS VIEW

$2\frac{1}{2}$ $1\frac{1}{4}$

10.

COMPLETE THIS VIEW

5 $1\frac{1}{2}$ $\frac{1}{2}$ $\frac{1}{2}$

$2\frac{1}{2}$ $\frac{1}{2}$ $\frac{3}{4}$

11.

COMPLETE THIS VIEW

5 $2\frac{1}{2}$ $1\frac{1}{2}$

$2\frac{1}{2}$ $\frac{3}{4}$

12.

5 $\frac{3}{4}$ $1\frac{1}{2}$

$\frac{3}{4}$ $\frac{3}{4}$

45°

$1\frac{1}{4}$ $2\frac{1}{2}$ $\frac{5}{8}$

COMPLETE THIS VIEW

$\varnothing\frac{5}{8}$

PROBLEM SHEET 8-2. MULTIVIEW DRAWINGS. Draw each problem on a separate sheet and complete as indicated.

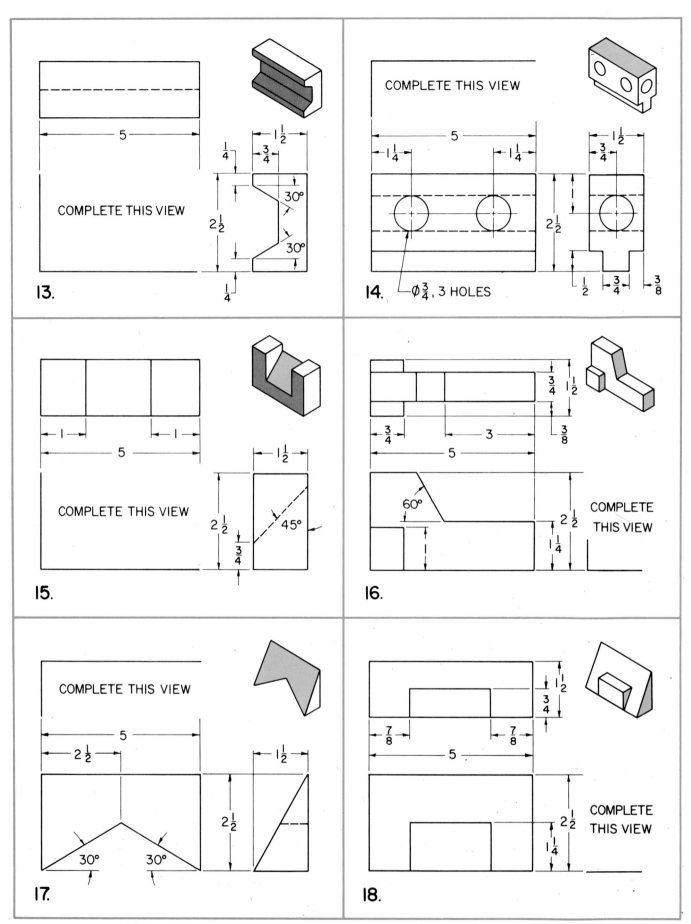

13.

14. ⌀ 3/4, 3 HOLES — COMPLETE THIS VIEW

15.

16.

17.

18.

PROBLEM SHEET 8-3. MULTIVIEW DRAWINGS. Draw each problem on a separate sheet.
Draw as many views as necessary to fully describe each problem.

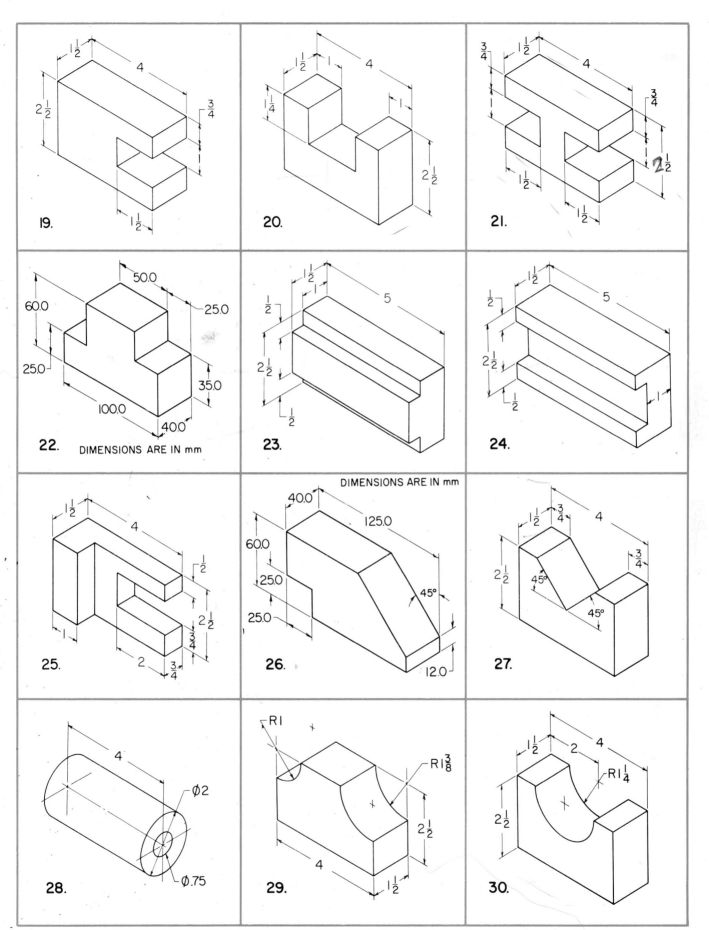

PROBLEM SHEET 8-4. MULTIVIEW DRAWINGS. Draw each problem on a separate sheet. Draw as many views as necessary to fully describe each problem.

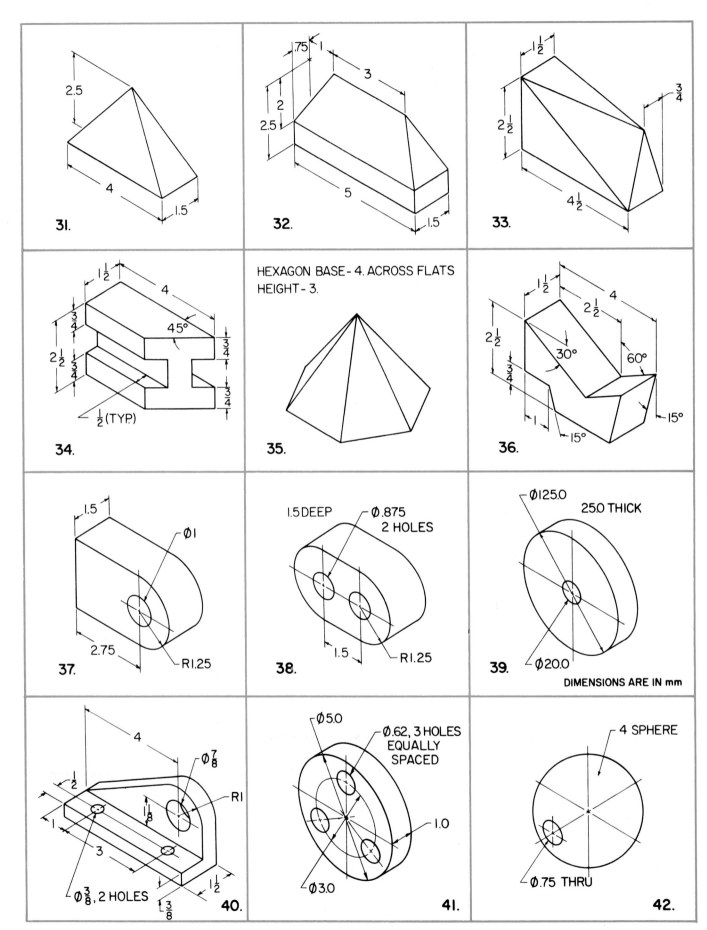

31.

32.

33.

34.

HEXAGON BASE - 4. ACROSS FLATS
HEIGHT - 3.

35.

36.

37.

38.

39.

DIMENSIONS ARE IN mm

40.

41.

42.

PROBLEM SHEET 8-5. MULTIVIEW DRAWINGS. Draw each problem on a separate sheet.
Draw as many views as necessary to fully describe each problem.

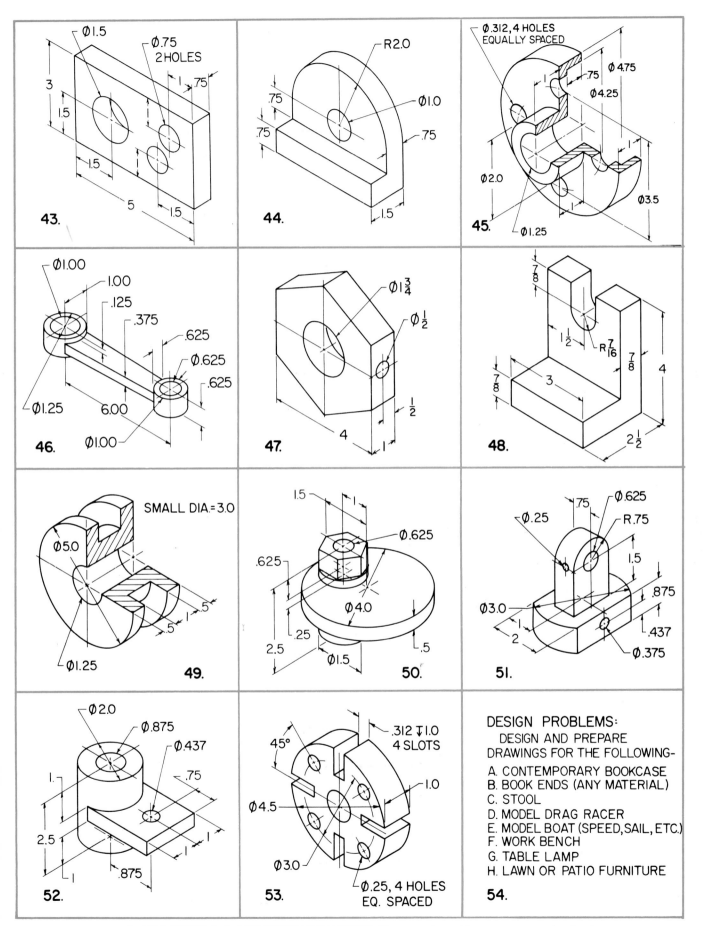

43.

44.

45.

46.

47.

48.

49.

50.

51.

52.

53.

54.

DESIGN PROBLEMS:
DESIGN AND PREPARE
DRAWINGS FOR THE FOLLOWING-

A. CONTEMPORARY BOOKCASE
B. BOOK ENDS (ANY MATERIAL)
C. STOOL
D. MODEL DRAG RACER
E. MODEL BOAT (SPEED, SAIL, ETC.)
F. WORK BENCH
G. TABLE LAMP
H. LAWN OR PATIO FURNITURE

PROBLEM SHEET 8-6. MULTIVIEW DRAWINGS. Draw each problem on a separate sheet.
Draw as many views as necessary to fully describe each problem.

Unit 9

DIMENSIONING AND NOTES

After reading this unit, you will be able to state how dimensions give a size description of an object. You will be able to explain the methods of placing dimensions and notes on a drawing. You will be able to apply general rules for dimensioning. You will be able to dimension circles, holes, and arcs. You will be able to describe five methods used to dimension drawings in metrics.

If an object is to be manufactured according to the designer's specifications, the person crafting the product usually needs more information than that furnished by a scale drawing of its shape. DIMENSIONS and SHOP NOTES are needed, Fig. 9-1.

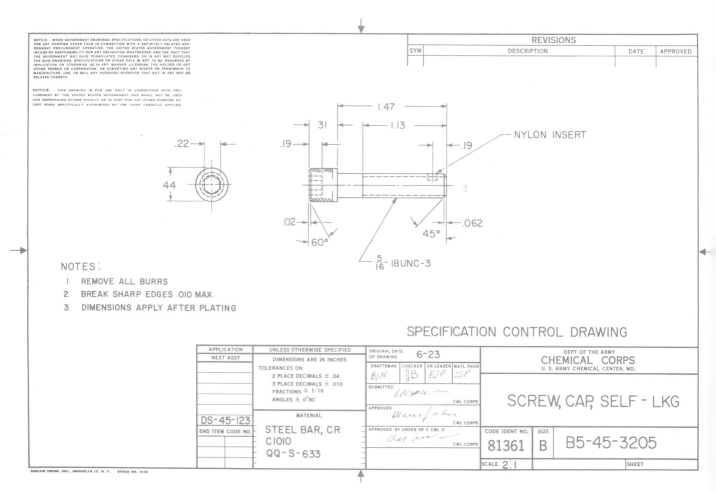

Fig. 9-1. Industry drawing showing dimensions and shop notes.

DIMENSIONS define the sizes of the geometrical features of an object. Dimensions and notes are added by hand or by using the keyboard in CAD applications. They are presented in appropriate units of measure—inches, millimeters, etc. The latest drafting standards recommend decimal inch dimensioning. Decimals are easier to add, subtract, multiply, and divide. However, fractional inch dimensioning is still widely employed. Both types will be found in this text.

Metric drawings are usually dimensioned in millimeters (mm).

NOTES provide additional information not found in the dimensions.

READING DIRECTION FOR DIMENSIONS

Dimensions are placed on the drawing in either an UNIDIRECTIONAL or an ALIGNED manner, Fig. 9-2. Unidirectional dimensioning is preferred.

Unidirectional dimensions are placed to read from the bottom of the drawing.

Aligned dimensions are placed parallel to the dimension line. The numerals are read from the bottom and from the right side of the drawing.

Regardless of the method used, dimensions shown with leaders and all notes are lettered parallel to the bottom of the drawing.

DIMENSIONING A DRAWING

From your study of the ALPHABET OF LINES, you will remember that special lines are used for dimensioning, Fig. 9-3. The dimension line is a fine solid line used to indicate distance and location. It should be fine enough to contrast with the object lines. It is broken for the insertion of the dimension. The line is capped with arrowheads.

The line that extends out from the drawing is called an extension line. It projects about 1/8 in. (3.0 mm) beyond the last dimension line. It should not touch the drawing. The smaller or detail dimensions are nearest the view, while the larger or overall dimensions are farthest from the view.

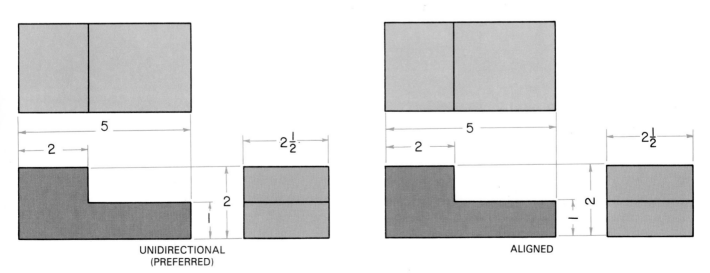

Fig. 9-2. The two accepted methods for dimensioning drawings. The UNIDIRECTIONAL method is preferred.

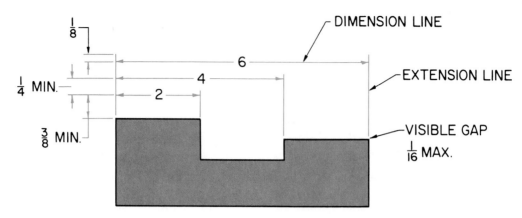

Fig. 9-3. The lines used for dimensioning. Note how they contrast with the object lines.

Fig. 9-4. Arrowheads are drawn freehand. The solid arrowhead is generally preferred.

Arrowheads are drawn freehand and should be carefully made, Fig. 9-4. The solid arrowhead is generally preferred. It is made narrower and slightly longer than the open arrowhead. For most applications, arrowheads 1/8 in. (3.0 mm) long are satisfactory. Rub-on arrowheads are often used by the drafter to save time.

GENERAL RULES FOR DIMENSIONING

To be easily understood, dimensions should conform to the following general rules:
1. Place dimensions on the views that show the true shape of the object, Fig. 9-5.
2. Unless absolutely necessary, dimensions should not be placed within the views.
3. If possible, dimensions should be grouped together rather than scattered about the drawing, Fig. 9-6.
4. Dimensions must be complete, so no scaling of the drawing is required. It should be possible to determine sizes and shapes without assuming any measurements.
5. Draw dimension lines parallel to the direction of measurement. If there are several parallel dimension lines, the numerals should be staggered to make them easier to read, Fig. 9-7.

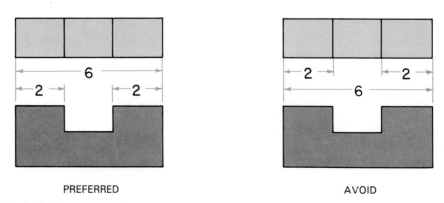

PREFERRED   AVOID

Fig. 9-5. Keep the dimensions on the view that shows the true shape of the object.

136

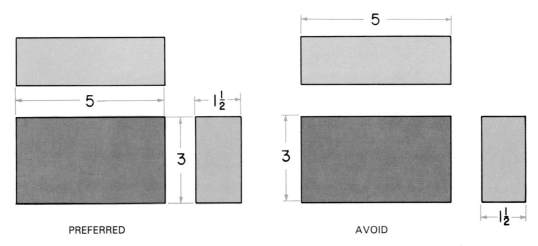

Fig. 9-6. Keep dimensions grouped for easier understanding of the drawing.

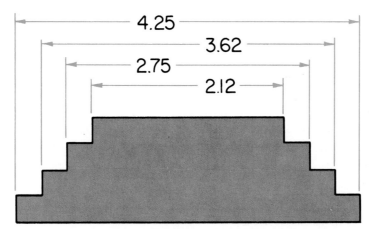

Fig. 9-7. When there are several parallel dimension lines, numerals should be staggered to make them easier to read.

6. Dimensions should not be duplicated unless they are absolutely necessary to the understanding of the drawing. Omit all unnecessary dimensions, Fig. 9-8.
7. Plan your work carefully so the dimension lines do not cross extension lines, Fig. 9-9.
8. When all dimensions on a drawing are in inches, the inch symbol (") should not be used. Dimensions on metric drawings are in millimeters unless otherwise noted.
9. Numerals and fractions must be drawn in proper relation to one another. See Fig. 9-10.

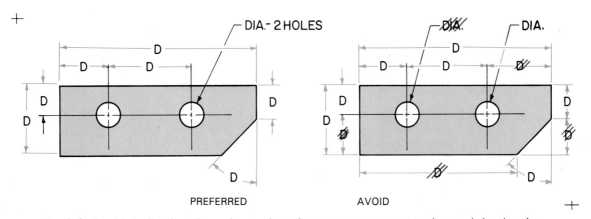

Fig. 9-8. Avoid duplicating dimensions unless they are necessary to understand the drawing.

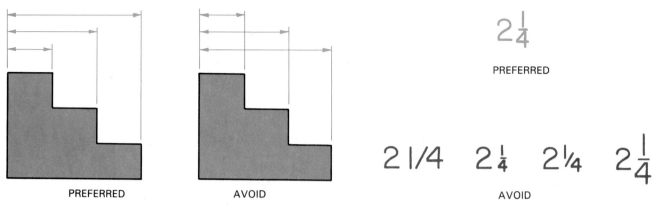

PREFERRED

AVOID

Fig. 9-9. Line crossing can be kept to a minimum if the shortest dimension lines are placed next to the object outline.

Fig. 9-10. Numerals and fractions must be drawn in proper size relation to one another.

DIMENSIONING CIRCLES, HOLES, AND ARCS

Many products are manufactured from parts that contain circles, round holes, or arcs in their design, Fig. 9-11. They are often produced by turning, boring, drilling, reaming, counterboring, spotfacing, and countersinking, Fig. 9-12.

Fig. 9-11. These jet engine castings are typical of manufactured parts that contain circles, round holes, and arcs in their design. (Precision Castparts Corp.)

Fig. 9-12. Machining operations which produce circles or holes. A—Reaming produces a very accurately sized hole. The hole is first drilled slightly undersize before reaming. B—Boring is an internal machining operation. C—Turning work on a lathe. D—Spotfacing machines a surface that permits a bolt head or nut to bear uniformly over its entire surface. E—Drilling is an operation often performed on a drill press. F—Counterboring prepares a hole to receive a fillister or socket head screw. (Clausing)

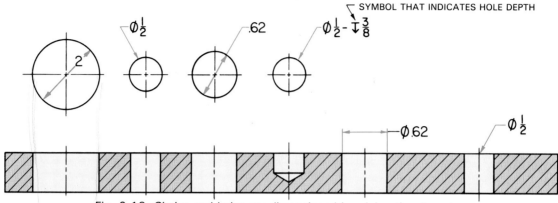

Fig. 9-13. Circles and holes are dimensioned by giving the diameter.

The American National Standards Institute (ANSI) has set standards on how to call out a diameter or a radius. The Greek letter ϕ (phi, pronounced fi) indicates that the dimension is a diameter. The symbol is placed BEFORE the dimension. Circles and round holes are dimensioned as shown in Fig. 9-13.

Where it is not clear that a hole goes through the part, the abbreviation THRU follows the dimension. Symbols are recommended to show hole depth, and whether the hole is to be spotfaced, counterbored, or countersunk. See Fig. 9-14. Also refer to page 347 in Useful Information.

If the diameters of several concentric circles must be dimensioned on a drawing, it may be more convenient to show them on the front view, Fig. 9-15.

The correct way to use a leader to indicate a diameter is shown in Fig. 9-16. The leader ALWAYS points to the center of the diameter.

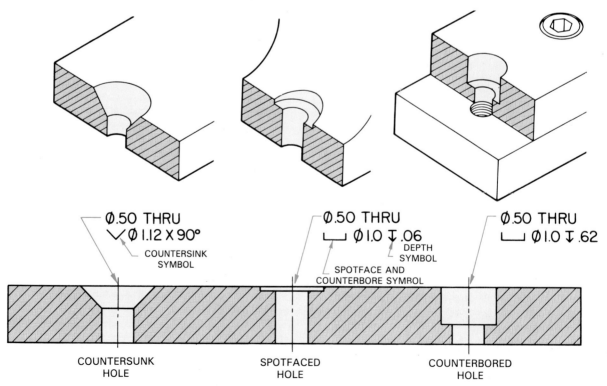

Fig. 9-14. Where it is not clear that a hole goes through the part, the abbreviation THRU follows the dimension. Symbols indicate hole depth, and whether the hole is spotfaced, counterbored, or countersunk.

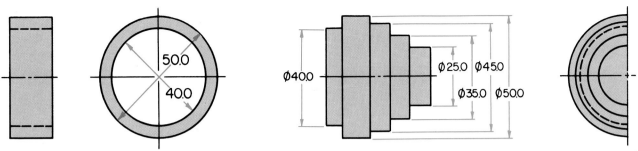

Fig. 9-15. Recommended ways to dimension concentric circles. Note that when the method to the right is used, only half a right side view is needed.

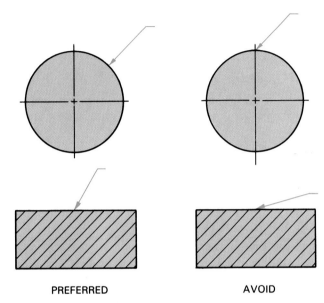

PREFERRED   AVOID

Fig. 9-16. A leader is employed to direct attention to a note or to indicate sizes of arcs and circles. When used with arcs and circles, the leader should radiate from their centers.

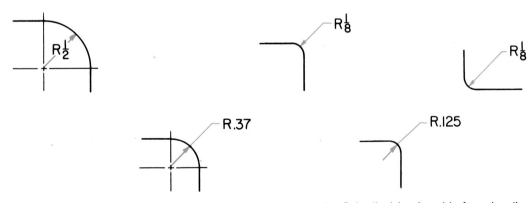

Fig. 9-17. Recommended ways for dimensioning arcs. Note that the R (radius) is placed before the dimension.

Arcs are dimensioned by giving the radius, Fig. 9-17. The capital letter R indicates that the dimension is a radius. It is placed BEFORE the dimension.

Round holes and cylindrical parts are dimensioned from centers, never from the edges, Fig. 9-18.

When it is necessary to dimension a series of holes around a circle, use a note to designate the number of holes, their size, and the diameter of the circle on which they are located, Fig. 9-19.

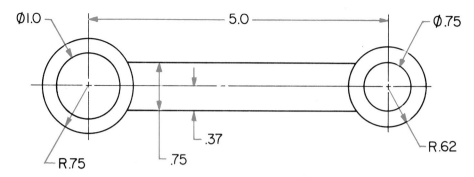

Fig. 9-18. Where feasible, round holes and cylindrical parts are dimensioned from centers.

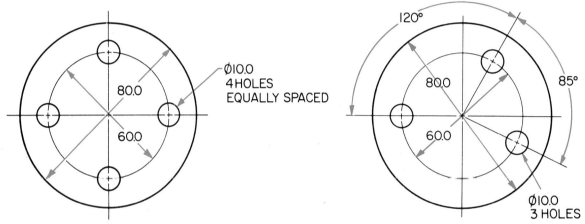

Fig. 9-19. Left. The correct way to dimension equally spaced holes. Right. Holes that are not equally spaced are dimensioned this way.

DIMENSIONING ANGLES

Angular dimensions are expressed in degrees (°), minutes ('), and seconds ("). Angles are dimensioned as shown in Fig. 9-20.

DIMENSIONING SMALL PORTIONS OF AN OBJECT

When the space between extension lines is too small for both the numerals and the arrowheads, dimensions are indicated as shown in Fig. 9-21.

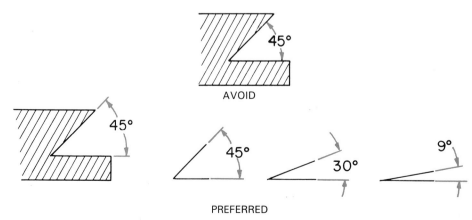

Fig. 9-20. Dimensioning angles.

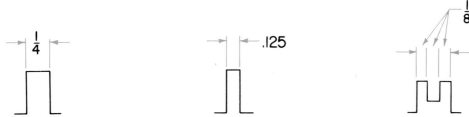

Fig. 9-21. When the space between extension lines is too small to place both arrowheads and numerals, dimensions may be indicated as shown.

THE CHANGEOVER FROM CONVENTIONAL MEASUREMENT TO METRIC MEASUREMENT

The changeover from conventional measurement (inch, foot, yard, etc.) to metric measurement has been very expensive and no one is sure when it will be complete. One problem has been at the INTER-FACE (surfaces where parts come together) of a part designed to metric standards that must fit an existing part that was designed to inch standards.

Five methods have been devised for drafters to provide the information necessary to make such a part.
1. Dual dimensioning.
2. Dimensioning with letters and tabular chart.
3. Metric dimensioning with readout chart.
4. Dimensioning with metric units only.
5. Undimensioned master drawings.

DUAL DIMENSIONING

The use of DUAL DIMENSIONING was the first method devised to dimension engineering drawings with both inch and metric units, Fig. 9-22.

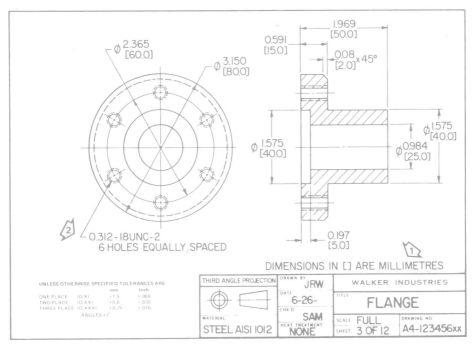

Fig. 9-22. DUAL DIMENSIONED drawing. Dual dimensioning has limited use. It was the first method devised to dimension engineering drawings with both inch and metric units. Notice the following important points on the drawing indicated by arrows. 1. Note indicating how metric dimensions are identified. 2. Thread size is not given in metric units because there is no metric thread this size.

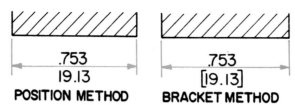

POSITION METHOD BRACKET METHOD

Fig. 9-23. Methods of indicating inches and millimeters on a dual dimensioned drawing.

The dual dimensions were presented by the POSITION METHOD or the BRACKET METHOD, Fig. 9-23. The inch dimension was placed first on drawings of products to be made in the United States. The metric dimension was placed first on drawings to be produced where metrics was the basic form of measurement.

Dual dimensioning is the most complicated dimensioning system. It is seldom used. However, large numbers of dual dimensioned drawings are still in use. It is presented in this text so you will be aware of its existence.

DIMENSIONING WITH LETTERS

Dimensioning with letters, Fig. 9-24, is a technique that is sometimes used. Letters (A, B, C, etc.) are used in place of either inch or millimeter dimensions. A TABULAR CHART is added to the drawing. The chart shows the metric and inch equivalents of each letter.

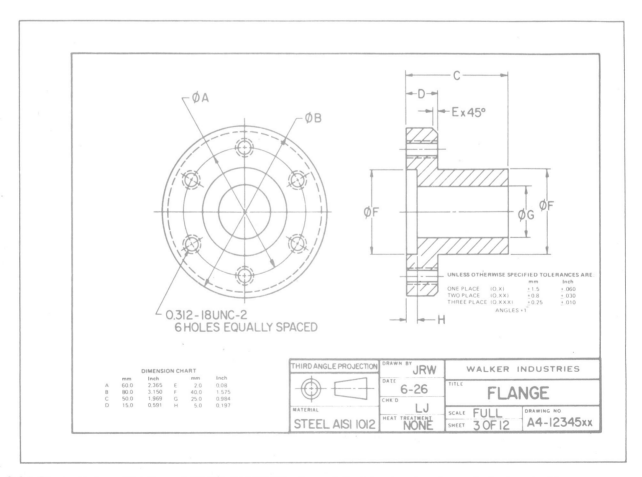

Fig. 9-24. Dimensioning with letters. A TABULAR CHART at lower left of the drawing shows the inch and millimeter equivalents for each letter.

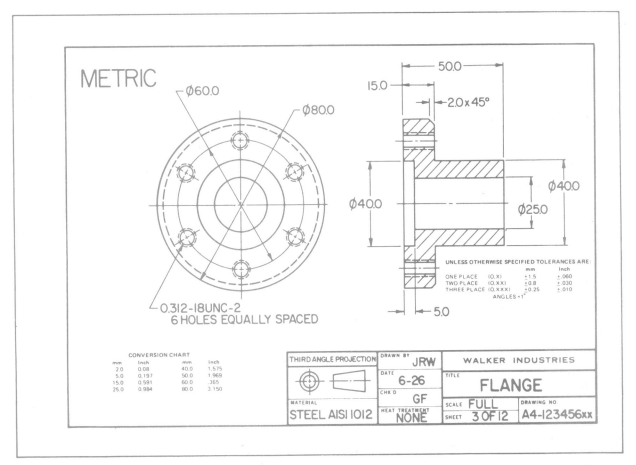

Fig. 9-25. Metric dimensioned drawing with READOUT CHART that shows inch equivalents to the metric dimensions.

METRIC DIMENSIONING WITH READOUT CHART

In doing metric dimensioning with a READOUT CHART, the part is designed to metric standards. Only metric dimensions are placed on the drawing, Fig. 9-25. A readout chart is added to the drawing. It shows metric dimensions in the left column and the inch equivalents in the right column. This technique permits a comparison of values.

DIMENSIONS WITH METRIC UNITS ONLY

Dimensioning with only metric units is the quickest way to get engineers, designers, drafters, and craftworkers to "think metric." Only metric dimensions are used on the drawing. See Fig. 9-26.

UNDIMENSIONED MASTER DRAWINGS

A MASTER DRAWING is first made without dimensions, Fig. 9-27. Next, prints are made. Metric dimensions may be added to one print, inch dimensions to another print. Notes and details can be added in whatever language is needed to produce the part—German, French, English, Japanese, etc.

Whatever dimensioning technique is used, drafting personnel must have both scales and templates on hand. They must also have a thorough knowledge of the metric system.

Drafters, engineers, or designers will find they cannot specify a 9/16 bolt by merely listing the metric equivalent of 9/16 in. (15.29 mm). There is no metric bolt that corresponds to this diameter. The same

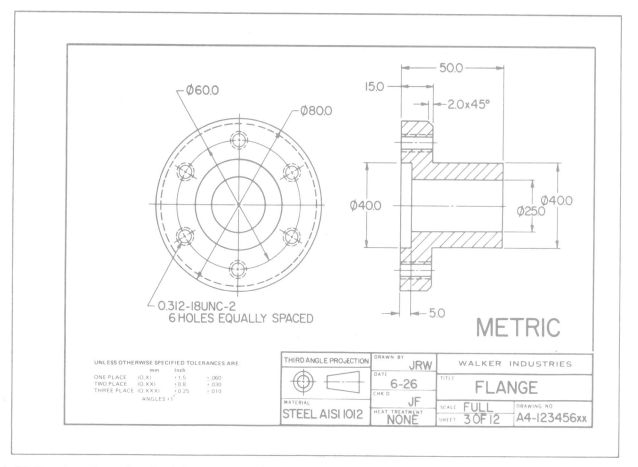

Fig. 9-26. Drawings dimensioned only in metric units. (Since there is NO metric equivalent for the thread indicated on the drawing, thread size is given in inch units.)

problem will occur if a 1.0 inch diameter is specified as a 25.4 mm diameter. There is no metric sized shaft 1.0 inch in diameter. The closest size would be 25.0 mm diameter. To get the 25.4 mm diameter shaft, an expensive machining operation would be necessary to turn a larger shaft to the required 25.4 mm diameter.

GENERAL RULES FOR METRIC DIMENSIONING

Metric dimensioning should conform to the following rules:
1. The millimeter is the standard metric unit for dimensioning engineering drawings.
2. Millimeter dimensioning on engineering drawings is based on the use of one-place decimals. All metric dimensioning should have one digit to the right of the decimal point. Two and three digits are used where critical tolerances are required. A zero will be shown to the right of the decimal point when full millimeter dimensions are shown.

125.0, not 125

3. When a dimension is less than one millimeter, a zero will be shown to the left of the decimal point.

0.5, not .5

4. All metric drawings should be clearly identified as such. The symbol for millimeter (mm) does not have to be added to every dimension if one of the notes shown in Fig. 9-28 is placed on the drawing.
5. When the millimeter symbol is used on a drawing, a space is placed between the dimension and the symbol.

125.0 mm, not 125.0mm

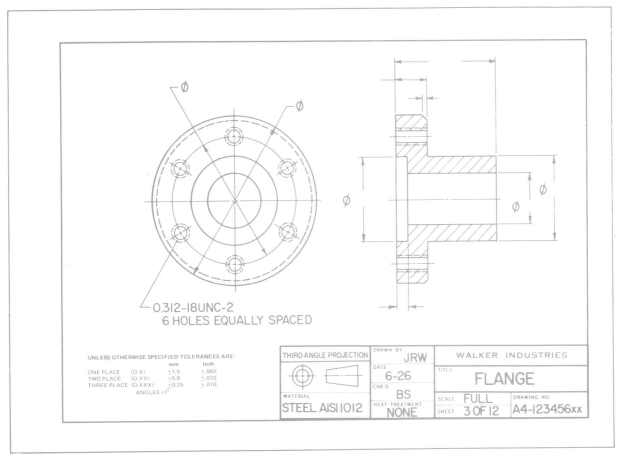

Fig. 9-27. No dimensions are shown on the MASTER DRAWING (except for thread size). After the print is made, dimensions in either inch or metric units are added.

ALL DIMENSIONS ARE IN MILLIMETERS.

Fig. 9-28. Metric drawings should be clearly identified by use of either of the above notes.

DRAFTING VOCABULARY

Aligned, ANSI, Appropriate, Arc, Arrowhead, Circle, Concentric, Counterboring, Countersinking, Cylindrical, Degrees, Diameter, Dimensions, Inch, Insertion, Interface, Leader, Millimeter, Minutes, Notes, Omit, Parallel, Radius, Scale drawing, Scaling, Seconds, Series, Specifications, Spotfacing, Staggered, Symbol, Unidirectional.

TEST YOUR KNOWLEDGE—UNIT 9

Please do not write in the book. Place your answers on a piece of notebook paper.
1. Dimensions and shop notes are needed if _____.
2. Dimensions define _____.

3. Notes provide _____.
4. Unidirectional dimensions are read from _____.
5. Aligned dimensions are read from _____.
6. The dimension line is a: (Check the correct answer/s.)
   ____ a. Heavy solid line.
   ____ b. Fine solid line.
   ____ c. Fine dotted line.
   ____ d. None of the above.
7. The dimension line is capped with _____.
8. Dimensions are placed on views that show _____.
9. Dimensions should be _____ _____ rather than _____ about the drawing.
10. Dimensions must be complete so that: (Check the correct answer/s.)
    ____ a. No scaling of the drawing is necessary.
    ____ b. Sizes and shapes can be determined without assuming any measurements.
    ____ c. Both of the above.
    ____ d. None of the above.
11. When all dimensions are in inches the _____ _____ is not used.
12. Circle diameter is indicated by the symbol _____. It is placed _____ the dimension.
13. Sketch the symbols used to indicate the following. Label each symbol on your answer sheet.
    a. Hole depth.

    b. Countersunk hole.

    c. Counterbored or spotfaced hole.

14. When a leader is used to indicate a diameter it always points to the _____ of the diameter.
15. The letter _____ indicates the radius of an arc. It is placed _____ the dimension.
16. Define a dual dimensioned drawing.
17. Some drawings are dimensioned with letters. Where does the person using the drawing get the inch/millimeter sizes needed to make the part?
18. What is a master drawing and how is it used?

OUTSIDE ACTIVITIES

1. Contact several industrial drafting departments. If possible, secure copies of the standards they use for dimensioning. Compare how similar or dissimilar they are. Do they vary by industry?
2. Secure copies of industrial prints. Study how they are dimensioned and how notes are used. Are there any provisions for foreign components to be used with the items on the prints? Will the parts be exported?

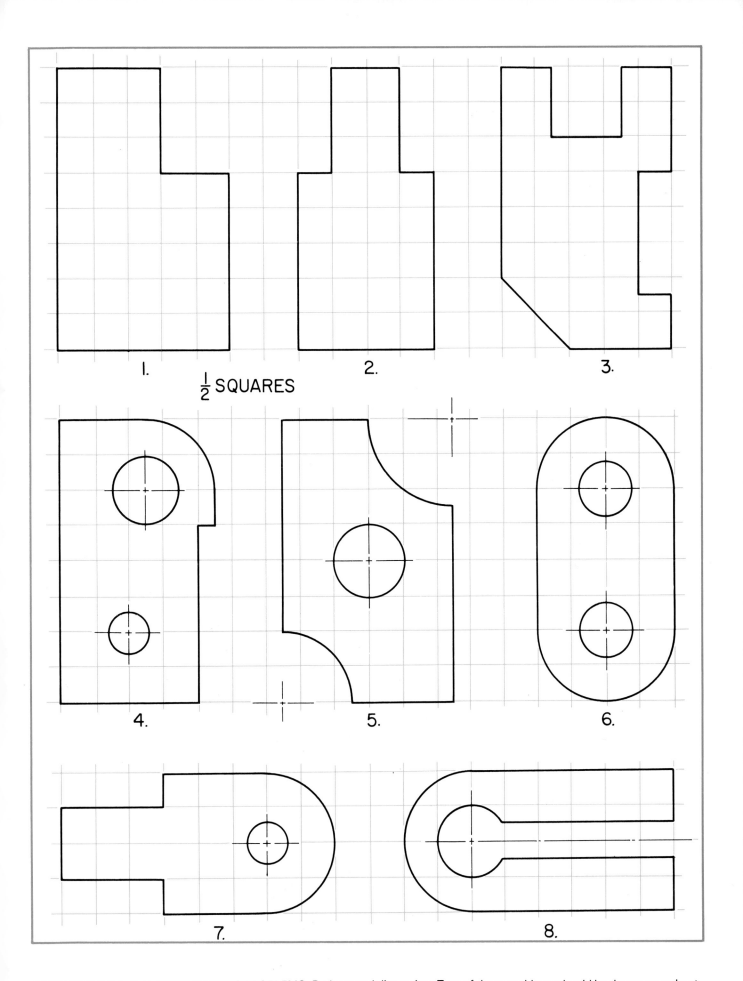

1.

$\frac{1}{2}$ SQUARES

2.

3.

4.

5.

6.

7.

8.

PROBLEMS 9-1 to 9-8. DIMENSIONING PROBLEMS. Redraw and dimension. Two of these problems should be drawn on a sheet.

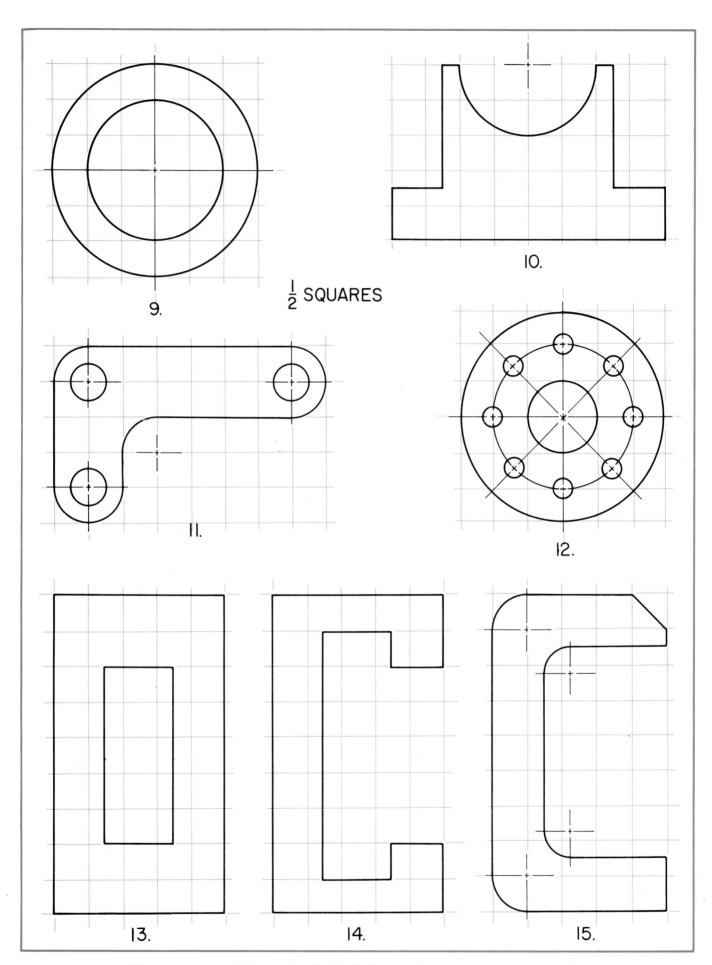

$\frac{1}{2}$ SQUARES

9.

10.

11.

12.

13.

14.

15.

PROBLEMS 9-9 to 9-15. DIMENSIONING PROBLEMS. Redraw and dimension problems.

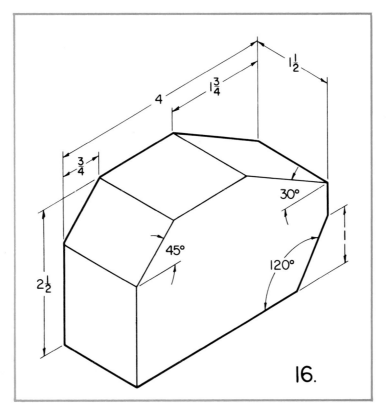

PROBLEM 9-16. SHIM. Prepare the necessary views and correctly dimension them.

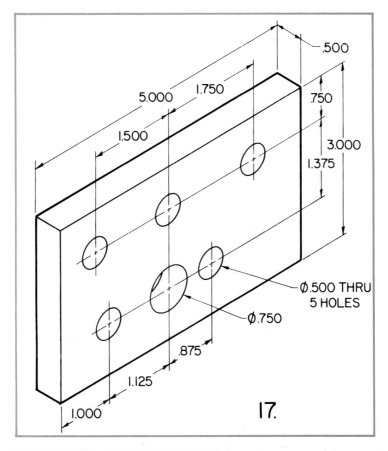

PROBLEM 9-17. ALIGNMENT PLATE. Prepare the necessary views and correctly dimension them.

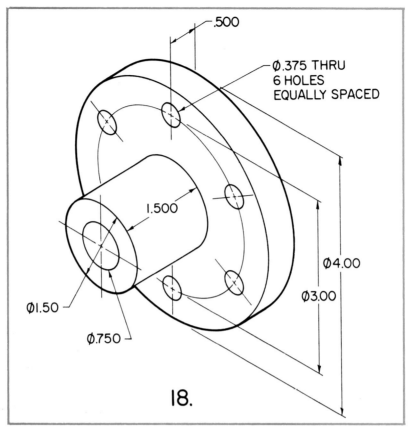

.500

Ø.375 THRU
6 HOLES
EQUALLY SPACED

1.500

Ø4.00

Ø3.00

Ø1.50

Ø.750

18.

PROBLEM 9-18. FACE PLATE. Prepare a two view drawing and
dimension correctly.

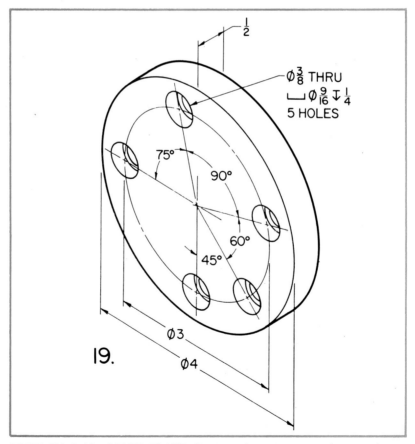

$\frac{1}{2}$

$\emptyset\frac{3}{8}$ THRU
$\sqcup\emptyset\frac{9}{16} \downarrow \frac{1}{4}$
5 HOLES

75°

90°

60°

45°

Ø3

Ø4

19.

PROBLEM 9-19. COVER PLATE. Prepare the necessary views and
correctly dimension them.

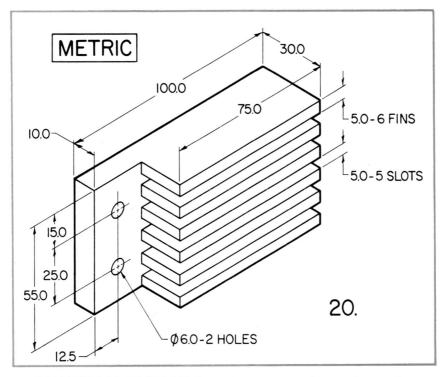

PROBLEM 9-20. HEAT SINK. Draw the necessary views and correctly
dimension them.

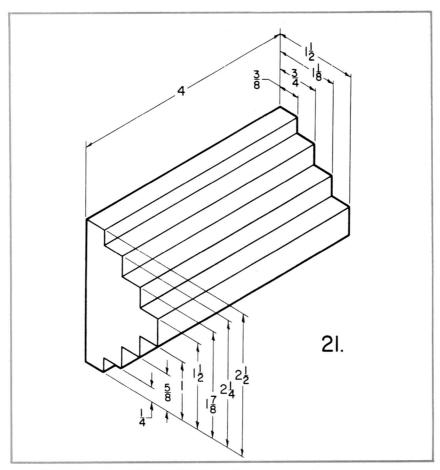

PROBLEM 9-21. STEP BLOCK. Draw the necessary views and correctly
dimension them.

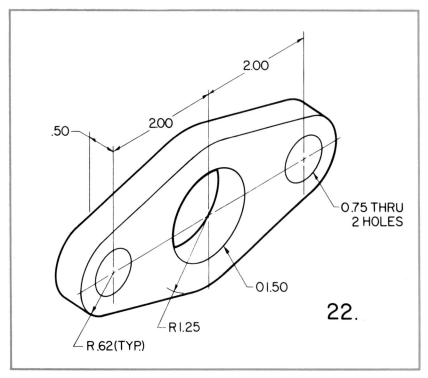

PROBLEM 9-22. GASKET. Prepare a two view drawing and dimension correctly.

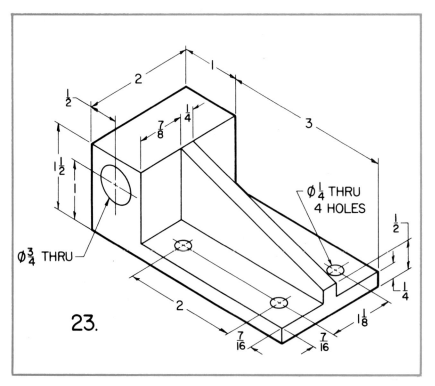

PROBLEM 9-23. MOTOR BRACKET. Draw the necessary views and correctly dimension them.

Unit 10

SECTIONAL VIEWS

After studying this unit, you will be able to identify when sectional views should be used. You will recognize designs with difficult internal features which would lead to a confusing jumble of hidden object lines. You will construct section lines. You will apply seven sectional views to various drawings: full sections, half sections, revolved sections, aligned sections, removed sections, offset sections, and broken art sections.

Sectional views permit complicated interior features to be shown on a drawing without a large number of confusing hidden object lines. Sectional views are also useful when explaining interior construction or hidden features which cannot be shown clearly by various outside views and hidden lines. This unit will provide a working knowledge of the principle types of sectional views. It will also explain when each type is to be used by the skilled drafter.

Details on objects that are simple in design can be shown in regular multiview drawings. When an object has some of its design features hidden from view, as in Fig. 10-1, it is another matter. It is not possible to show the interior structure without using a "jumble" of hidden lines. Sectional views are employed to make drawings of this type less confusing and easier to understand or visualize, Fig. 10-2.

SECTIONAL VIEWS show how an object would look if a cut were made through it perpendicular (at an exact right angle) to the direction of sight, Fig. 10-3. The section drawing should be placed behind the arrows. Sectional view drawings are necessary for a clear understanding of the shape of complicated parts.

CUTTING PLANE LINE

The CUTTING PLANE LINE indicates the location of an imaginary cut made through the object to reveal its interior structure, Fig. 10-4. The extended lines are capped with arrowheads that show the direction of sight to view the section. Some drafters prefer the use of a simplified cutting plane line that employs only the ends.

Letters A-A, B-B, etc., identify the section if it is moved to another position on the drawing. The same technique is used if several sections are used on a single drawing.

SECTION LINES

General purpose section lining is often employed on drawings where the material specifications ("specs") are shown elsewhere on the drawing.

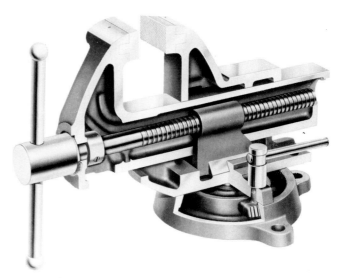

Fig. 10-1. Pictorial sectional view showing the interior details of a machinist's vise. (Columbia Vise and Mfg. Co.)

155

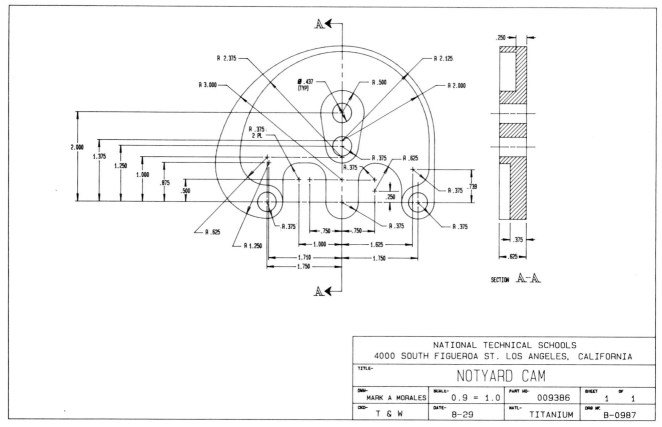

SECTION A-A

NATIONAL TECHNICAL SCHOOLS
4000 SOUTH FIGUEROA ST. LOS ANGELES, CALIFORNIA

TITLE-					
		NOTYARD CAM			

DWN- MARK A MORALES	SCALE- 0.9 = 1.0	PART NO- 009386	SHEET 1	OF 1
CKD- T & W	DATE- 8-29	MATL- TITANIUM	DRB NO. B-0987	

Fig. 10-2. Computer generated sectional view. The student developing this drawing had to have a good working knowledge of basic drafting procedures and techniques. (CADAPPLE, T & W Systems.)

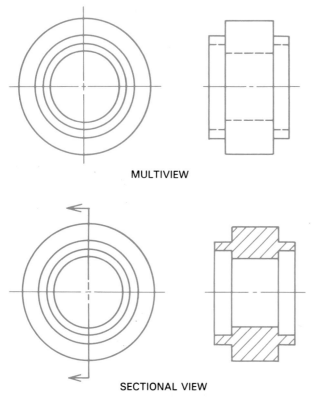

MULTIVIEW

SECTIONAL VIEW

Fig. 10-3. Comparison of a conventional multiview drawing and a drawing in section. The sectional view is always placed behind the arrows.

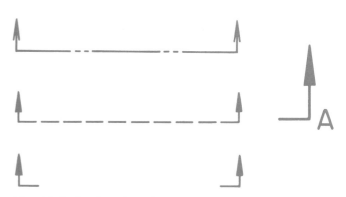

Fig. 10-4. Cutting plane lines. To save time, some drafting departments prefer to use the simplified cutting plane line that employs only the ends.

Spacing of general purpose section lining is by eye and usually at 45 degrees, Fig. 10-5. Line spacing is somewhat dependent on the drawing size or the area to be sectioned. Spacing of 1/8 in. (3.0 mm) is commonly used. However, larger spacing is recommended when large areas must be sectioned. General purpose section lining is the same as the symbol used for cast iron, refer to Fig. 10-21.

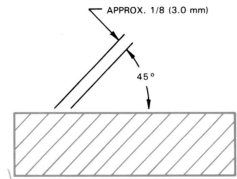

← APPROX. 1/8 (3.0 mm)

45°

Fig. 10-5. General purpose section lining is spaced by eye and usually drawn at 45 degrees. Spacing of 1/8 in. (3.0 mm) is commonly used.

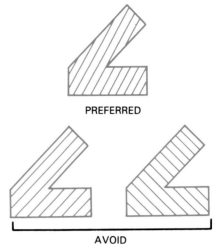

PREFERRED

AVOID

Fig. 10-7. The outline shape of the section may require the section lining to be drawn at other than 45 degrees.

Do not use section lines that are too thick or spaced too closely. Also avoid lines that are not uniformly spaced or are drawn in different directions. See Fig. 10-6.

When section lines will be parallel, or nearly parallel with the outline of an object, they should be drawn at some other angle, Fig. 10-7.

Should two or more pieces be shown in section, the section lines should be drawn in opposite directions and/or angles, to provide contrast, Fig. 10-8.

OUTLINE SECTIONING is a permissable and time saving technique for indicating large sections, Fig. 10-9. Section lines are only drawn near the boundary of the sectioned area. The interior portion is left clear.

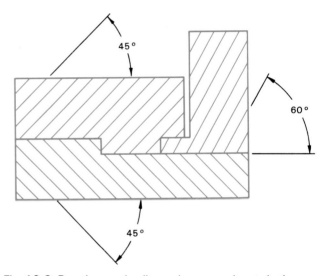

45°

60°

45°

Fig. 10-8. Drawing section lines when several parts in the same section are adjacent (next to each other).

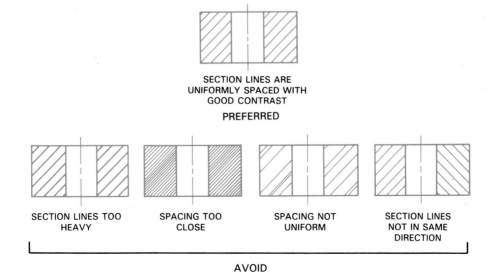

SECTION LINES ARE
UNIFORMLY SPACED WITH
GOOD CONTRAST

PREFERRED

SECTION LINES TOO
HEAVY

SPACING TOO
CLOSE

SPACING NOT
UNIFORM

SECTION LINES
NOT IN SAME
DIRECTION

AVOID

Fig. 10-6. Section line spacing.

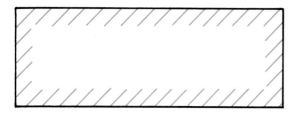

Fig. 10-9. Outline sectioning saves time when large areas must be sectioned.

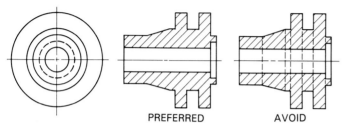

Fig. 10-10. Unless absolutely necessary for a complete understanding of a view, hidden lines behind the cutting plane are omitted.

Unless absolutely necessary for a complete understanding of a view, or for dimensioning purposes, hidden lines behind the cutting plane are omitted, Fig. 10-10.

FULL SECTION

A FULL SECTION is developed by imagining that the cut has been made through the entire object, Fig. 10-11. The part of the object between your eye and the cut is removed to reveal the interior features of the object. The resulting features are drawn as part of the regular multiview projection.

HALF SECTION

The shape of one-half of the interior features and one-half of the exterior features are shown in a HALF SECTION, Fig. 10-12. Half sections are best suited for symmetrical objects (those objects having the same size, shape, and relative position on opposite sides of a dividing line or plane). Cutting plane lines are passed through the piece at right angles to each other. One-quarter of the object is considered removed to show a half section of the interior structure. Unless needed for clarity or dimensioning, hidden lines are not used on half sections.

REVOLVED SECTION

REVOLVED SECTIONS are primarily utilized to show the shape of such objects as spokes, ribs, and stock metal shapes, Fig. 10-13. A revolved section is a drawing within a drawing. To prepare a revolved section, imagine that a section of the part to be shown is cut out and revolved 90 degrees in the same view. Do not draw the lines of a normal view through the revolved section, Fig. 10-14.

ALIGNED SECTION

It is not considered good drafting practice to make a full section of a symmetrical object that has an odd number of holes, webs, or ribs. An ALIGNED SECTION shows two of the holes, webs, or ribs depicted on the view, one of them

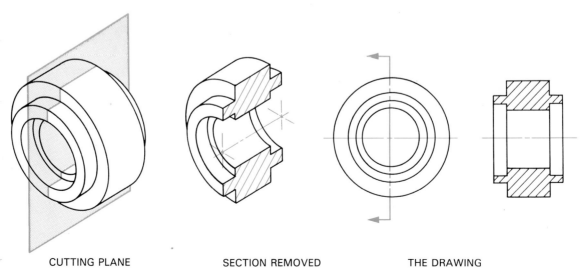

CUTTING PLANE   SECTION REMOVED   THE DRAWING

Fig. 10-11. The full section.

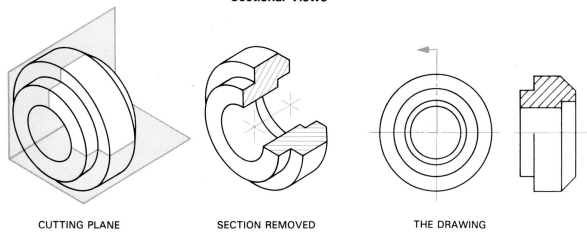

CUTTING PLANE   SECTION REMOVED   THE DRAWING

Fig. 10-12. The half section.

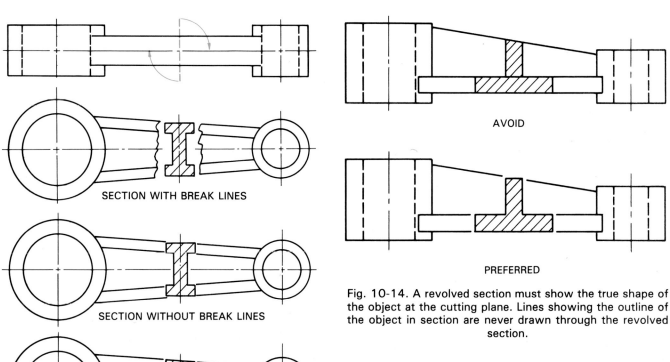

SECTION WITH BREAK LINES

SECTION WITHOUT BREAK LINES

AVOID

Fig. 10-13. A revolved section may be presented with break lines or without break lines. Note that object lines do not pass through the revolved section when the break lines are omitted.

AVOID

PREFERRED

Fig. 10-14. A revolved section must show the true shape of the object at the cutting plane. Lines showing the outline of the object in section are never drawn through the revolved section.

being revolved into the plane of the other, Fig. 10-15. The actual or true projection in a full section may be misleading or confusing.

REMOVED SECTION

There are times when it is not possible to draw a needed sectional view on one of the regular

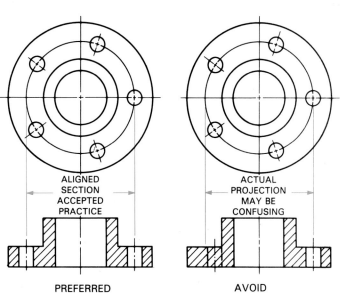

ALIGNED SECTION ACCEPTED PRACTICE

ACTUAL PROJECTION MAY BE CONFUSING

PREFERRED   AVOID

Fig. 10-15. An aligned section.

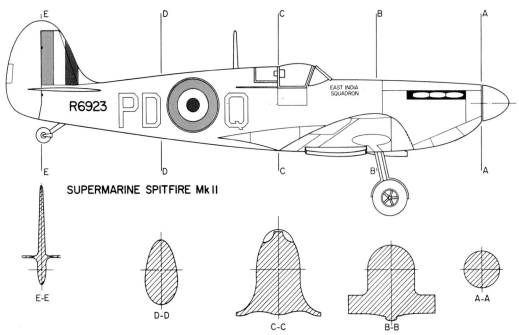

Fig. 10-16. A drawing showing the use of removed sections.

views. When this occurs, a REMOVED SECTION is generally used, Fig. 10-16. The section (or sections) are placed elsewhere on the drafting sheet.

A removed section is also employed when the section must be enlarged for better understanding of the drawing.

OFFSET SECTION

While the cutting plane is ordinarily taken straight through an object, it may be necessary to show features not located on such a straight cutting plane. An OFFSET SECTION uses a cutting plane that is stepped or offset to pass through the required features, Fig. 10-17. The features are drawn as if they were in the same plane.

Offsets in the cutting plane are not shown in the sectional view. The section view appears as one complete drawing. Use reference letters A-A, B-B, etc., at the ends of cutting plane lines.

BROKEN OUT SECTION

The BROKEN OUT SECTION is employed when a small portion of a sectional view will provide the required information. Break lines define the section and are shown on one of the regular views, Fig. 10-18.

CONVENTIONAL BREAKS

When an object with a small cross section but of some length must be drawn, the drafter is faced

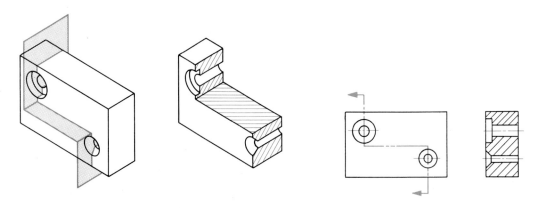

Fig. 10-17. An offset section.

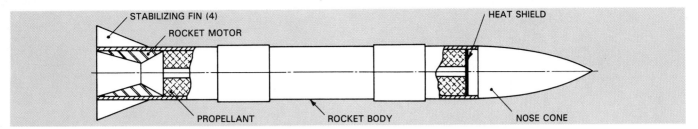

Fig. 10-18. Broken out section.

with many choices. If the drawing is made full size, it may be too long to fit on the sheet. If the scale used is small enough to fit the part on a sheet, the details may be too small to give the required information or to be dimensioned.

In such cases, the object can be fitted to the sheet by reducing its length by means of a CONVENTIONAL BREAK, Fig. 10-19. This method permits a portion of the object to be deleted from the drawing. A conventional break can only be

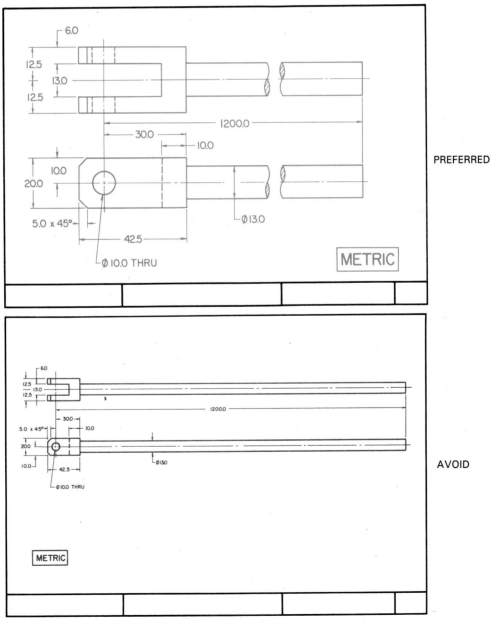

Fig. 10-19. Conventional break in use.

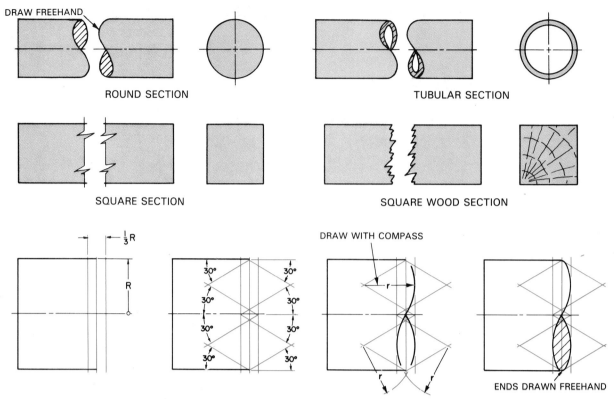

Fig. 10-20. Conventional breaks. By using breaks, an object with a small cross section but some length may be drawn full size (or in a larger scale) on a standard size drawing sheet.

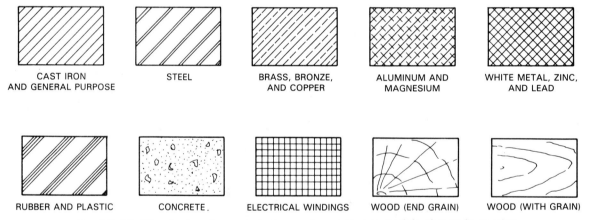

Fig. 10-21. Standard code symbols to use for various materials shown in section.

used when the cross section of a part is uniform its entire length.

Conventional breaks for round and tubular shapes may be drawn freehand, or with instruments as shown in Fig. 10-20.

SYMBOLS TO REPRESENT MATERIALS

Fig. 10-21 shows symbols which may be employed on sectional views to indicate a number of different materials. The lines should be drawn dark and thin to contrast with the heavier object lines.

DRAFTING VOCABULARY

Clarity, Complicated part, Confusing, Contrast, Cutting plane line, Depicted, Details, Exterior, Features, General purpose, Interior, Interior feature, Interior structure, Offset, Omitted, Parallel, Perpendicular, Reference, Sectional view,

Specifications, Stepped, Symmetrical object, Uniformly spaced.

TEST YOUR KNOWLEDGE—UNIT 10

Please do not write in the book. Place your answers on a piece of notebook paper.

1. What are sectional views?
2. When are section views used?
3. The _____ _____ _____ indicates location of imaginary cut made through object to show its interior details.
4. Sections are identified by the use of _____.
5. The _____ section is used when the cut has been made through the entire object.
6. The _____ section shows half of the interior and half of the exterior of the object.
7. _____ sections are primarily used to show the shape of such things as spokes, ribs, and stock metal shapes.

8. The broken out section is used when _____ _____.
9. Objects with small cross sections but of considerable length can be fitted on a standard size sheet by making use of the _____.
10. _____ lines may be used to indicate the material that has been cut.

OUTSIDE ACTIVITIES

1. Obtain an industry print using a sectional view. With color pencils, highlight the different component sections in the plan to demonstrate how the various parts fit together.
2. Using a cutaway object, draft the sectional view showing the internal features. Use both the object and the completed drawing as a classroom teaching aid.
3. Create sectional view drawings of common drafting tools such as a drawing pencil, a T-square head, or a bow compass.

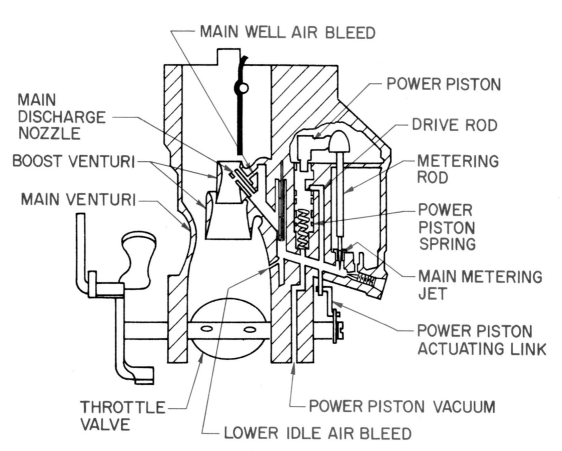

Sectional view showing the interior structure of an automobile engine carburetor. It would be very confusing if only a conventional multiview drawing were used.

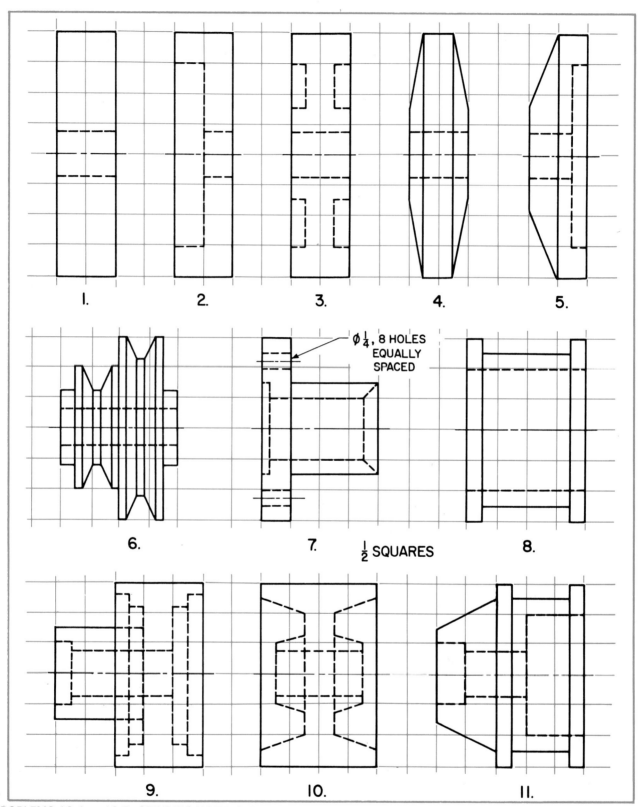

1.

2.

3.

4.

5.

6.

7. ½ SQUARES

8.

Ø ¼, 8 HOLES EQUALLY SPACED

9.

10.

11.

PROBLEMS 10-1 to 10-5. GRINDING WHEELS. Draw the views necessary to show the shape of each wheel. Make one view a full section or a half section as directed by your instructor. PROBLEM 10-6. PULLEY. Draw the views necessary to show the shape of this pulley. Draw one view as a half section. PROBLEM 10-7. COUPLING. Draw the views necessary to show the shape of this coupling. Draw one view as a full section. PROBLEM 10-8. BUSHING. Draw the views necessary to show the shape of this bushing. Draw one view as a half section. PROBLEM 10-9. SPACER. Draw the views necessary to show the shape of this spacer. Draw one view as a half section. PROBLEM 10-10. FLAT BELT PULLEY. Draw the views necessary to show the shape of this pulley. Draw one view as a full section. PROBLEM 10-11. ADAPTER BEARING. Draw the views necessary to show the shape of this bearing. Draw one view as a half section.

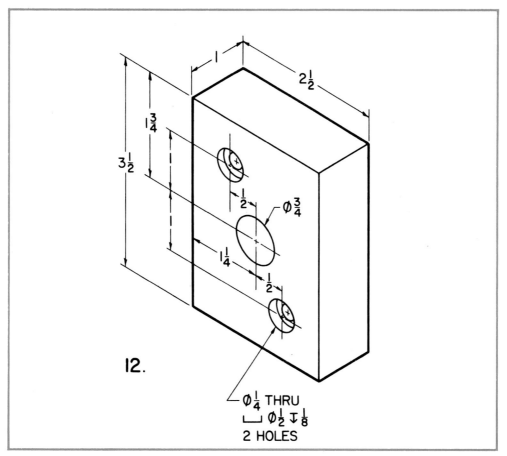

12.

$\varnothing\frac{1}{4}$ THRU
$\sqcup$ $\varnothing\frac{1}{2}$ $\top$ $\frac{1}{8}$
2 HOLES

PROBLEM 10-12. SPACER. Draw the views necessary to show the shape of the spacer. Draw one view as an offset section through the three holes.

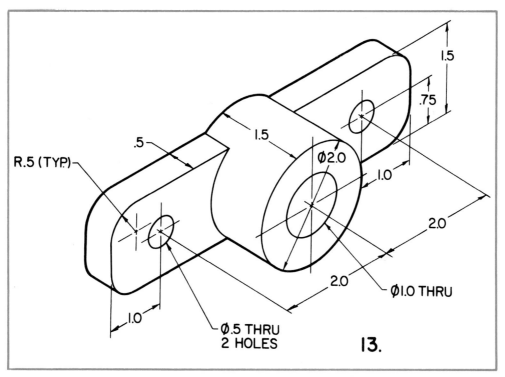

R.5 (TYP.)

$\varnothing$2.0

$\varnothing$1.0 THRU

$\varnothing$.5 THRU
2 HOLES

13.

PROBLEM 10-13. SPECIAL BUSHING. Draw the views necessary to show the shape of the bushing. Draw one view as a full section along the horizontal plane.

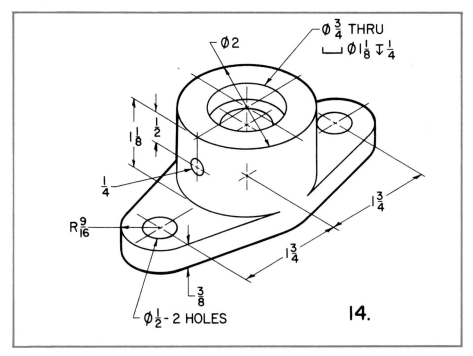

PROBLEM 10-14. BRACKET. Draw the views necessary to show the shape of the bracket. Include a broken out section through the 1/4 in. diameter hole on one view.

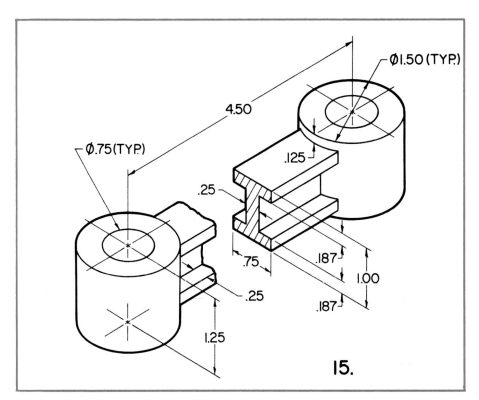

PROBLEM 10-15. CONNECTING ROD. Draw the views necessary to show the shape of the rod. The cross section of the rod may be shown as a revolved section or as a removed section.

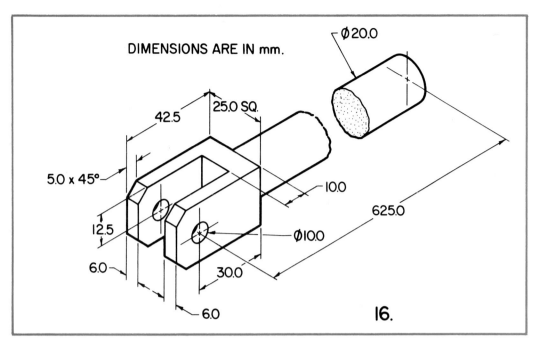

DIMENSIONS ARE IN mm.

Ø20.0

25.0 SQ.

42.5

5.0 x 45°

12.5

6.0

6.0

30.0

Ø10.0

10.0

625.0

16.

PROBLEM 10-16. TORQUE ROD. Draw the views necessary to show the shape of the rod. Use the conventional break to fit the object on the drawing sheet.

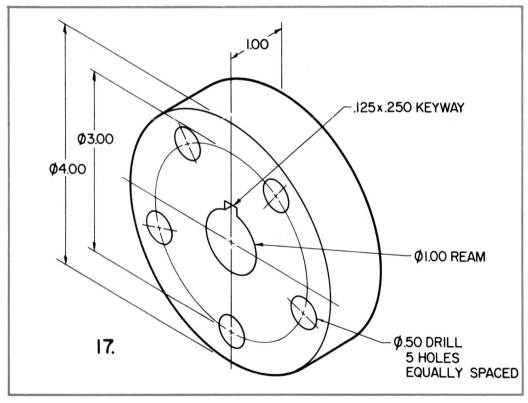

1.00

.125 x .250 KEYWAY

Ø3.00

Ø4.00

Ø1.00 REAM

17.

Ø.50 DRILL
5 HOLES
EQUALLY SPACED

PROBLEM 10-17. ADAPTER PLATE. Draw the views necessary to show the shape of the plate. Draw one view as a full section.

Unit 11

AUXILIARY VIEWS

After studying this unit, you will recognize that some objects may have a surface that is not parallel to any of the six regular planes of projection. You will comprehend that such a surface must be projected to another plane parallel to the inclined plane in order to show its true shape and size. You will define the new plane as an auxiliary plane and the new view as an auxiliary view. You will understand why auxiliary construction lines are not projected at right angles to the inclined surface. You will develop and draw auxiliary views.

The true shape and size of objects having angular or slanted surfaces cannot be drawn using the regular top, front, and right side views. The LATCH PLATE shown in Fig. 11-1 is such an object. The true length of the angular surface is shown on the front view, but this view does not show its width. The true width of the surface is shown on the top and right side views but neither view shows its true length.

An additional or AUXILIARY VIEW is needed to show the true length and true width of the angular surface, Fig. 11-2.

When drawing an auxiliary view, remember that the view is ALWAYS projected from the regular

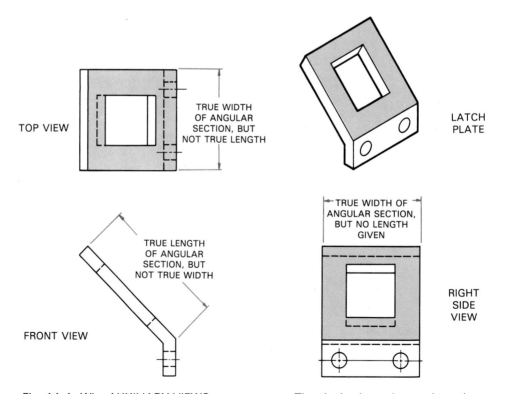

TOP VIEW

TRUE WIDTH OF ANGULAR SECTION, BUT NOT TRUE LENGTH

TRUE LENGTH OF ANGULAR SECTION, BUT NOT TRUE WIDTH

FRONT VIEW

LATCH PLATE

TRUE WIDTH OF ANGULAR SECTION, BUT NO LENGTH GIVEN

RIGHT SIDE VIEW

Fig. 11-1. Why AUXILIARY VIEWS are necessary. The single views do not show the true shape (length and width) of the cut off portion of the object.

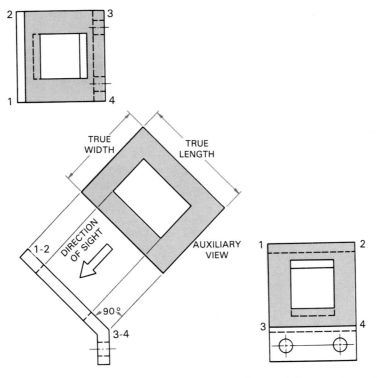

Fig. 11-2. The AUXILIARY VIEW shows the true shape of the angular surface.

view on which the true length of the inclined surface is shown. Also, the construction lines projecting from the angled surface are ALWAYS at right angles to that surface. In other words, the auxiliary view is constructed perpendicular to the inclined line.

When drawing auxiliary views, the usual practice is to show only the angled portion of the view. It is seldom necessary to draw a full projection of the object. See Fig. 11-3. It is also often possible to eliminate one of the conventional views when using an auxiliary view, Fig. 11-4.

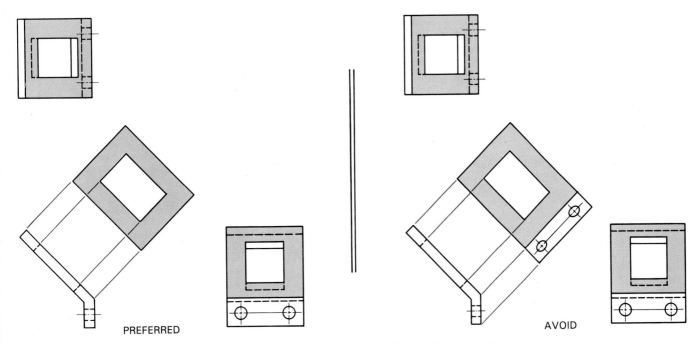

Fig. 11-3. It is not necessary to draw a full projection of the object.

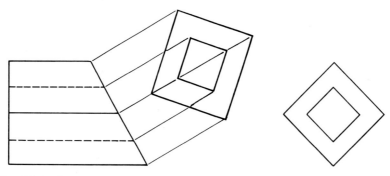

Fig. 11-4. One view may often be eliminated when using an auxiliary view.

An auxiliary view can be projected from any view that shows the angled surface as a line. It would be called a FRONT AUXILIARY if the view is projected from the front view. A TOP AUXILIARY is projected from the top view. A RIGHT SIDE AUXILIARY is projected from the right side view.

Considerable time can be saved when drawing auxiliary views of symmetrical objects by drawing one half of the view, Fig. 11-5.

Auxiliary views that include rounded surfaces or circular openings may cause minor problems. Fig. 11-6 shows how a circular surface would be drawn. Proceed as follows:

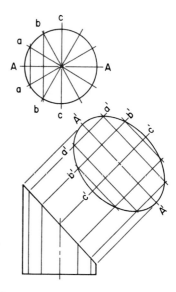

Fig. 11-6. Drawing an auxiliary view of a circular object.

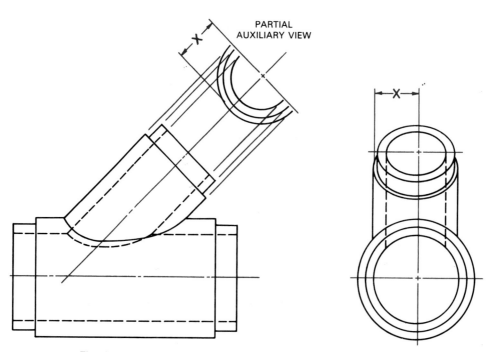

Fig. 11-5. A half auxiliary view can be used to describe some symmetrical objects.

1. Draw the needed front, top, or right side views. Divide the circular view into 12 equal parts. This is done with a 30-60 degree triangle. Project the divisions from the circular view to the other view(s). Identify each division with letters as shown. Use A-A for the horizontal center line.
2. At any convenient distance from the inclined surface, draw the center line A'-A' for the auxiliary view. The center line will be parallel to the the inclined face.
3. Project the necessary points from the inclined surface through center line A'-A'. These lines are at right angles (90 degrees) to the inclined face.
4. Using dividers or a compass, transfer measurements from horizontal center line A-A to circumference points a, b, c, etc., to the proper projection lines to the right and left of inclined center line A'-A'. This constructs points a', b', c'.
5. Complete the auxiliary view by connecting the points with a French curve.

DRAFTING VOCABULARY

Angular surface, Auxiliary view, Convenient distance, Inclined surface, Project, Right angle, Symmetrical, Transfer, True length, True width.

TEST YOUR KNOWLEDGE—UNIT 11

Please do not write in the book. Place your answers on a piece of notebook paper.
1. Why are auxiliary views needed?
2. The auxiliary view is always projected from the view that shows the inclined surface as a _____. The construction lines projecting from the inclined surface are always at _____ or _____ to the cut.
3. The auxiliary view when projected from the front view is called a _____ _____.
4. When drawing an auxiliary view of a symmetrical object, much time can be saved by drawing _____ _____ of the view.
5. Connect the points of rounded surfaces or circular openings by using a _____ _____.

OUTSIDE ACTIVITIES

1. Make a sketch of an object that would require an auxiliary view.
2. Make a collection of items from your school that need auxiliary views to show true size and shape. An example would be starting blocks from the track. Draw the views needed to construct such a part.
3. Make a sketch of a home or garage peaked roof as an auxiliary view. Add appropriate dimensions. What is the relationship between the area of roof or the number of shingles used as compared to the steepness of the roof?

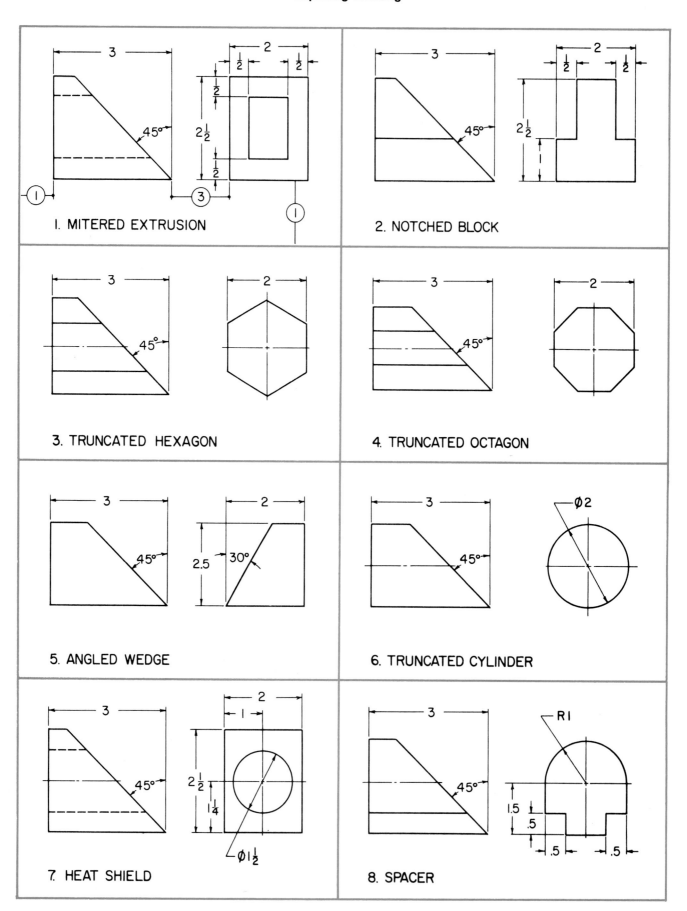

1. MITERED EXTRUSION

2. NOTCHED BLOCK

3. TRUNCATED HEXAGON

4. TRUNCATED OCTAGON

5. ANGLED WEDGE

6. TRUNCATED CYLINDER

7. HEAT SHIELD

8. SPACER

PROBLEMS 11-1 to 11-8. Space drawings according to the circled dimensions. The top views may be eliminated.

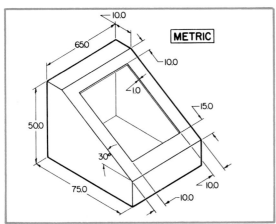

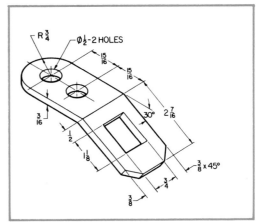

PROBLEM 11-9. Left. INSTRUMENT CASE. 1—Use a vertical drawing sheet format. 2—Allow 4 in. between front and top views. 3—Locate the auxiliary view 1 1/2 in. from the front view. 4—The front view is 1 in. from the left border. PROBLEM 11-10. Right. BRACKET. 1—Use a horizontal drawing sheet format. 2—Allow 3 in. between the front and top views. 3—Locate the auxiliary view 1 in. from the front view. 4—The front view is 2 3/4 in. from the left border.

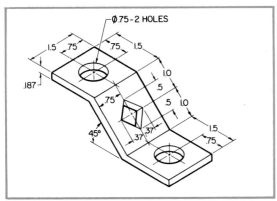

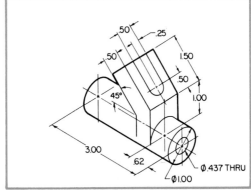

PROBLEM 11-11. Left. HANGER CLAMP. 1—Use a horizontal drawing sheet format. 2—Allow 3 in. between front and top views. 3—Locate the auxiliary view 1 1/2 in. from the front view. 4—Allow 2 1/2 in. between the front and right side view. 5—The front view is 3/4 in. from the left border. PROBLEM 11-12. Right. SHIFTER BAR. 1—Use a horizontal drawing sheet format. 2—Allow 2 in. between the front and top views. 3—Locate the auxiliary view 1 in. from the front view. 4—Allow 2 in. between the front and right side view. 5—The front view is 2 in. from the left border.

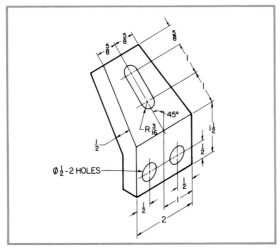

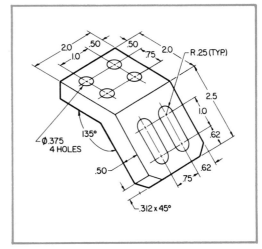

PROBLEM 11-13. Left. SUPPORT. 1—Use a vertical sheet format. 2—Allow 2 1/2 in. between front and top views. 3—Locate the auxiliary view 1 1/2 in. from the front view. 4—The front view is 1 1/4 in. from the left border. PROBLEM 11-14. Right. ADJUSTABLE BRACKET. 1—Use a horizontal sheet format. 2—Allow 2 1/8 in. between the front and top views. 3—Locate the auxiliary view 1 in. fron the front view. 4—Allow 2 3/4 in. between the front and right side views. 5—The front view is 3/4 in. from the left border.

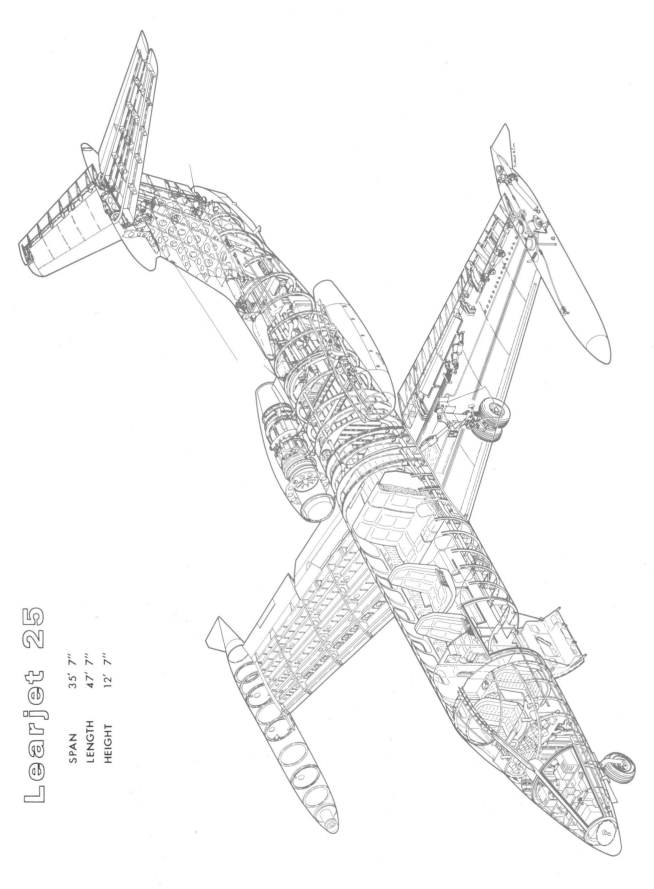

Learjet 25

SPAN	35' 7"
LENGTH	47' 7"
HEIGHT	12' 7"

Fig. 12-1. Pictorials have many uses. This one shows the structural details of a business jet. (Gates-Learjet Corp.)

Unit 12

PICTORIALS

After studying this unit, you will comprehend the use of pictorial drawings. From multiview drawings, you will be able to construct five types of pictorial drawings: isometric, cavalier oblique, cabinet oblique, parallel perspective, and angular perspective. You will be able to illustrate using exploded assembly drawings or cutaway pictorial drawings. You will determine how to center a pictorial drawing on a drawing sheet.

A PICTORIAL DRAWING shows a likeness (shape) of an object as viewed by the eye. The pictorial of the jet aircraft in Fig. 12-1 shows many structural details of the craft.

Pictorials are also an important part in some computer aided design (CAD) projects. They range from basic "wire frame" pictorials in Fig. 12-2, to moving, full color three dimensional renditions as in Fig. 12-3. The complexity of the computer generated pictorial is determined by the program level and computer capacity.

If you have worked with radio or electronic kits, you are familiar with pictorial drawings. These

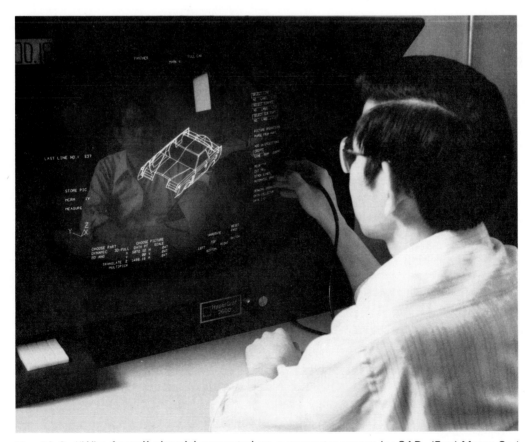

Fig. 12-2. "Wire frame" pictorial generated on a computer screen by CAD. (Ford Motor Co.)

175

Fig. 12-3. Full color three-dimensional pictorial generated by computer graphics. This pictorial requires a higher level program and larger capacity computer than the "wire frame" pictorial shown in Fig. 12-2. (Evans & Sutherland)

drawings show the builder what needs to be done and how to do it to complete the project, Fig. 12-4.

You should become familiar with several types of pictorial drawings, Fig. 12-5. Only the basics of pictorial drawing will be covered in this text.

ISOMETRIC DRAWINGS

All ISOMETRIC DRAWING lines which show the width, length, and depth are drawn full size (or in some proportionate scale). An isometric drawing shows an object as it is. Edges which are upright are shown as vertical lines. The object is assumed to be in position with its corners toward you, and its horizontal edges sloping away at angles of 30 degrees to the right and to the left. Refer to Fig. 12-5.

An isometric drawing is made entirely with instruments, using three base lines, Fig. 12-6. One of the lines is vertical, the other two are drawn at an angle of 30 degrees to the horizontal. The base lines may be reversed if more information can be shown with the object in this position, Fig. 12-7.

Lines which are not parallel to the three base lines are called NON-ISOMETRIC LINES. They are drawn by transferring reference points from a multiview drawing to the isometric drawing, Fig. 12-8. The length of non-isometric lines cannot be measured directly on the isometric drawing.

Isometric circles are shown in Fig. 12-9. Note that isometric circles are circles which are not true ellipses. Isometric circles may be drawn using a compass and a 30-60 degree triangle.

Pictorials

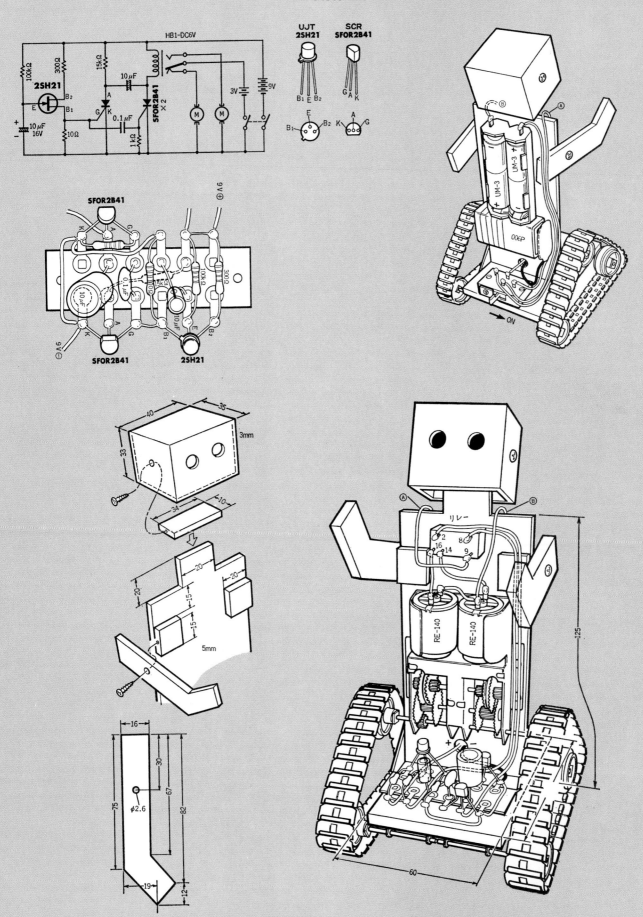

Fig. 12-4. Pictorials are used to give instructions in many hobby areas. Shown is a model robot that can be programmed to follow a specified travel pattern.

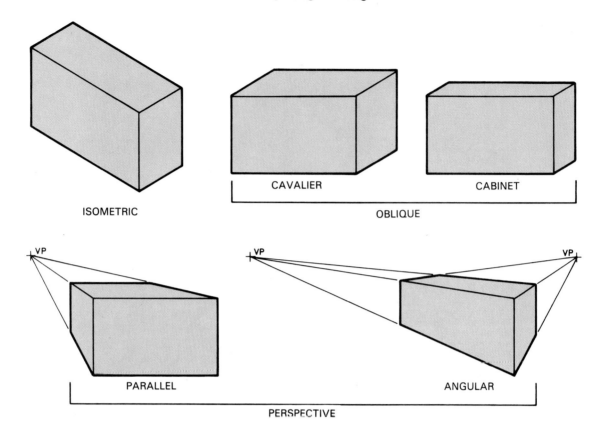

CAVALIER

CABINET

ISOMETRIC

OBLIQUE

PARALLEL

ANGULAR

PERSPECTIVE

Fig. 12-5. Types of pictorial drawings.

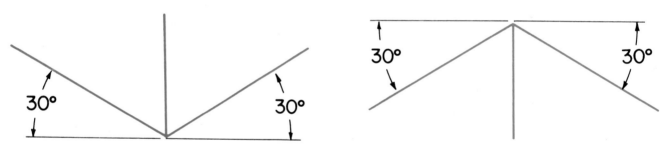

30°

30°

30°

30°

Fig. 12-6. Base lines required to make isometric drawings.

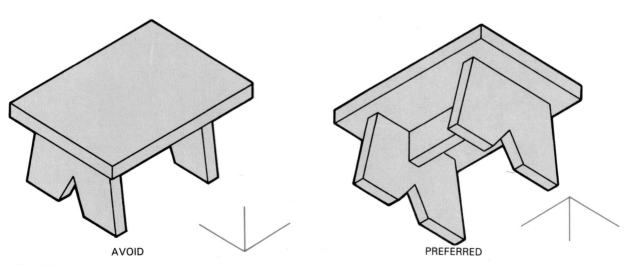

AVOID

PREFERRED

Fig. 12-7. Reverse the base lines if more information can be shown with the object in that position. Note the details in the underside view.

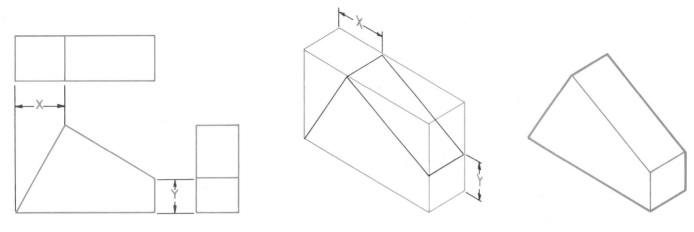

Fig. 12-8. Non-isometric lines are made by transferring reference points from a multiview drawing of the object.

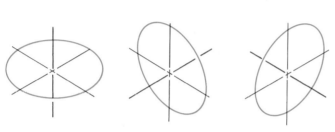

Fig. 12-9. Isometric circles. Note true ellipse for comparison.

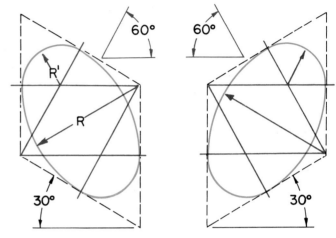

Fig. 12-11. Drawing an isometric circle on a vertical plane with a compass.

Figs. 12-10 and 12-11 show how to draw an isometric circle using the compass and triangle technique. Proceed as follows:

1. Using 30 degree angles, construct an isometric square the same size as the circle. The side of the square is the same length as the circle diameter, as shown to the left of Fig. 12-10.
2. Using 60 degree angles, draw lines from the corners of the isometric squares as shown. The intersection points of these lines provide the centers for drawing radius R^1. Construct arcs tangent to the sides of the isometric square

between the 60 degree construction lines as shown in Figs. 12-10 and 12-11.

It is handy to know how to draw isometric circles using drafting tools. An easier way to draw such circles and arcs is to use an ellipse template of appropriate size, Fig. 12-12.

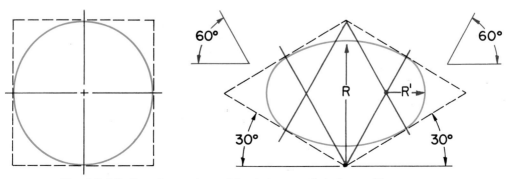

Fig. 12-10. Drawing an isometric circle on a flat plane with a compass.

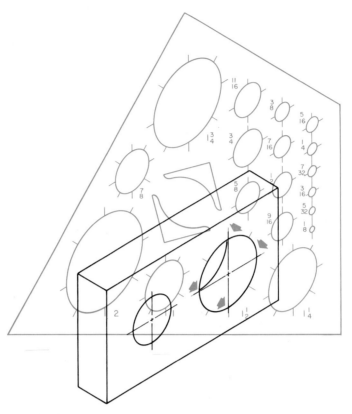

Fig. 12-12. A template is a quick way to draw isometric circles.

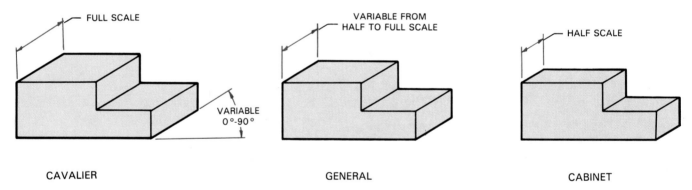

CAVALIER

GENERAL

CABINET

Fig. 12-13. Three types of oblique drawings. The CAVALIER and CABINET drawings are most frequently used.

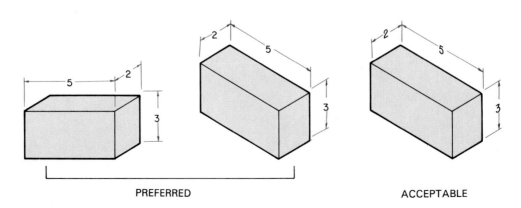

PREFERRED

ACCEPTABLE

Fig. 12-14. Dimensioning pictorials.

180

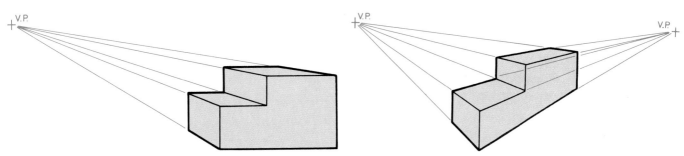

Fig. 12-15. Perspective drawings, with vanishing points indicated.

OBLIQUE PICTORIAL DRAWINGS

An OBLIQUE DRAWING is a pictorial drawing with the longest dimension or front parallel to the picture plane; refer to Fig. 12-5. The front view is shown in true shape and size. The other views are similar to isometric drawings. Three types of oblique drawings are shown in Fig. 12-13. Each type shows the front of the object as it appears in multiview form. The angle of the depth axis may be any angle, but 15, 30, or 45 degrees are generally used.

CAVALIER OBLIQUE. An oblique drawing in which the depth axis lines are full scale (full size).

CABINET OBLIQUE. Depth axis lines are drawn one-half scale.

GENERAL OBLIQUE. Depth axis lines vary from one-half to full scale.

DIMENSIONING PICTORIAL DRAWINGS

To dimension isometric and oblique drawings, draw the dimension lines parallel to the corresponding planes of the object (lines of the drawing). See Fig. 12-14 for examples.

It is preferred that dimensions be located outside the outline of the object. They should be placed to read in a horizontal direction, regardless of the direction of the dimension line. Acceptable alternate dimensioning is also shown in Fig. 12-14.

PERSPECTIVE DRAWINGS

A PERSPECTIVE DRAWING is a drawing used by an architect, artist, or drafter to show an object as it would appear to the eye when viewed from a certain position, Fig. 12-15. Compare the perspective drawings to the isometric drawing and the oblique drawings in Fig. 12-5.

A perspective drawing can best be described by imagining that you are landing a jet aircraft. As you approach the runway, you notice that the parallel sides of the runway appear to converge (meet) on the horizon, Fig. 12-16. The point where

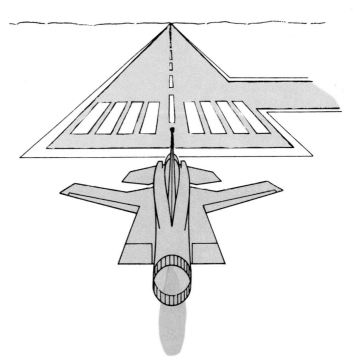

Fig. 12-16. Objects appear to become smaller the further away they are from the eye. Note how the parallel sides of the runway seem to converge or run together in the distance.

the runway edges meet is called the VANISHING POINT (V.P. or VP). In making a perspective drawing, lines used to show the depth of the object (receding lines), if continued, will converge to form a V.P. Perspective drawings are based on the principle of parallel lines converging in the distance. Refer again to Fig. 12-15.

To make a simplified ONE-POINT or PARALLEL PERSPECTIVE DRAWING, the front view is drawn in its true shape in full or scale size, Fig. 12-17. Draw the horizon line and select the VANISHING POINT at random. If at first you do not obtain a pleasing effect, try another vanishing point location. Information on locating

vanishing points more accurately may be obtained from a text covering advanced drafting procedures. A perspective may be located in any position relative to the V.P., Fig. 12-18. Project construction lines from each point on the front view to the vanishing point, Fig. 12-19.

Simple ANGULAR or TWO-POINT PERSPECTIVES may be drawn as shown in Fig. 12-20. Follow these steps: First — Locate the posi-

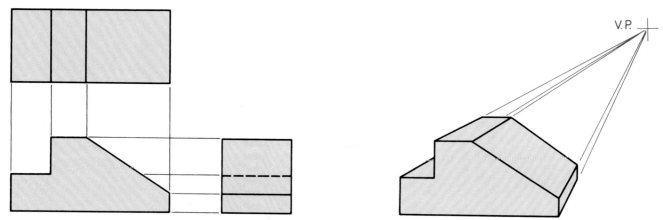

Fig. 12-17. On PARALLEL or ONE-POINT PERSPECTIVE DRAWINGS, the front view is shown in its true shape and in full or scale size.

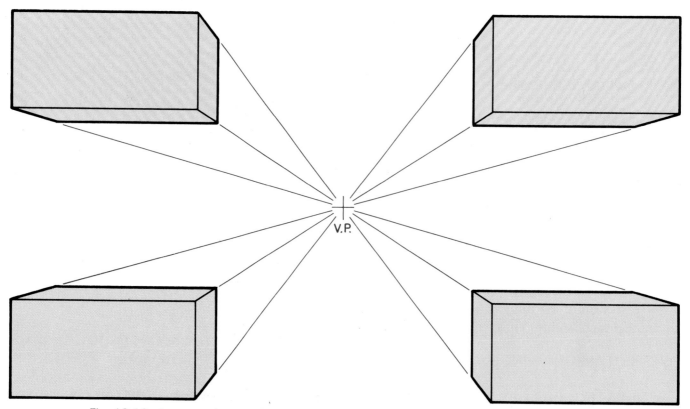

Fig. 12-18. A perspective may be located in any position relative to the vanishing point (V.P.).

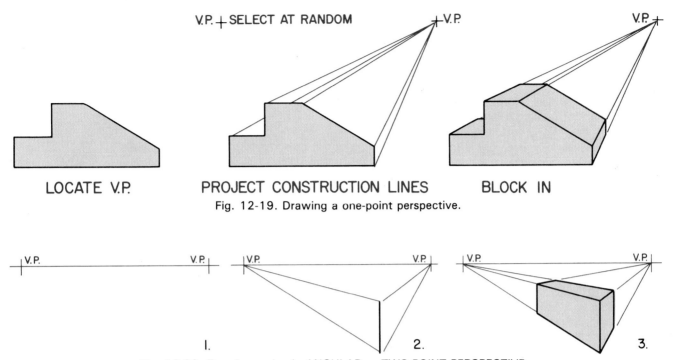

V.P. + SELECT AT RANDOM + V.P. V.P. +

LOCATE V.P.   PROJECT CONSTRUCTION LINES   BLOCK IN

Fig. 12-19. Drawing a one-point perspective.

V.P. V.P. V.P. V.P. V.P. V.P.

1.   2.   3.

Fig. 12-20. Drawing a simple ANGULAR or TWO-POINT PERSPECTIVE.

tion of the horizon and two vanishing points. Second—Locate and draw a vertical line the full scale height of the object to be drawn. Third—Draw construction lines from the vertical line to the vanishing points. For simple two-point perspectives like this, it is permissible to locate the length and depth in a position that gives the most pleasing visual effect. Darken the necessary lines with visible object lines.

In drawing perspectives, considerable saving of time plus greater accuracy will result if a perspective drawing board is used, Fig. 12-21. This board is available from firms that sell drafting supplies.

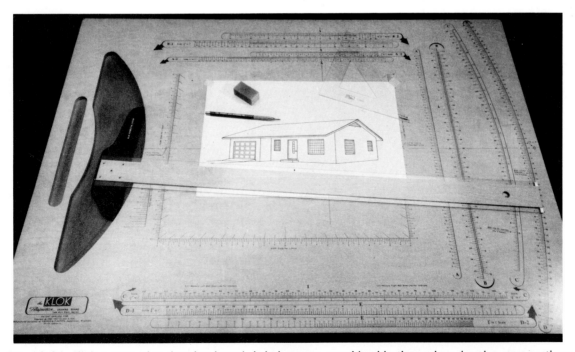

Fig. 12-21. Klok perspective drawing board. It helps save considerable time when drawing perspectives.

EXPLODED ASSEMBLY DRAWINGS

EXPLODED ASSEMBLY DRAWINGS are nothing more than a series of pictorial drawings (usually isometrics). They show the parts that make up the object in proper location to one another, Fig. 12-22.

Exploded assembly drawings are easy to read and can be understood without having an extensive knowledge of print reading. Industry makes extensive use of exploded assembly drawings, especially in areas where semi-skilled workers are employed.

Such drawings are also used in many hobby areas. You are probably familiar with them if you build model cars, planes, boats, or rockets.

CUT-AWAY PICTORIAL DRAWINGS

CUT-AWAY PICTORIAL DRAWINGS have been developed to show the interior details of a product, Fig. 12-23. They are usually employed in instructional manuals where the interior details of the product are important in understanding the theory of operation.

HOW TO CENTER PICTORIAL DRAWINGS

There are several ways to center pictorial drawings on a drawing sheet.

Fig. 12-23. Cut-away pictorial drawing of a modern autofocus camera. (Minolta Corp.)

One method, which is widely used, is called "eyeballing it." You estimate, by sight, the approximate location of the starting point and develop your drawing from this point.

Another method requires the drawing of the pictorial on a second sheet of paper. When it has been checked for accuracy, a piece of tracing vellum is centered over it and the drawing traced. Or, if regular drawing paper is used, the total width and height of the pictorial is measured. With this information as the starting point, center the drawing on the sheet.

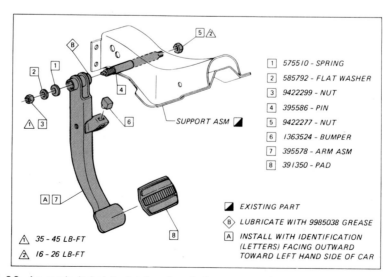

1	575510	SPRING
2	585792	FLAT WASHER
3	9422299	NUT
4	395586	PIN
5	9422277	NUT
6	1363524	BUMPER
7	395578	ARM ASM
8	391350	PAD

◢ EXISTING PART

Ⓑ LUBRICATE WITH 9985038 GREASE

Ⓐ INSTALL WITH IDENTIFICATION (LETTERS) FACING OUTWARD TOWARD LEFT HAND SIDE OF CAR

⚠ 35 - 45 LB-FT

⚠ 16 - 26 LB-FT

Fig. 12-22. An exploded pictorial drawing. Industry uses drawings of this type in areas where the workers have little training in print reading. (General Motors Corp.)

A third method used to center pictorial drawings which works well with most isometric and oblique problems is shown in Figs. 12-24 and 12-25. Follow the steps shown.

DRAFTING VOCABULARY

Alternate, Angular perspective, Appropriate size, Approximate location, Depth axis, Exploded

1. DESCRIBING AN ISOMETRIC VIEW.

2. LOCATE CENTER OF DRAWING AREA.

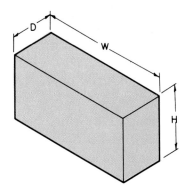
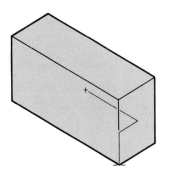
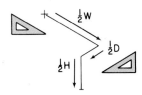

3. PLOT STARTING POINT OF DRAWING.

4. COMPLETE DRAWING.

Fig. 12-24. An easy way to center isometric drawings.

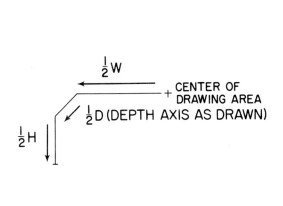
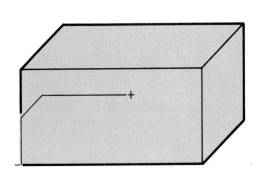

1. PLOT STARTING POINT OF DRAWING.

2. COMPLETE DRAWING.

Fig. 12-25. An easy way to center oblique drawings.

assembly drawing, Isometric drawing, Non-isometric lines, Oblique drawing, One-point perspective, Parallel perspective, Perspective drawing, Pictorial drawing, Proportionate scale, Receding lines, Reversed, Simplified, Structural details, Template, Two-point perspective, Vanishing point, Wire frame.

TEST YOUR KNOWLEDGE—UNIT 12

Please do not write in the text. Place your answer on a sheet of notebook paper.

1. Pictorial drawing is a method of showing an object as:
   a. It would look on the drawing.
   b. It would appear to the eye.
   c. Another form of multiview drawing.
   d. All of the above.
   e. None of the above.
2. Why are pictorials often used?
3. List 4 types of pictorial drawings:
   a. _____
   b. _____
   c. _____
   d. _____
4. Isometric pictorial drawings are drawn about three base lines. Sketch these base lines.
5. The lines that represent angles in an isometric drawing are known as _____ lines.
6. Prepare a sketch that shows the difference between a cavalier and a cabinet pictorial drawing.
7. Oblique drawings may be drawn at any angle, but _____ angles are usually used.
8. What is an "exploded assembly drawing"?

OUTSIDE ACTIVITIES

1. Collect pictorial drawings used to advertise real estate. Use both residential and commercial illustrations. Describe to the class why you think pictorials are used to "sell" real estate.
2. Make a bulletin board display of pictorial drawings such as exploded assembly views of models that you and your classmates have built. Explain why the pictorial views assisted in constructing the models.
3. Search through old magazines (not library copies) and find examples of cutaway pictorial drawings. Make a comparison of drawings using full color as compared to only printed in black and white. Try to make sharp black and white photocopies of the full color cutaways. Does color help explain or "sell"?

Pictorials

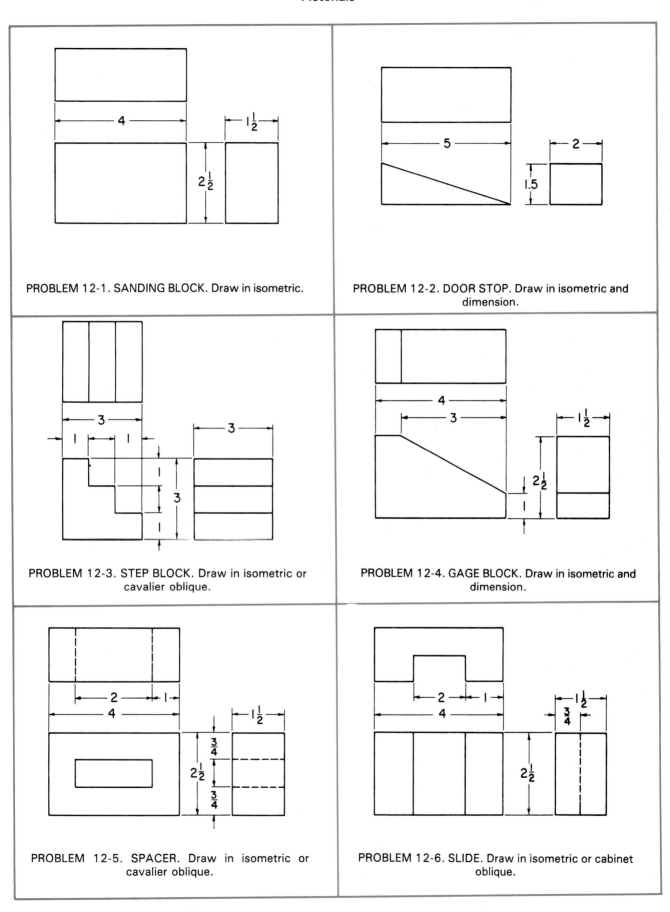

PROBLEM 12-1. SANDING BLOCK. Draw in isometric.

PROBLEM 12-2. DOOR STOP. Draw in isometric and dimension.

PROBLEM 12-3. STEP BLOCK. Draw in isometric or cavalier oblique.

PROBLEM 12-4. GAGE BLOCK. Draw in isometric and dimension.

PROBLEM 12-5. SPACER. Draw in isometric or cavalier oblique.

PROBLEM 12-6. SLIDE. Draw in isometric or cabinet oblique.

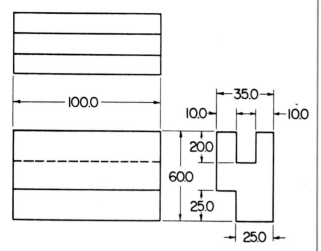

PROBLEM 12-7. GUIDE. Draw in isometric.

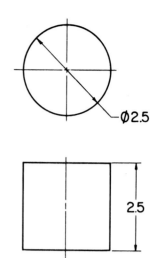

PROBLEM 12-8. CANISTER. Draw in isometric.

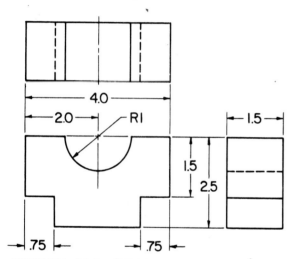

PROBLEM 12-9. SUPPORT. Draw in cabinet or cavalier oblique and dimension.

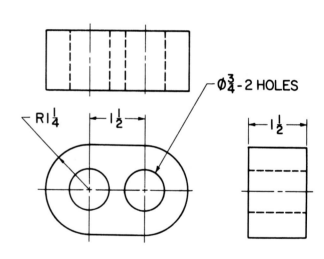

PROBLEM 12-10. LINK. Draw in cabinet oblique.

DIMENSIONS ARE IN mm.

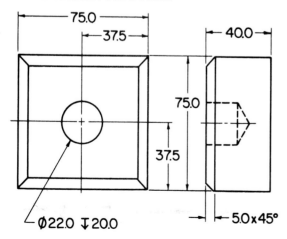

PROBLEM 12-11. CANDLESTICK HOLDER. Draw in isometric.

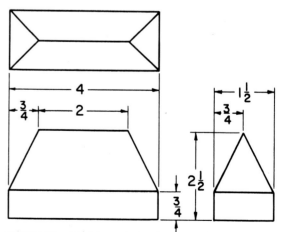

PROBLEM 12-12. BALANCE WEDGE. Draw in isometric and dimension.

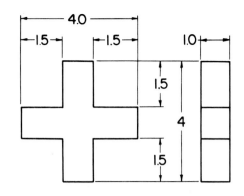

PROBLEM 12-13. CROSS. Draw in one-point perspective.

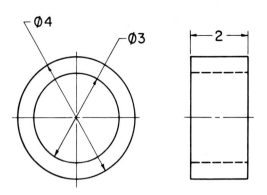

PROBLEM 12-14. BEARING. Draw in isometric or one-point perspective.

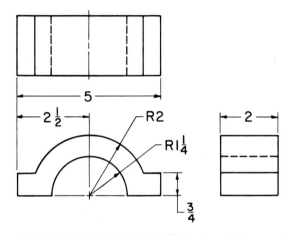

PROBLEM 12-15. BEARING CAP. Draw in isometric and dimension.

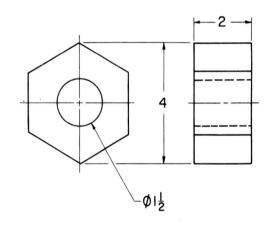

PROBLEM 12-16. HEXAGONAL BASE. Draw in isometric or cabinet oblique.

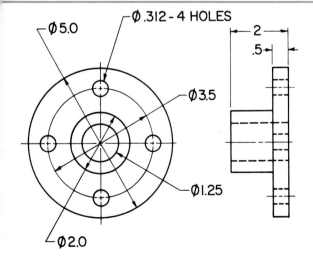

PROBLEM 12-17. FACE PLATE. Draw in isometric.

PROBLEMS:
Design and draw pictorial form—
1. Book rack
2. Shoe shine box
3. Modern bird house
4. Coffee table
5. End table
Design and draw as an exploded pictorial drawing—
1. Picture frame
2. Book case
3. Jewelry box
4. Wall storage cabinet
5. Desk, computer table, or stereo equipment stand

PROBLEM 12-18. SPECIAL DESIGN PROBLEMS. These problems should be drawn on a B or C size drawing sheet.

Unit 13

PATTERN DEVELOPMENT

After studying this chapter, you will be able to describe and define pattern development. You will be able to demonstrate and prepare patterns using the parallel line development method and the radial line development. You will develop patterns for a cylinder, rectangular prism, truncated prism, truncated cylinder, pyramid, right rectangular pyramid, and cone.

A PATTERN is a full-size drawing of the various outside surfaces of an object stretched out on a flat plane, Fig. 13-1. A pattern is frequently called a STRETCHOUT. The pattern can be bent or folded into three dimensional shapes.

Patterns or stretchouts are produced by utilizing a form of drafting called PATTERN DEVELOPMENT. The method is also known as SURFACE DEVELOPMENT and SHEET METAL DRAFTING.

Pattern development is important to many occupations and hobbies.

Patterns were required to make the clothing and shoes you wear. Wallets and handbags are cut to shape using patterns as guides. Pattern development plays an important part in the fabrication of sheet metal ducts and pipes needed in the installation of heating and air conditioning units. Stoves, refrigerators, and other appliances are fabricated from many sheet metal parts. Accurate patterns had to be developed for the parts before the appliance could be put into production.

Drafters in the aerospace industry must be familiar with pattern development techniques.

Manufacturing space-age materials into components for airplanes and rockets requires many patterns, Fig. 13-2. The technique used is called LOFTING. The computer generated patterns are developed directly on materials by a photographic process using lasers.

Every plate on a ship's hull and superstructure started with a pattern. Otherwise, they would have required extensive cutting and fitting (very costly) before they could be used.

If you have ever built a model boat or flying model airplane, Fig. 13-3, you know that many patterns are needed to cut the parts to shape for assembly.

Many pattern developments are prepared by CAD (Computer Aided Design), See Fig. 13-1. However, before the drafter can use CAD for this type of work, he or she must have a thorough knowledge of pattern development techniques.

As you can see, patterns play an important part in the manufacture of many products. How many items in the drafting room can you name that use pattern development in their manufacture?

HOW TO DRAW PATTERNS

Regular drafting techniques, as described elsewhere in this text, are used to draw patterns. Basic pattern development falls into the two categories of parallel line development and radial line development. Refer to Fig. 13-4.

PARALLEL LINE DEVELOPMENT. The technique employed to make patterns for prisms and

Pattern Development

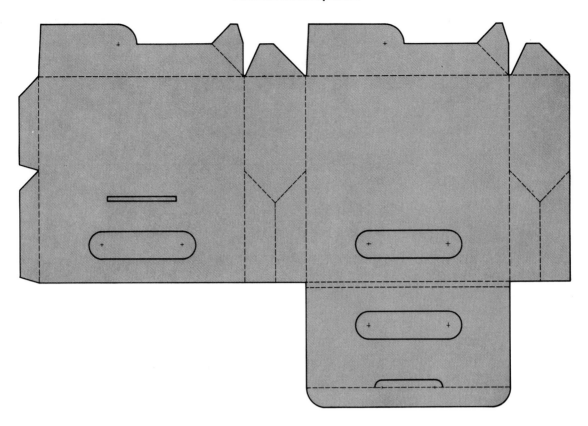

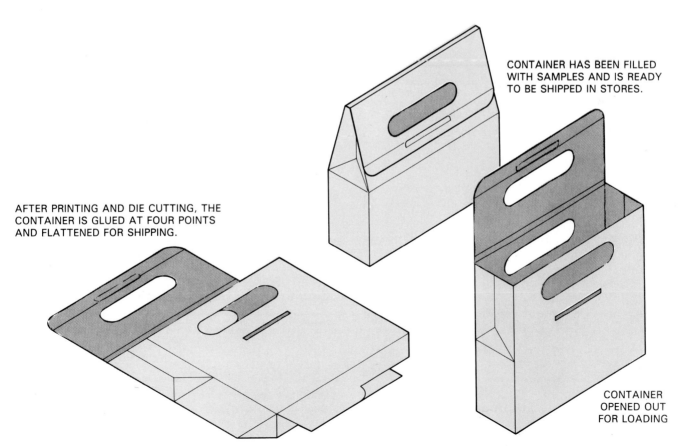

CONTAINER HAS BEEN FILLED WITH SAMPLES AND IS READY TO BE SHIPPED IN STORES.

AFTER PRINTING AND DIE CUTTING, THE CONTAINER IS GLUED AT FOUR POINTS AND FLATTENED FOR SHIPPING.

CONTAINER OPENED OUT FOR LOADING

Fig. 13-1. A computer generated pattern of a container used to introduce samples of a new product. The design reduces waste to a minimum, only four points are glued, and it can be folded for easy shipment.

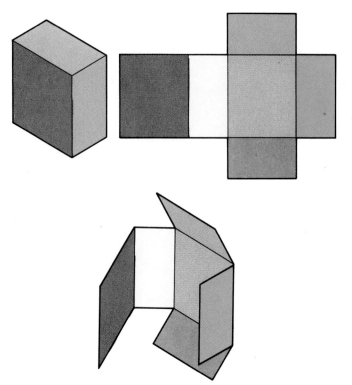

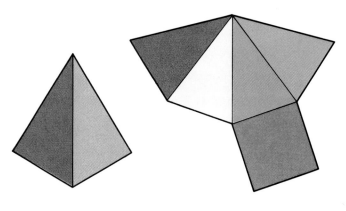

PARALLEL LINE DEVELOPMENT

Fig. 13-2. Very accurate patterns are needed to cut and shape the materials used to manufacture this aircraft. (Grumman Aerospace Corp.)

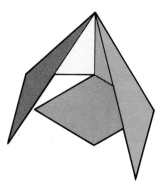

RADIAL LINE DEVELOPMENT

Fig. 13-3. A flying model airplane depends upon accurate patterns to construct wing ribs, fuselage, formers, and tail sections.

Fig. 13-4. PARALLEL LINE DEVELOPMENT is used to draw patterns for prisms and cylinders. (The lines used are parallel or at right angles to each other.) Patterns for regular tapering shapes like cones and pyramids are developed using RADIAL LINE DEVELOPMENT (the lines used to radiate out from a single point).

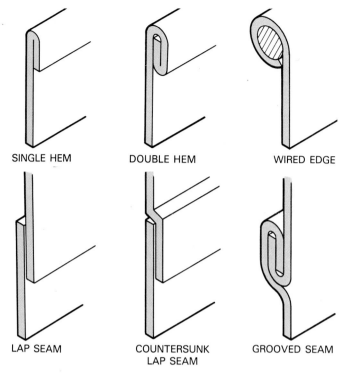

SINGLE HEM   DOUBLE HEM   WIRED EDGE

LAP SEAM   COUNTERSUNK LAP SEAM   GROOVED SEAM

Fig. 13-5. Typical hems, edges, and seams used to join and give rigidity to sheet metal. Extra material is required and must be included in the design.

meanings. Sharp folds or bends are indicated on the stretchout by a visible object line. See PATTERN DEVELOPMENT OF A RECTANGULAR PRISM, page 196, for an example of this type.

Curved surfaces are shown on the pattern with construction lines or center line. An example of this is shown in PATTERN DEVELOPMENT OF A CYLINDER, page 195.

Stretchouts are seldom dimensioned.

When developing a pattern or stretchout for a sheet metal product, it is often necessary to allow additional material for HEMS, EDGES, and SEAMS, Fig. 13-5.

HEMS strengthen the lips of sheet metal objects. They are made in standard fractional sizes, 3/16 in., 1/4 in., 3/8 in., etc. Standard metric sizes are 4.0 mm, 6.0 mm, 10.0 mm, etc.

A WIRE EDGE provides extra strength and rigidity to sheet metal edges.

SEAMS make it possible to join sheet metal sections. They are usually finished by soldering, spot welding, or riveting.

REFERENCE LINES and REFERENCE POINTS are needed when developing the stretchout of a circular object. To provide these, the circle is divided into twelve (12) equal parts. Fig. 13-6 shows how to divide a circle to produce the reference points.

Irregular curves are drawn using a FRENCH CURVE. The points of the irregular curve are first

cylinders. The lines used to develop the pattern are parallel or at right angles to each other.

RADIAL LINE DEVELOPMENT. Patterns for regular tapering forms such as cones, pyramids, etc. are developed by this technique. The lines used to develop the pattern radiate out from a single point.

Combinations and variations of the basic developments are used to draw patterns for more complex geometric shapes.

While regular drafting techniques are utilized in pattern development, some lines have additional

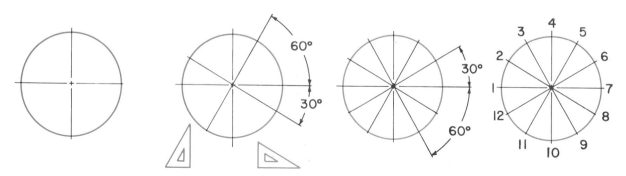

Fig. 13-6. How to divide a circle into twelve (12) equal parts.

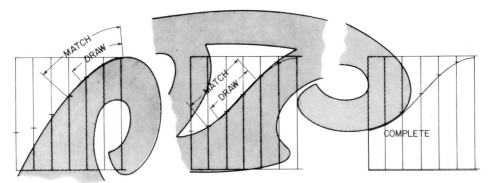

Fig. 13-7. Using a French Curve to draw an irregular curve. When drawing curved lines take care to make the curve of the line flow smoothly to produce one continuous line.

plotted. The points may then be connected with lightly sketched lines. Match the French Curve to the sketched line or points, taking care to make the curve of the line flow smoothly. Refer to Fig. 13-7 to see how this is done.

DRAFTING VOCABULARY

Extensive, Fabrication, Fitting, Hem, Irregular curve, Pattern, Pattern development, Parallel line development, Prism, Radial line development, Seam, Stretchout, Surface development, Wire edge.

UNIT 13—TEST YOUR KNOWLEDGE

Please do not write in the book. Place your answers on a sheet of note paper.

1. Pattern development is defined as a form of drafting that _____
_____.

2. Pattern development is also known as:
   a. _____.
   b. _____.

3. Patterns are also known as _____.

4. List four uses of patterns.
   a. _____.
   b. _____.
   c. _____.
   d. _____.

5. In pattern development, a heavy solid line (visible object line) indicates that _____
_____.

6. Very light lines (construction lines) and center lines indicate that _____.

7. Irregular curves are drawn with a _____
_____.

8. The lines used to develop patterns for prisms and cylinders use lines that are _____ or at _____ angles to each other.

9. The lines used to develop patterns for cones or pyramids _____ from a single _____.

OUTSIDE ACTIVITIES

1. Secure illustrations of products that require pattern development in their manufacture. They may be cut from discarded magazines and newspapers. Label each product as to how the pattern was developed, by parallel line development or by radial line development.

2. Using a discarded product such as a paper cup, milk carton, or snack food package, carefully unfold the product. Draw a full size pattern for the product.

PATTERN DEVELOPMENT
OF A CYLINDER

1. Draw front and top views of required cylinder. Divide top view into twelve (12) equal parts and number as shown.
2. The height of the pattern or stretchout is the same as the height of the front view. Project construction lines from the top and bottom of the front view.
3. Allow sufficient space (one-inch is adequate) between the front view and the pattern, and draw a vertical line. This will locate line 1 of the pattern.

4. Set your compass or divider from 1 to 2 (the points where the division lines intersect the circle) on the top view. Transfer this distance to the extended lines of the pattern to locate reference lines 1, 2, 3, - 12, 1.
5. Draw the top and bottom tangent to the extended lines.
6. Allow 1/4 in. for seams, and go over all outlines with visible object lines. The lines that represent the curves or circular lines are drawn in color, or, are left as construction lines.

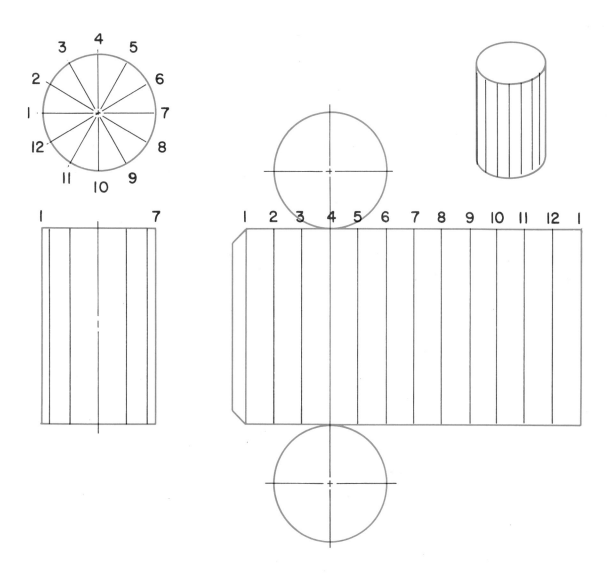

PATTERN DEVELOPMENT OF A
RECTANGULAR PRISM

1. Draw the front and top views.
2. The height of the pattern is the same as the height of the front view. Project construction lines from the top and bottom of the front view.
3. Measure over one inch from the front view and draw a vertical line between the extended lines to locate line 1.
4. Set your compass or divider from 1 to 2 on the

top view, and transfer this distance to the extended lines. Locate the other distances in the same manner.
5. Construct the top and bottom as shown.
6. Allow 1/4 in. for seams, and go over all outlines and folds with visible object lines. The pattern may be cut out, folded to shape, and cemented together using rubber cement.

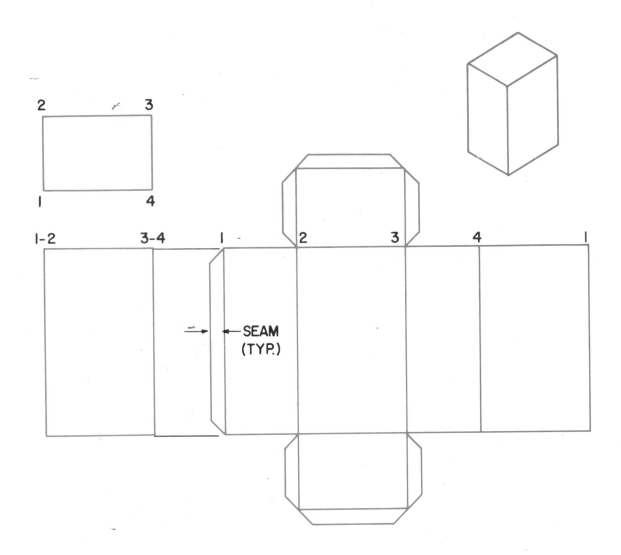

PATTERN DEVELOPMENT OF A
TRUNCATED PRISM

1. Draw front and top views. Number the points as shown.
2. Proceed as in previous examples of pattern development.
3. Mark off and number the folding points. Project point 1 on the front view, to line 1 of the stretchout. Repeat with points 2, 3, and 4 to lines 2, 3, and 4.
4. Connect the points 1 to 2, 2 to 3, 3 to 4, and 4 to 1.
5. Draw the top and bottom in position.
6. Allow material for seams, and go over the outline and fold lines with visible object lines. The pattern produces a truncated prism.

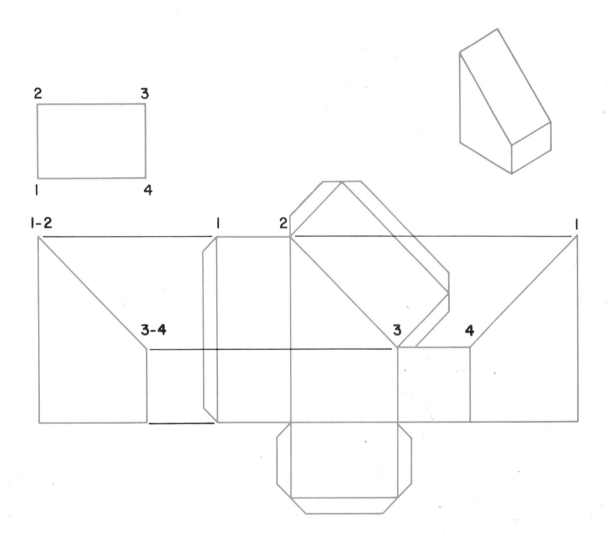

PATTERN DEVELOPMENT OF A
TRUNCATED CYLINDER

1. Draw the front and top views. Divide and number as shown.
2. Extend lines from the top and bottom of the front view.
3. Allow one inch between the front view and the pattern and draw line 1.
4. Set your compass or dividers from 1 to 2 on the top view and step off twelve (12) equal divisions on the extended lines of the pattern. Number them.
5. Draw vertical construction lines at each of the above divisions.
6. The curve of the pattern is developed by projecting lines from the points on the front view. Point 1 is projected over until it intersects line 1 on the pattern; points 2-12 intersect lines 2 and 12; etc. When all of the points are located they are connected with a curved line drawn with a French Curve.
7. Complete by adding the top and bottom. The top is developed as an auxiliary view. The pattern produces a truncated prism.

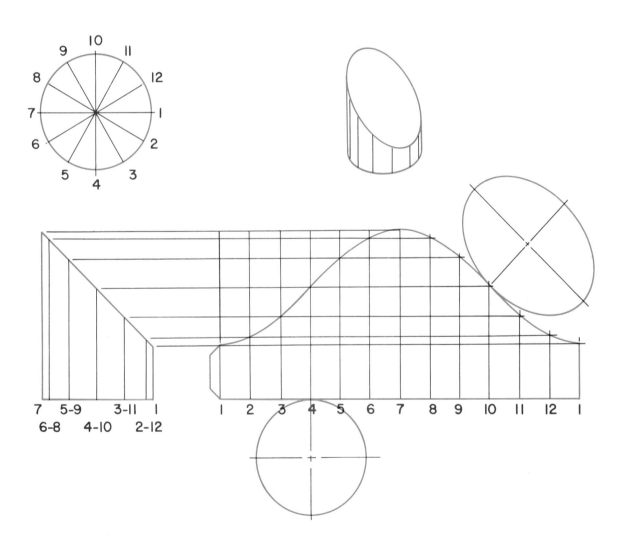

PATTERN DEVELOPMENT
OF A PYRAMID

1. Draw the front and top views. Number as shown.
2. Locate center line X of the stretchout.
3. Set your compass to a radius equal to X-1 on the front view and using the above center, draw arc A-B.
4. Draw a vertical line through center X and arc A-B.

5. Set compass from 1 to 2 on top view and at point where vertical line intersects the arc as the starting point, step off two (2) divisions on each side of the line. (Points 1-2-3-4-1 on the stretchout.)
6. Connect the points, draw the bottom in place, and go over the outline and folds with object lines. The completed pattern will produce a pyramid.

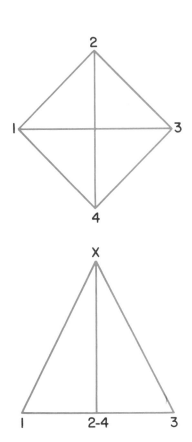

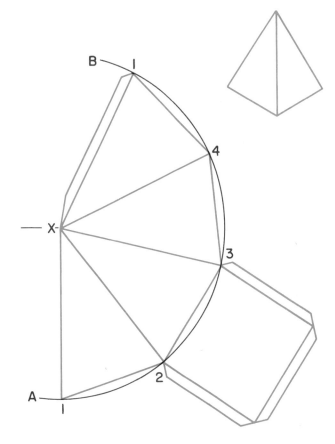

PATTERN DEVELOPMENT
OF A CONE

1. Draw the front and top views. Divide the top view into twelve (12) equal parts. Number as shown.
2. Locate center line X of the stretchout.
3. Set your compass from X to 1 on the front view and with X of the stretchout as the center, draw arc A-B.
4. Draw a vertical construction line through center line X and the arc.

5. Set your compass from 1 to 2 on the top view, and the point where the vertical line intersects the arc as the starting point, step off six (6) divisions on each side of the line. (Points 1-2-3-4 etc. on the pattern.)
6. Go over the outline carefully with object lines.
7. The lines that represent the curved portion may be drawn in color, or, they may be left as construction lines.

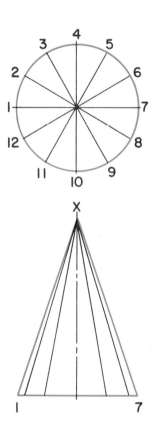

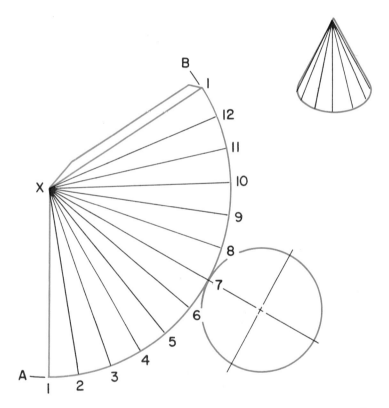

PATTERN DEVELOPMENT FOR A
RIGHT RECTANGULAR PYRAMID

1. Draw the front and top views. Number as shown.
2. Neither view shows the true length of the pyramid edges. To find the true length, rotate one edge as shown and project it to the front view where it appears in true length when the projected point is connected with point X.
3. Locate center line X of the stretchout.
4. Set compass to true length of the pyramid's edge. With X as the center, draw arc A-B.
5. Draw a vertical line through center X and arc A-B.
6. Set compass from 1-2 on the top view. With the point where the vertical line intersects arc A-B as the starting point, step off pyramid side 1-2.
7. Reset compass to distance 2-3 on the top view and step off side 2-3 on arc A-B.
8. Repeat the sequence and step off sides 3-4 and 4-1 on arc A-B.
9. Connect the points. Draw the bottom in place. Go over the outline and folds with object lines. The pattern produces a right pyramid.

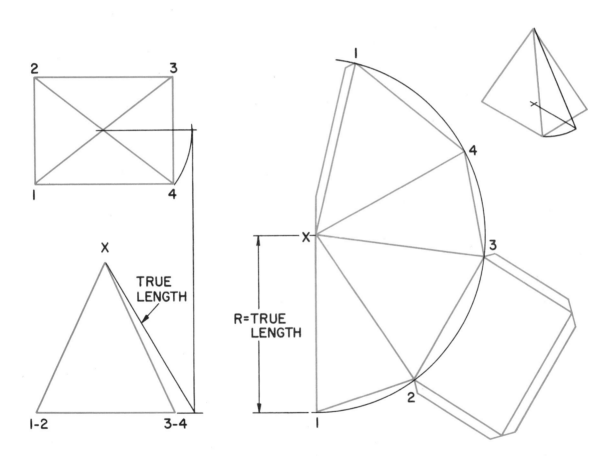

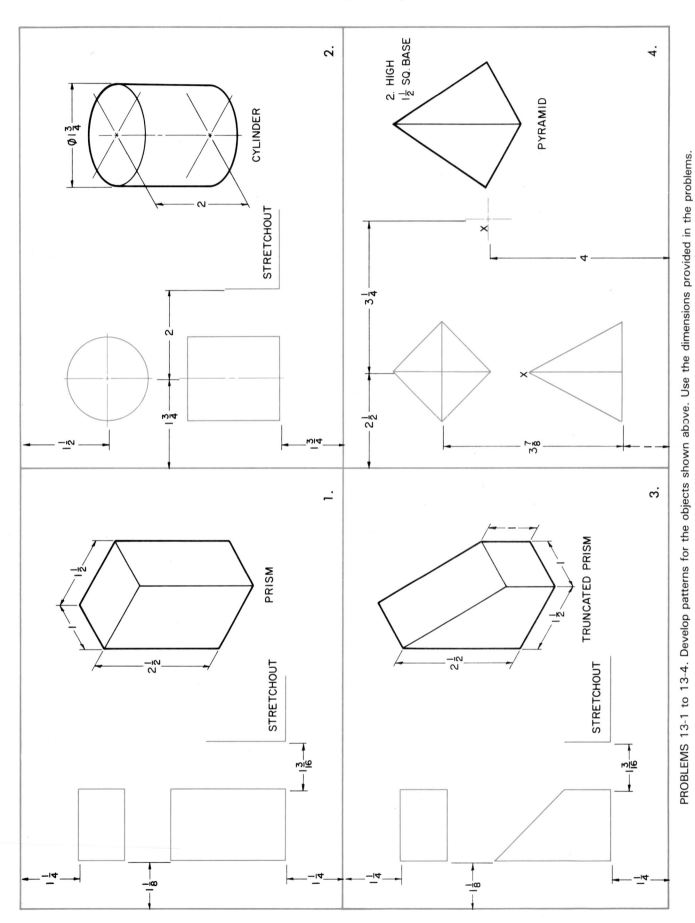

PROBLEMS 13-1 to 13-4. Develop patterns for the objects shown above. Use the dimensions provided in the problems.

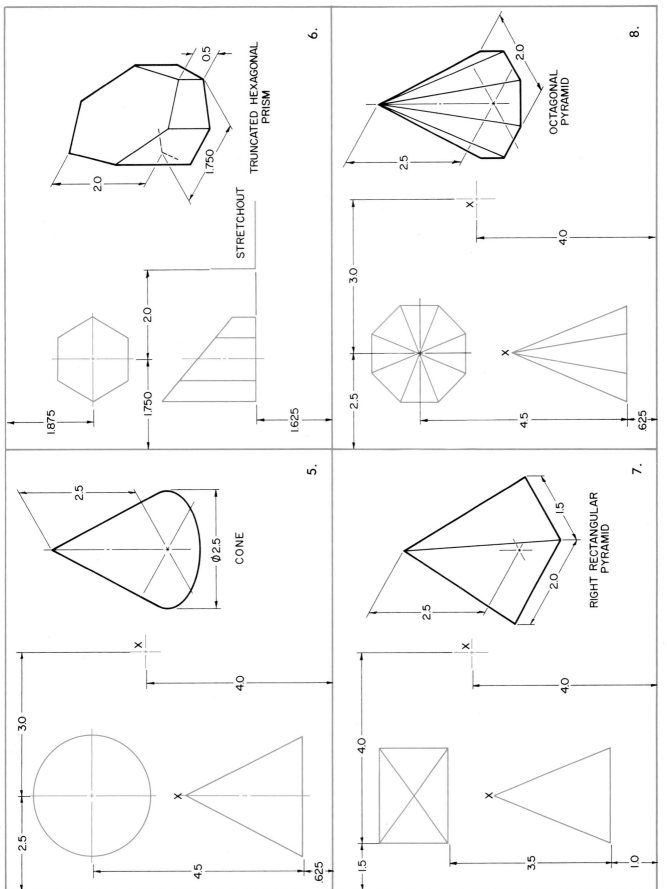

PROBLEMS 13-5 to 13-8. Develop patterns for the objects shown above. Use the dimensions provided in the problems.

203

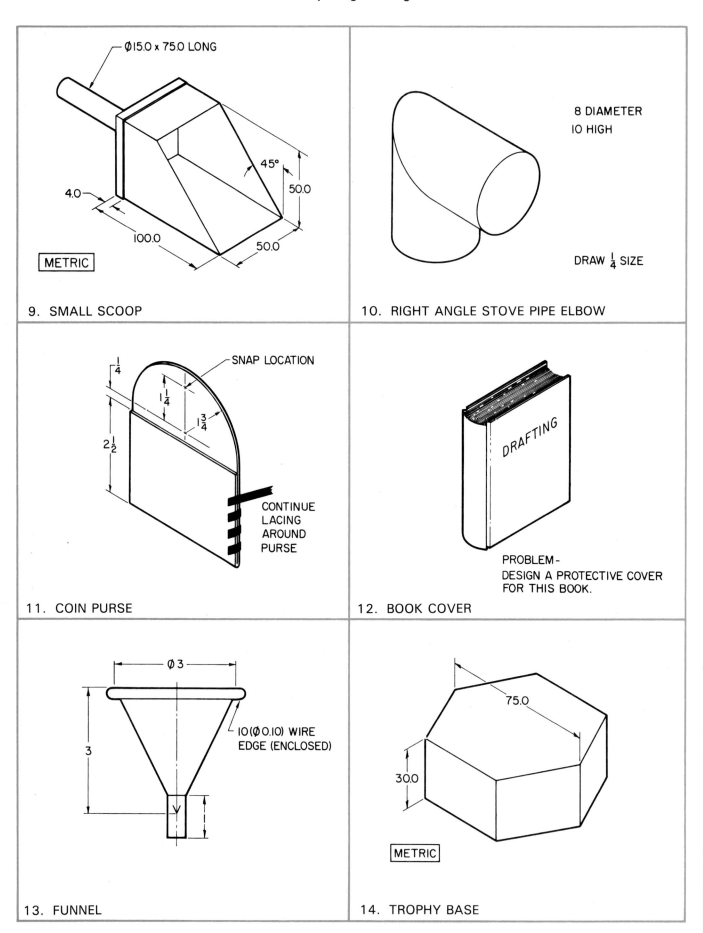

Ø15.0 x 75.0 LONG

45°

50.0

4.0

100.0

50.0

METRIC

9. SMALL SCOOP

8 DIAMETER
10 HIGH

DRAW ¼ SIZE

10. RIGHT ANGLE STOVE PIPE ELBOW

SNAP LOCATION

¼

1¼

1¾

2½

CONTINUE
LACING
AROUND
PURSE

11. COIN PURSE

DRAFTING

PROBLEM -
DESIGN A PROTECTIVE COVER
FOR THIS BOOK.

12. BOOK COVER

Ø3

10 (Ø 0.10) WIRE
EDGE (ENCLOSED)

3

13. FUNNEL

75.0

30.0

METRIC

14. TROPHY BASE

PROBLEMS 13-9 to 13-14. Develop patterns for the objects shown above. Use the dimensions provided in the problems.

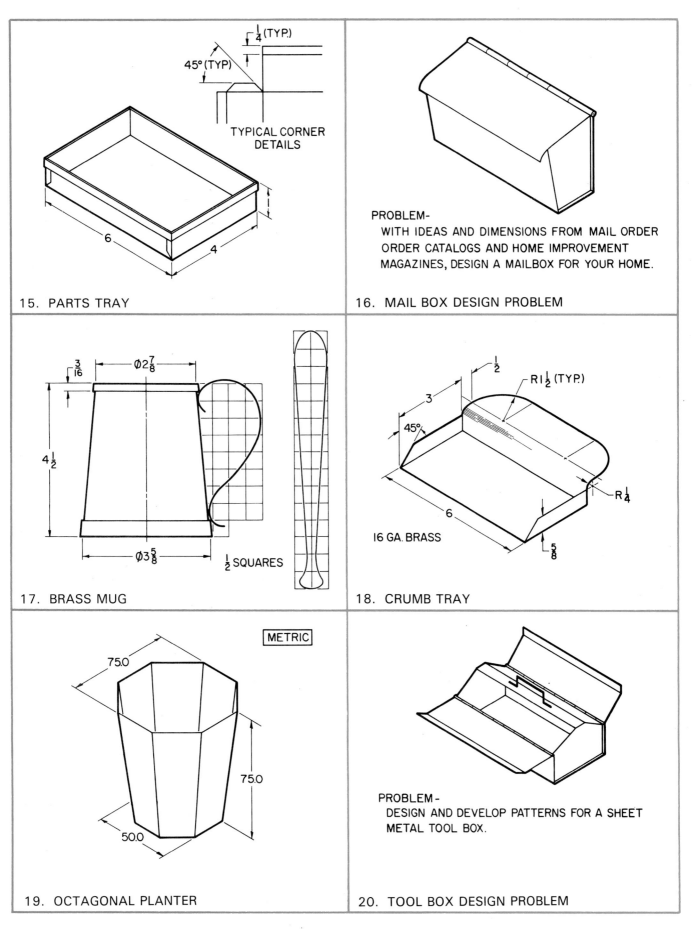

$\frac{1}{4}$ (TYP.)

45° (TYP.)

TYPICAL CORNER
DETAILS

6

4

15. PARTS TRAY

PROBLEM-
WITH IDEAS AND DIMENSIONS FROM MAIL ORDER
ORDER CATALOGS AND HOME IMPROVEMENT
MAGAZINES, DESIGN A MAILBOX FOR YOUR HOME.

16. MAIL BOX DESIGN PROBLEM

$\frac{3}{16}$

$\emptyset 2\frac{7}{8}$

$4\frac{1}{2}$

$\emptyset 3\frac{5}{8}$

$\frac{1}{2}$ SQUARES

17. BRASS MUG

$\frac{1}{2}$

R1$\frac{1}{2}$ (TYP.)

3

45°

6

R$\frac{1}{4}$

16 GA. BRASS

$\frac{5}{8}$

18. CRUMB TRAY

METRIC

75.0

75.0

50.0

19. OCTAGONAL PLANTER

PROBLEM-
DESIGN AND DEVELOP PATTERNS FOR A SHEET
METAL TOOL BOX.

20. TOOL BOX DESIGN PROBLEM

PROBLEMS 13-15 to 13-20. Develop patterns for the objects shown above. Use the dimensions provided in the problems.

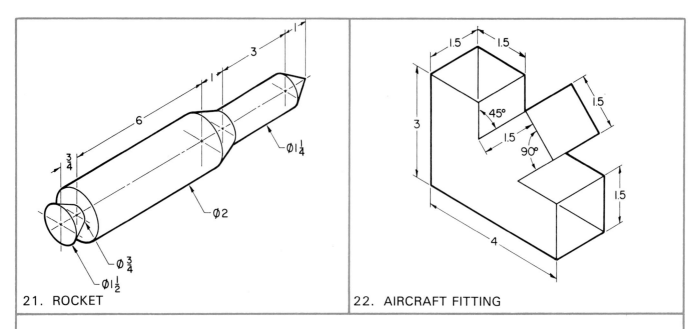

21. ROCKET

22. AIRCRAFT FITTING

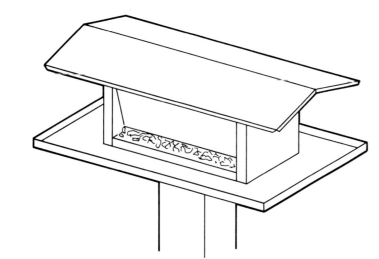

PROBLEM:

DESIGN AND DEVELOP PAT-
TERNS FOR A BIRD FEEDER.
THE ROOF MUST BE HINGED
FOR EASY LOADING OF FEED.
CONSTRUCT A CARDBOARD
MODEL OF YOUR DESIGN.

23. BIRD FEEDER DESIGN PROBLEM

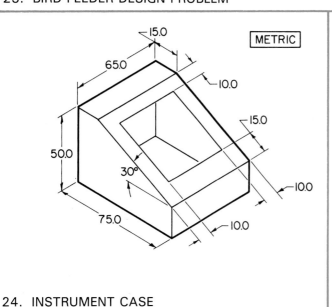

METRIC

24. INSTRUMENT CASE

DESIGN PROBLEMS-

DESIGN AND DEVELOP PATTERNS FOR THE
FOLLOWING PROJECTS. MAKE A CARDBOARD
MODEL OF YOUR CHOICE.

- WASTE BASKET
- PLANTER
- DUST PAN
- CHARCOAL SCOOP
- BOOK ENDS
- LEATHER HAND OR WALLET
- JEWEL BOX

25. PROBLEMS

PROBLEMS 13-21 to 13-25. Develop patterns for the objects shown or described above. Use the dimensions provided in the problems or design appropriate sized products for the end use.

Unit 14

WORKING DRAWINGS

After studying this chapter, you will be able to explain the use of working drawings. You will be able to distinguish detail drawings and assembly drawings. You will be able to interpret the information provided on working drawings. You will be able to design and draw various working drawings.

It would not be practical for industry to manufacture a product without using drawings which provide complete manufacturing details.

Required drawings range from a single freehand sketch for a simple part, to many thousands of drawings needed to manufacture a complex product like the Space Shuttle. Working drawings tell the craftworker what to make, and establish the standards to which he or she must work, Fig. 14-1.

Working drawings fall into two categories, DETAIL DRAWINGS and ASSEMBLY DRAWINGS. A DETAIL DRAWING includes a view (or views) of the product, with dimensions and other pertinent information required to make the part.

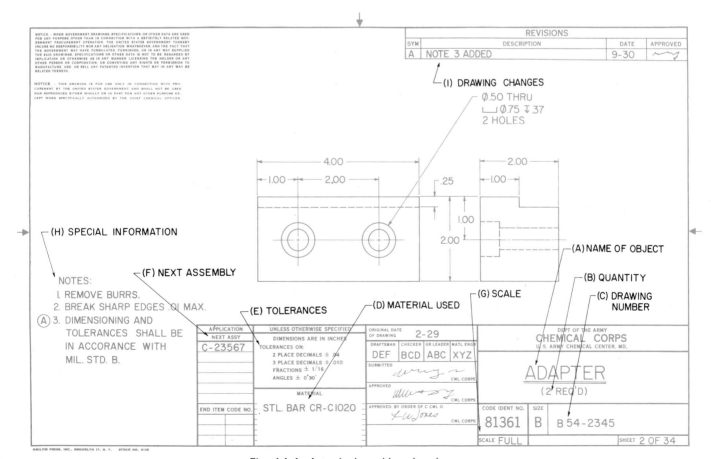

Fig. 14-1. A typical working drawing.

Refer to Fig. 14-1 and note the following information on a detail drawing.

A. NAME OF THE OBJECT. The name assigned to the part. This should appear on all drawings making reference to the piece.

B. QUANTITY. The number of units needed in each assembly.

C. DRAWING NUMBER. Number assigned to the drawing for filing and reference purposes.

D. MATERIAL USED. The exact material specified to make the part.

E. TOLERANCES. The permissible deviation, either oversize or undersize, from a basic dimension.

F. NEXT ASSEMBLY. The name given to the major assembly on which the part is to be used.

G. SCALE. Drawings made other than full size are called scale drawings.

H. SPECIAL INFORMATION. Information pertinent to the correct manufacture of the part that is not included on the various views of the object.

I. REVISIONS. Changes that have been made on the original drawing.

In most instances, the detail drawing provides information on a single part. See Fig. 14-2. Can you identify all the information points in Fig. 14-2? It is also permissible to draw all of the parts and the assembly of a small or simple mechanism on the same sheet, Fig. 14-3.

An ASSEMBLY DRAWING contains views that show where and how the various parts fit into the assembled product. Fig. 14-4 is an example of a typical assembly drawing. They also show how the complete object will look.

When the object is large or complex it may not be possible to present all of the information on one sheet. A SUB-ASSEMBLY DRAWING illustrates the assembly of only a small portion of the complete object, Fig. 14-5. A sub-assembly drawing solves the problem of not fitting all the drawing information on a single sheet.

Assembly and sub-assembly drawings may also be in pictorial form, Fig. 14-6.

Under certain situations, a drawing may include

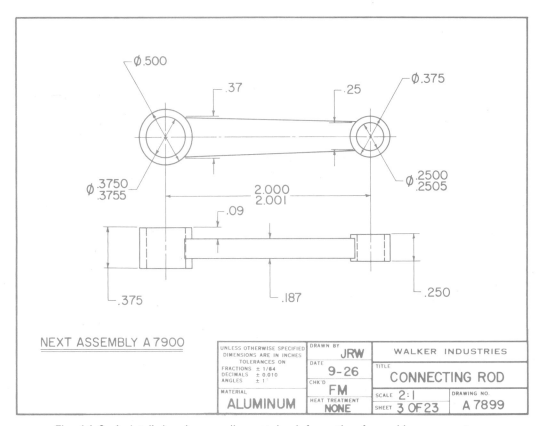

Fig. 14-2. A detail drawing usually contains information for making one part.

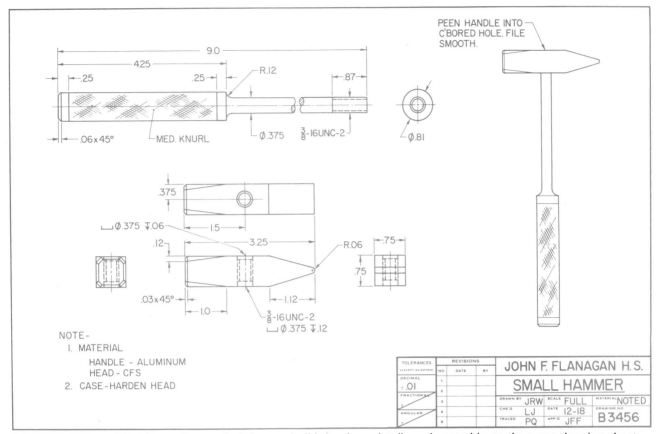

Fig. 14-3. If the object is small enough, it is permissible to draw details and assembly on the same drawing sheet.

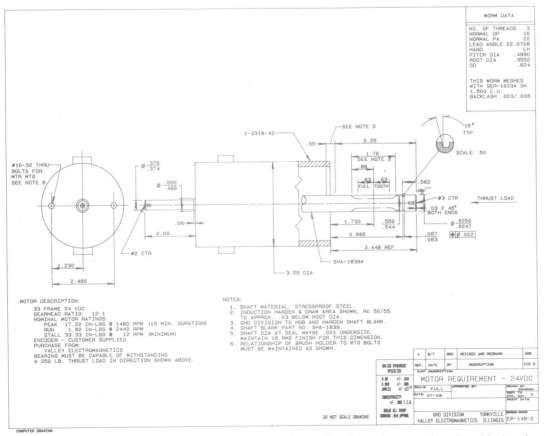

Fig. 14-4. An assembly drawing generated by computer aided drafting. The dimensions are placed on this drawing for reference purposes for quality control. (T & W Systems, Inc.)

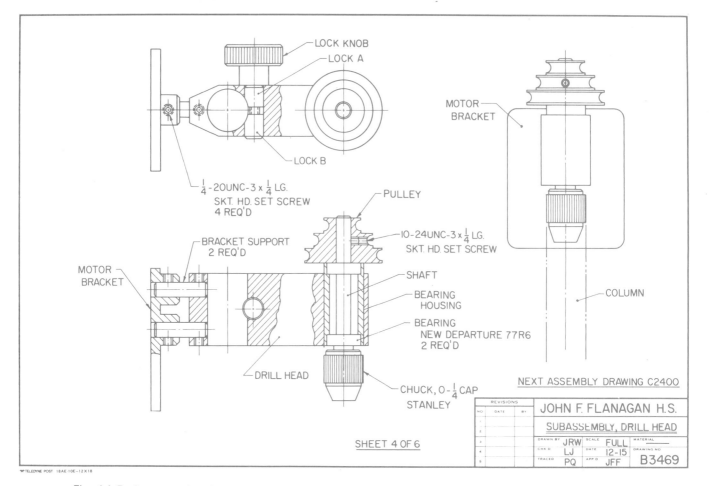

Fig. 14-5. A conventional type sub-assembly drawing. It shows the assembly of a small drill press head.

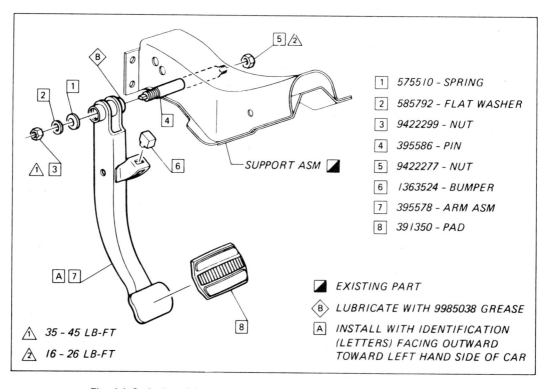

1	575510 - SPRING
2	585792 - FLAT WASHER
3	9422299 - NUT
4	395586 - PIN
5	9422277 - NUT
6	1363524 - BUMPER
7	395578 - ARM ASM
8	391350 - PAD

◨ EXISTING PART

B LUBRICATE WITH 9985038 GREASE

A INSTALL WITH IDENTIFICATION
(LETTERS) FACING OUTWARD
TOWARD LEFT HAND SIDE OF CAR

△1 35 - 45 LB-FT

△2 16 - 26 LB-FT

Fig. 14-6. A pictorial type sub-assembly drawing. This type of drawing
is often used with semiskilled workers who have received a minimum
of training in print reading. (General Motors Corp.)

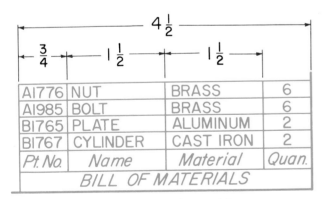

Fig. 14-7. Parts list.

a PARTS LIST, Fig. 14-7, and/or a BILL OF MATERIALS, Fig. 14-8. The information on the parts list is read from top down when the list is located at the upper right corner of the drawing sheet, or, read from the bottom up when located above the title block. Either position is acceptable. The parts are usually listed according to size or in their order of importance.

An IDENTIFYING NUMBER is placed on each sheet for convenience in filing or rapid location of the drawing, Fig. 14-9. The drawing number is also usually the part number stamped or printed on the object after it is manufactured.

Pt. No.	Name	Material	Quan.
A1776	NUT	BRASS	6
A1985	BOLT	BRASS	6
B1765	PLATE	ALUMINUM	2
B1767	CYLINDER	CAST IRON	2

BILL OF MATERIALS

Fig. 14-8. Bill of materials.

DRAFTING VOCABULARY

Assembly drawing, Categories, Complex product, Detail drawing, Deviation, Manufacturing details, Mechanism, Oversize, Permissible, Pertinent, Revision, Standards, Sub-assembly, Tolerances, Undersize, Working Drawing.

TEST YOUR KNOWLEDGE—UNIT 14

Please do not write in the text. Place your answers on a sheet of notebook paper.

1. Why are working drawings used?
2. The _____ drawing has a picture of the part with dimensions and other information needed to make the object.
3. The _____ drawing shows where and how the object described in the above drawing fits into the complete assembly of the object.
4. On large or complex objects, _____-_____ drawings are frequently used. These drawings show the assembly of only a small portion of the complete object.

The drawing described in question #2 includes much information necessary to manufacture a part. Some of this information is described below. Place the letter of the correct description in the blank before the words below.

5. ____ QUANTITY
6. ____ TOLERANCES
7. ____ MATERIAL USED
8. ____ SCALE
9. ____ REVISIONS

A. The exact material specified to make the part.
B. Drawings made other than to full size.
C. The number of units needed in each assembly.
D. The permissible deviation, either oversize or undersize, from a basic dimension.

UNLESS OTHERWISE SPECIFIED DIMENSIONS ARE IN INCHES TOLERANCES ON FRACTIONS ± 1/64 DECIMALS ± 0.010 ANGLES ± 1°	DRAWN BY JRW	WALKER INDUSTRIES	
	DATE 9-23	TITLE BRACKET	
	CHK'D FM		
MATERIAL STEEL AISI-1020	HEAT TREATMENT NONE	SCALE FULL	DRAWING NO. A 123456-1
		SHEET 1 OF 5	

Fig. 14-9. Each plate is given an identifying number for convenience in filing and locating drawings.

E. Changes that have been made on the original drawing.

10. Why are drawings given identifying numbers?

OUTSIDE ACTIVITIES

1. From a local industry, try to obtain samples of detail drawings. Include samples of assembly drawings and samples of sub-assembly drawings. Prepare a bulletin board display using the theme WORKING DRAWINGS.

2. Design a suitable title block for the drawing sheets of a company that you are planning to start. Make sure there are blanks to include all pertinent information required.

3. Secure the piston and rod assembly from a small single cylinder gasoline engine. Prepare detail drawings for these parts. Indicate all dimensions as decimals. It may be necessary for you to learn how to read a micrometer to make accurate measurements.

4. Design a product with sales appeal. Prepare the drawings necessary to mass-produce it in the school laboratory. Make your drawings on tracing vellum so that a number of prints can be made.

5. Research the topic of Geometric Dimensioning and Tolerancing. Report back to the class how industry uses this topic in preparing their assembly drawings. Show some examples.

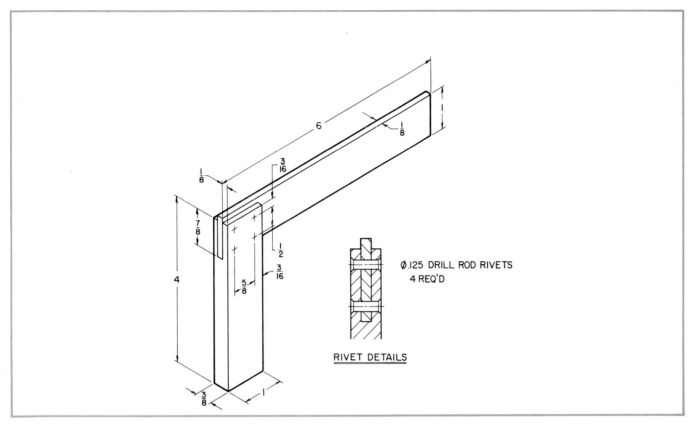

PROBLEM 14-1. MACHINIST'S SQUARE. Make detail and assembly drawings.

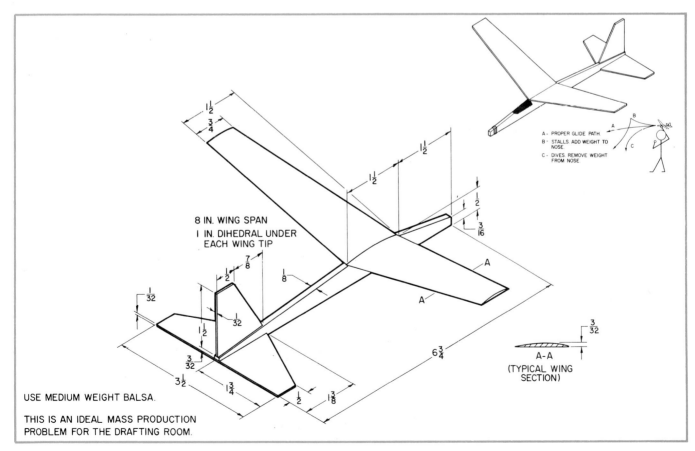

PROBLEM 14-2. MODEL GLIDER. Prepare detail and assembly drawings. Design the necessary patterns, fixtures, etc., needed to mass-produce the glider.

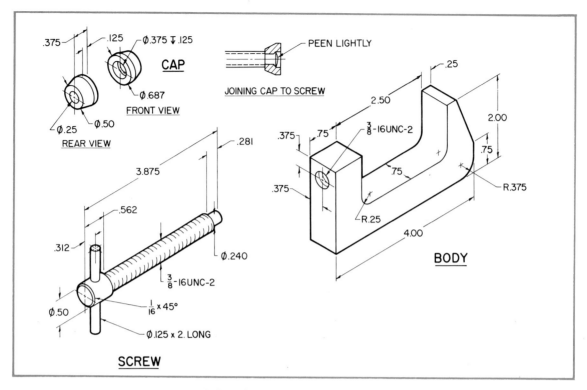

PROBLEM 14-3. C-CLAMP. Prepare detail and assembly drawings.

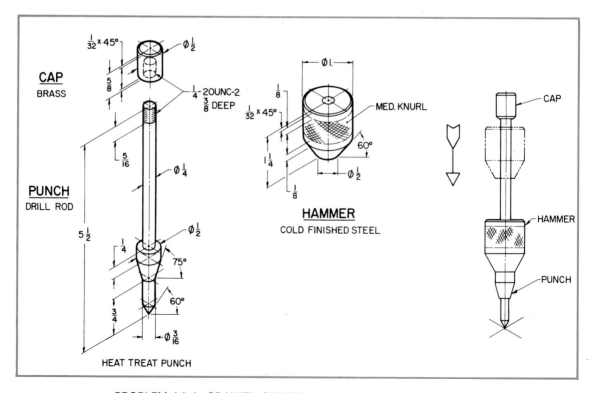

PROBLEM 14-4. GRAVITY CENTER PUNCH. Make detail drawings.

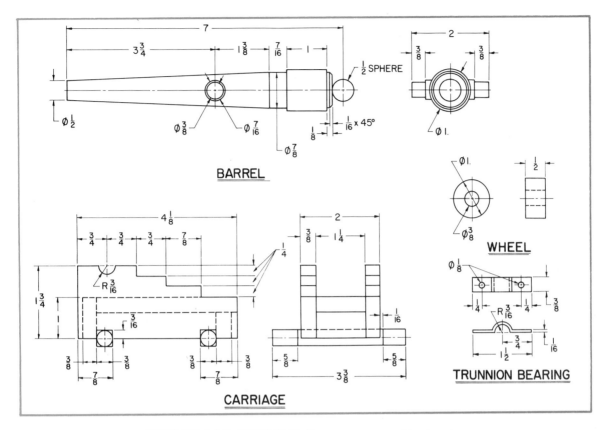

PROBLEM 14-5. DECK GUN. Prepare an assembly drawing.

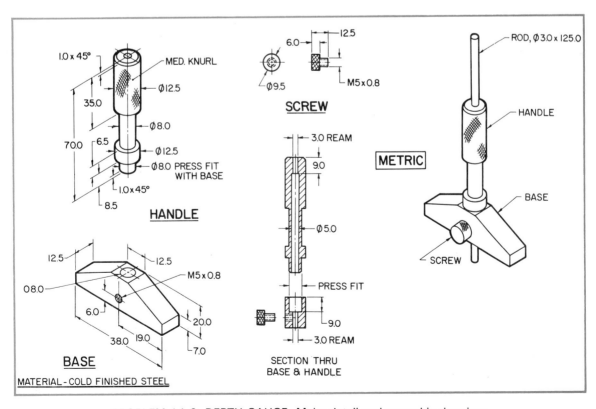

PROBLEM 14-6. DEPTH GAUGE. Make detail and assembly drawings.

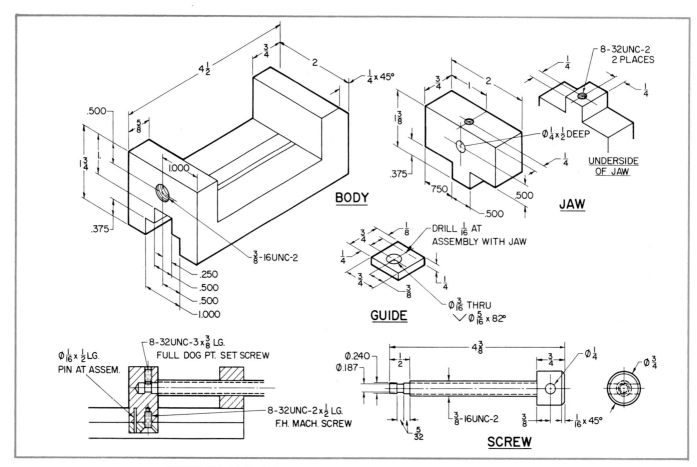

PROBLEM 14-7. SMALL VISE. Prepare detail and assembly drawings.

DESIGN PROBLEM—

USE THE BASIC DIMENSIONS GIVEN
AND DESIGN A DESK. IT DOES NOT
HAVE TO LOOK LIKE EITHER
DESK SHOWN.

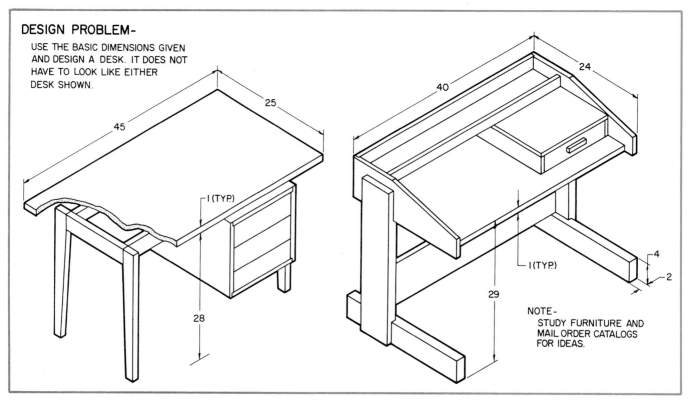

NOTE—
STUDY FURNITURE AND
MAIL ORDER CATALOGS
FOR IDEAS.

PROBLEM 14-8. CONTEMPORARY DESK. Design a desk using the dimensions given. Prepare the drawings necessary to make your design.

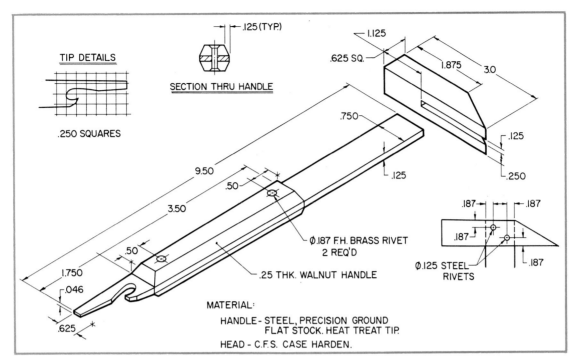

PROBLEM 14-9. HANDY HELPER. Prepare detail and assembly drawings.

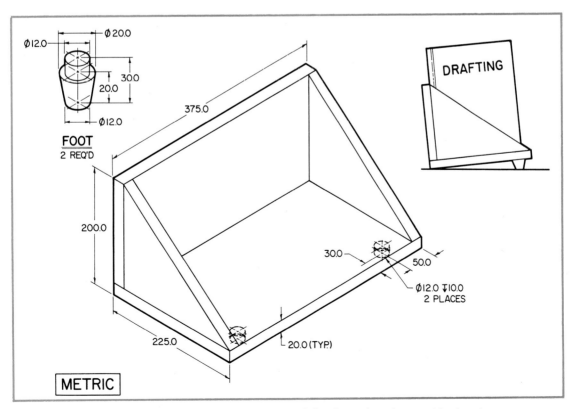

PROBLEM 14-10. BOOK RACK. Make a fully dimensioned assembly drawing.

Unit 15

PRINT MAKING

After studying this unit, you will be able to recognize when prints are used in place of original drawings. You will be able to demonstrate how to make a pencil or an ink tracing. You will be able to explain several print making techniques.

Original drawings are seldom found on a job because they would soon become worn or soiled, and difficult to read. Also, several sets of identical drawings are frequently needed at the same time since craftworkers at different locations are working on the product or structure. High cost makes it impractical to draw up a set of plans for each person who needs them. It is costly to create new originals to replace plans that are damaged or ruined. Reproductions, called PRINTS, of the original drawings are used in these situations.

When several sets of prints are needed, the original drawings are reproduced or duplicated. The technique used must produce accurate copies of the originals, must not destroy the original drawings, and must be speedy and reasonable in cost.

Several of the more important duplicating processes will be described in this unit. The original usually must be made as a tracing, Fig. 15-1. It is made on semitransparent (translucent) material such as tracing paper, cloth, or film.

HOW TO MAKE A TRACING

The first step in making a tracing for reproduction is to draw it on a semitransparent or translucent material. There are many types of this material available in sheet or roll form.

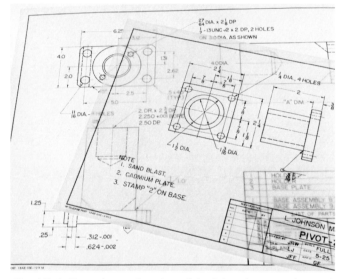

Fig. 15-1. To make a print, the original drawing must be made on tracing media (vellum, film, or cloth). This material is translucent (semitransparent).

TRACING PAPER is frequently used because it is less expensive than film or cloth. Tracing paper is employed extensively for making preliminary plans or sketches.

TRACING CLOTH is the most expensive of the tracing materials. Tracing cloth has been replaced almost entirely by tracing films. It is still used when the original tracings must withstand severe handling. The cloth is made from specially treated, fine grade linen which can take pencil or ink.

TRACING FILMS made from acetate, mylar, or polyester are ideally suited for work requiring print stability (it will not expand or contract in heat, cold, or high humidity). Film has high transparency and produces prints with maximum contrast. One surface has a matte finish that will take pencil or ink. Corrections are made with an

218

eraser or damp cloth. Tracing films are tough and tear resistant. They do not become discolored or brittle with age.

PENCIL TRACINGS

Sharp, clear reproductions can be made from pencil tracings with modern reproduction equipment and materials. At present, most drawings used by industry are done in pencil.

Many drafters make original drawings on the tracing material that will be used to produce the copies. For more complex jobs, the layout is usually traced from the original drawing onto other sheets which are used to make copies. This is where we get the term TRACING.

Tracings for reproduction are made the same way as the conventional drawings you have been making. Block in the needed views with a 4H or 5H pencil. If they are drawn very lightly, they will not need to be erased. Complete the drawing, add notes, and dimension with a 2H pencil. Keep the lines UNIFORMLY SHARP and DENSE as this type line reproduces best.

Every effort must be made to keep the tracing clean. Working with clean hands and a piece of clean paper under your hand when you letter, dimension, or add notes will also help to make sharp reproductions.

INKED TRACINGS

Inked tracings are more difficult to prepare than tracings in pencil. A knowledge of inking tech-

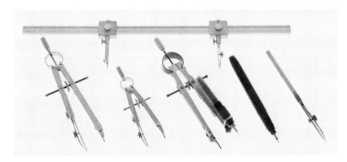

Fig. 15-2. Drafting equipment used in inking: Top—Beam compass with inking attachment. Left to right—Bow compass, small bow compass, bow compass with technical pen attachment, technical pen, ruling pen.

niques, much patience, and lots of practice are required to produce acceptable inked tracings.

A dense black, waterproof ink called INDIA INK is used. Lines are drawn with a RULING PEN or a TECHNICAL PEN, Fig. 15-2. The ruling pen can be adjusted to make different widths. The ruling pen is held in much the same manner as you would hold a pencil. Fill a ruling pen with the quill on the ink bottle cap, Fig. 15-3. NEVER dip the pen in the ink.

Fig. 15-3. The correct way to fill a ruling pen. NEVER dip the pen in the ink bottle.

Technical pens with different size points are needed for each line width desired. However, the technical pen is preferred because the line width is uniform each time that particular size line weight is drawn. A technical pen must be held vertical.

Practice inking lines until they are uniform in width. Make sure the ink does not run under the edge of your drawing instruments.

Most of the difficulties you will encounter when starting to ink are shown in Fig. 15-4. Many of these difficulties can be avoided or their occurance greatly reduced if you keep the pen clean. Use great care so you do not accidentally slide your instruments or hand into freshly inked lines.

When inking a tracing, the experienced drafter follows a definite order of working. The order preferred by the author is to ink:
1. All circles, arcs, and tangents, starting with the smallest and working to the largest.
2. Hidden circles and arcs.
3. Irregular curved lines.

CORRECTLY INKED LINE

INSTRUMENT ACCIDENTALLY PUSHED
INTO FRESHLY INKED LINE

PEN NOT HELD VERTICALLY

INK RAN UNDER EDGE OF
INSTRUMENT. PEN NOT HELD VERTICAL

NOT ENOUGH INK TO COMPLETE LINE *

TOO MUCH INK IN PEN *

Fig. 15-4. Common mistakes made in inking. (*Applies only to ruling pen.)

4. Straight lines in this order:
   a. Horizontal.
   b. Vertical.
   c. Inclined.
5. Hidden straight lines in the same order listed in 4.
6. All center, extension, and dimension lines.
7. Border and section lines.
8. Lettering, dimensions, and arrowheads. The title block should be done last.

BLUEPRINTS

The BLUEPRINT PROCESS is the oldest of the techniques employed to duplicate drawings. The print consists of white lines on a blue background, Fig. 15-5.

Since this process requires the print to be developed, "fixed" in a chemical solution, washed, and then dried, blueprints are seldom used today. The newer print making processes are dry and much faster.

DIAZO PROCESS

The DIAZO PROCESS is a copying technique for making direct positive copies (dark lines on a white background) of almost anything that is drawn, written, or typed on translucent material, Fig. 15-6. The copies are produced quickly and inexpensively.

A print is made by placing the tracing in contact with sensitized material (paper or film) and

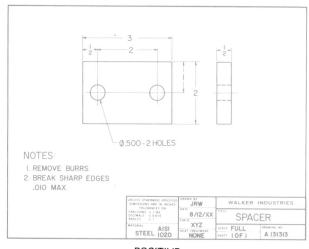

POSITIVE

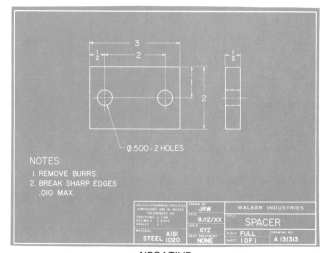

NEGATIVE

Fig. 15-5. The true blueprint is a print with white lines and a blue background. The term "blueprint" is commonly used when referring to all types of prints, regardless of the color of the lines or background.

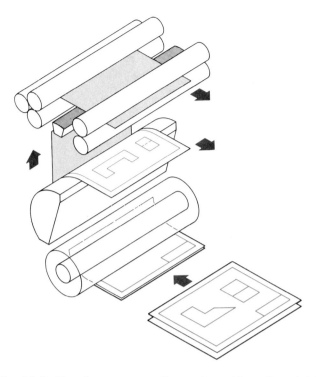

Fig. 15-6. The diazo process. On most machines the print is made in one continuous operation.

Fig. 15-7. A modern combination printer and developer type diazo printer. (Blu-Ray Inc.)

exposing them to light. Light cannot penetrate the opaque lines drawn on the tracing. Exposure takes place when light strikes the light sensitive coating of the print paper. After exposure, the print paper is developed by passing it through ammonia vapors.

With most modern diazo copying machines, Fig. 15-7, the print is made in one continuous operation. In addition to turning out the finished copy in seconds, it is clean, odorless, and quiet. The copies produced (depending on the type of sensitized paper used) can be black-on-white, color-on-white, or color-on-color.

XEROGRAPHY (ELECTROSTATIC) PROCESS

XEROGRAPHY (pronounced ze-rog'-ra-fee) is a print making process that uses an electrostatic charge (stationary electrical charge) to duplicate an original written, drawn, or printed image, Fig. 15-8. It is based on the principle that like electrical charges repel and unlike charges attract, Fig. 15-9.

The xerographic process makes a copy exactly duplicating the original. It is clean and dry. The copy can be enlarged or reduced in size if desired. The original drawing does not have to be on translucent material or on tracing media.

Full color copies can be made on color copying xerographic machines. This is very important in the electrical/electronics industry where wires are color coded.

THE MICROFILM PROCESS

The MICROFILM PROCESS was originally designed to reduce storage facilities and to protect prints from loss, Fig. 15-10. With this technique, the original drawing is reduced by photographic means to negative form. Finished negatives can be stored in roll form, or on cards adaptable to computerized storage and retrieval. Cards or film are easier to store than full-size tracings. To produce a working print, the microfilm image is retrieved from the files, and viewed or enlarged (called BLOW-BACKS) on photographic paper.. The print is often discarded or destroyed when it is no longer needed. Microfilms can also be viewed on a READER to check details, Fig. 15-11.

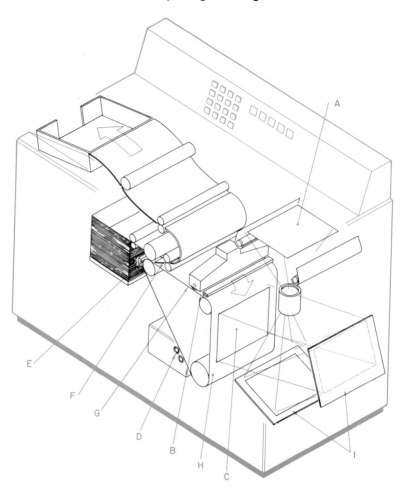

Fig. 15-8. Prints can be made on a xerography printer. Print size can be enlarged or reduced in size from the original drawing. A— Original drawing. B—A corona discharges a layer of positive charges on the photoconductor. C—White areas of the original drawing discharge the same areas on the photoconductor. D—Magnetic brushes disperse negatively charged toner (''ink'') over the remaining charged portion of the image. E—Positively charged paper presses against the image. It picks up the toner which is fused to the paper by pressure and heat. F—Rollers that apply the heat and pressure. G—Brushes and a vacuum cleaner remove the remaining toner and it is ready for the next cycle. H—Photoconductor belt. I—Mirrors and lens. The mirrors reverse the projected image so it will read correctly on the printed sheet. The lens is moved up or down to increase or decrease print size over the original drawing.

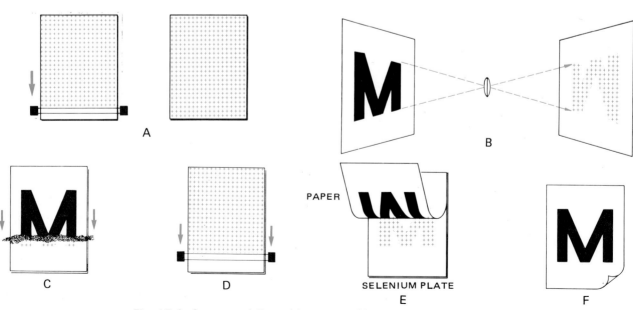

Fig. 15-9. Sequence followed in xerographic printing. (Xerox Corp.)

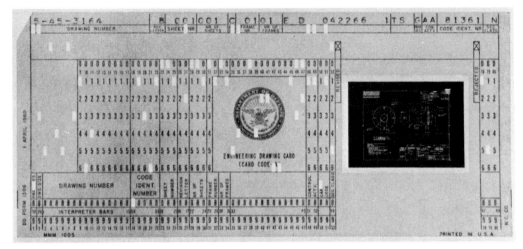

Fig. 15-10. The small negative on the microfilm aperture card is enlarged by a photographic process to the desired print size. When needed, the drawing can be retrieved by computerized techniques.

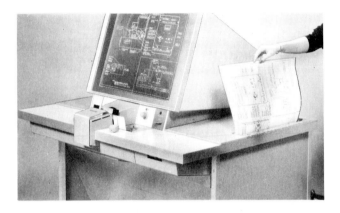

Fig. 15-11. A microfilm reader. The enlarged print can be viewed on a view screen in this unit and then made into a print of the required size. (RECORDAK)

COMPUTER GRAPHICS

Computer graphics is the newest way to produce original drawings and make prints from information stored electronically in computer memory. Refer to Unit 24 for a full explanation of this technique.

DRAFTING VOCABULARY

Accurate, Blow-backs, Computerized storage, Diazo, Duplicate, Microfilm, Opaque, Original drawings, Positive copies, Reproduction, Retrieval, Semitransparent, Sensitized, Stability, Tracing, Translucent, Xerography.

TEST YOUR KNOWLEDGE—UNIT 15

Please do not write in the book. Place your answers on a sheet of notebook paper.
1. Why are prints used instead of the original drawings?
2. To make reproductions, the original drawings are usually made in the form of _____.
3. Name the two most used basic types of tracing media.
4. A uniformly _____, _____ line reproduces best when making a tracing.
5. The ink used to make an ink tracing is called _____ ink.
6. List five methods for making prints.
7. The _____ method can make full color reproductions of the original drawings. Full color prints are used by the electrical/electronic industries where _____ are color coded.
8. Enlarged prints made from _____ are called blow-backs.

OUTSIDE ACTIVITIES

1. Make a pencil tracing of one of your drawings and duplicate it by one of the reproduction methods studied in this unit. Make an ink tracing of the same drawing and duplicate it by the same reproduction process. Compare the results. How do they differ? Were all the pencil lines uniformly sharp and dense? How do "weak" lines reproduce?

223

2. Secure an example of each of the reproduction processes described in this unit. Prepare a bulletin board display around them. Label each method.
3. Obtain an example of a microfilm aperture card and a print made from it. How does the quality compare to a full size reproduction?
4. Survey industries in your area. What type of reproductions are used by architects, contractors, and manufacturers?

Unit 16

DESIGN

After studying this unit, you will be able to describe the role of the industrial designer. You will be able to apply six basic design guidelines to problem solving situations and develop good designs for products. You will be able to demonstrate and prepare designs for your own projects.

Design is a plan for the simple and direct solution to a technical problem. Good design is the orderly and interesting arrangement of an idea to provide certain results and/or effects, Fig. 16-1.

A well designed product is functional (does what it is supposed to do and does it well), efficient, and dependable. Such a product is less expensive than a similar poorly designed product that does not function properly and must constantly be repaired.

The men and women doing this planning for industry are called INDUSTRIAL DESIGNERS, Fig. 16-2. In addition to a creative imagination, skilled designers must have a knowledge of engineering, production techniques, tools, machines, and materials to design a new product for manufacture, or to improve an existing product, Fig. 16-3.

DESIGN GUIDELINES

Product design requires much research and development. Many ideas and concepts must be studied, tried, refined, and then either used or discarded. The same technique is followed whether designing a simple trademark, Fig. 16-4, or a complex product like a ship, Fig. 16-5, or a student learning experience for this class.

Fig. 16-1. A design is a plan to solve a problem. The SOLAR CHALLENGER was designed to fly from Paris to London (about 200 miles or 320 km) using sunshine as its only power source. No batteries or other energy storage devices were employed. (The DuPont Company)

Fig. 16-2. This designer is developing an advanced automotive electrical system. As automotive electronics become more sophisticated (complex), systems must be designed that are more reliable and save valuable space and weight.
(United Technologies—Automotive Div.)

Fig. 16-4. While this trademark or logo appears to be simple, it was a challenging problem for its designer. The "mark" has great commercial value. People may not remember the company name but they recognize its trademark.

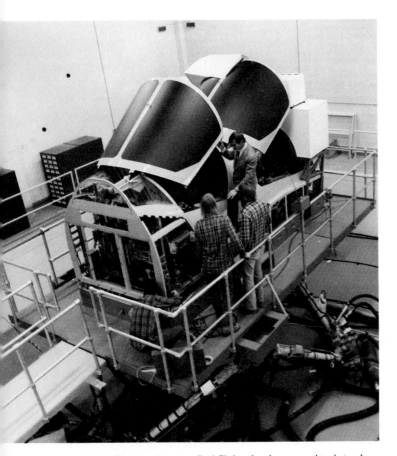

Fig. 16-3. Computer controlled flight simulator used to introduce flyers to new aircraft. Its movements, sounds, and visual simulation are so realistic that the aviators have no problem when they fly the real aircraft. Much creative imagination went into its design.
(Sperry Div., Sperry Corp.)

Fig. 16-5. Product design requires much research and development. Many people are responsible for the design of a complex object like an ocean going ship.

The design process is not an exact science. There is no special formula that if followed 1, 2, 3 . . . will guarantee a well designed product. However there are certain qualities that can help you identify good design.

FUNCTION. A well designed product "works." It does what it is designed to do and does it efficiently. Function often dictates product shape. Some design problems have several solutions, Fig. 16-6.

A

B

C

Fig. 16-6. A design problem may have several solutions. Here is how three companies designed a vertical take-off and landing (VTOL) aircraft. Each solved the problem. Can you explain how each aircraft operates? (A—Bell Helicopter. B—McDonnell Douglas. C—United Technologies, Sikorsky Helicopter.)

Fig. 16-7. A well designed product does not have one material masquerading as another material. One of the homes shown above has siding made from molded fiberglass, the other of real wood. Can you tell the one made with real wood?

HONESTY. The qualities of the materials (strength, weight, texture, etc.) used in the design are emphasized. One material is not made to look like another material, Fig. 16-7.

APPEARANCE. The elements of the design create interest. The product is visually pleasing, as shown in Fig. 16-8.

RELIABILITY. A well designed product is not always malfunctioning (breaking down) or failing. It should be easy to service and economical to maintain. See Fig. 16-9.

Fig. 16-9. A well designed product is dependable. It is not always in need of repairs. This truck section is being tested on a computer controlled road simulator. It tells the designers and engineers how the vehicle will react when subjected to severe road conditions. (Ford Motor Company)

Fig. 16-8. The elements of good design create interest and the product is visually pleasing. The PROBE IV is considered a very attractive automobile by many people. (Ford Motor Company)

SAFETY. When used properly, the product is safe. Protection from possible injury to the user is part of the basic design, Fig. 16-10.

QUALITY. Quality is a basic part of good design. It cannot be added after the product is made. Good design should be a guarantee of quality, Fig. 16-11. Quality is designed and built into items.

Fig. 16-10. A well designed product is safe. Tire fatigue (how fast it wears out or fails) and performance is being tested on this endurance machine. (Ford Motor Company)

Simply stated, good design is characterized by certain qualities. All of them are necessary. Overlooking or eliminating one can potentially destroy the entire design.

DESIGNING A PROJECT

How should YOU go about designing a project? One way is to employ a method followed by professional designers. Think of your project as a DESIGN PROBLEM:

1. THE IDEA. What kind of a project do you want to make? Keep your first few design problems simple. How will it be used? Will its use affect the shape of the project? Are there cost limitations?
2. DEVELOP YOUR IDEAS. What must your project do? How can it best be done? Research (study) similar products. How did other designers solve the problem? Make sketches of your ideas. Select the best of the ideas. When possible, submit your design for class critique (review).
3. MAKE MODELS. Construct models of your best ideas, Fig. 16-12. Evaluate them. Select the best. You may want to make refinements (changes) in your design when you see it in three dimensions.

Fig. 16-11. Quality is a basic part of good design. This hardwood (mahogany) stool is well designed and constructed, and it looks that way.

Fig. 16-12. Develop your ideas by making many sketches. Construct models of your best ideas to determine how they will look in three-dimensions. (The sketches shown were made with instruments and in ink so they would photograph well. Yours can be drawn in pencil.)

4. MAKE WORKING DRAWINGS. When satisfied that you have done the best you can to solve the problem, prepare working drawings for the project. If practical, make the drawings full size.
5. CONSTRUCT THE PROJECT. Make the project to the best of your ability. Do a job you will be proud to show to others, Fig. 16-13. Do not be afraid to ask for help from your instructor if you have a problem.

Developing a well designed project takes time. Do not be discouraged if your first attempts fall short of your goals. It takes practice and effort to acquire the skill and ability to solve design problems. The only way to succeed is by doing.

Keep a notebook of your design efforts. Include your sketches, drawings, and photos of projects you have designed. It will make it easier to evaluate your progress.

Fig. 16-13. The proof of a design is how it looks when it is constructed and put into use. This student-made night table is nicely designed and has been constructed using approved cabinet-making techniques. No nails or screws were used.

TECHNICAL VOCABULARY

Creative imagination, Dependable, Design, Design problem, Dictates, Efficient, Functional, Guidelines, Industrial designer, Limitations, Malfunctioning, Qualities, Refinements, Reliability, Research, Solution.

TEST YOUR KNOWLEDGE—UNIT 16

Please do not write in the book. Place your answers on a sheet of notepaper.
1. Good design is: (Check the correct answer.)
   a. _____ An orderly and interesting arrangement of an idea to produce certain results and/or effects.
   b. _____ A carefully thought out plan to produce a product that is functional and pleasing to the eye and/or touch.
   c. _____ A creative process.
   d. _____ All of the above.
   e. _____ None of the above.
2. The men and women who do design work in industry are called _____ _____.
3. They must have working knowledge of _____ _____ _____ _____ _____.
4. List the characteristics that make up good design.

5. List the steps you would follow to solve a design problem. Briefly describe each.
   1. _____ _____ _____
   2. _____ _____ _____
   3. _____ _____ _____
   4. _____ _____ _____
   5. _____ _____ _____

OUTSIDE ACTIVITIES

The design problems listed below are present to give you practice in design activities. You may also originate your own products to be designed.
a. Hand launched glider.
b. Jet propelled (CO_2 cartridge or Jetex Rocket engine) model automobile.

c. Book rack.
d. Box to store compact disks or tapes.
e. Tool box.
f. Turned bowl.
g. Clock.
h. Night stand table.
i. Stereo cabinet.
j. Bicycle rack.
k. Storage container for video cassettes.
l. Display cabinet for awards and trophies.
m. Fiberglass model boat.
n. Cutting board.
o. Metal base for a table lamp.

p. Coffee table.
q. Tray.
r. Turned wood lamp.
s. Plastic soft food spreader.
t. Salad server (laminated wood, wood, metal, plastic).
u. Wall shelf.
v. Shoeshine box.
w. Disposable waste basket (made from corrugated cardboard).
x. Desk top pencil holder.
y. Folder to store drafting plates.
z. Carrying case for drafting tools.

Unit 17

MODELS, MOCKUPS, AND PROTOTYPES

After studying this unit, you will comprehend why industry uses models, mockups, and prototypes. You will be able to identify and define the different uses of models, mockups, and prototypes. You will discover how various industries use them in design and preproduction functions. You will be able to recognize various model building materials.

To many people, modelmaking is an interesting hobby. You have probably made models of famous planes, boats, or cars from wood or plastic kits.

Industry makes extensive use of three modelmaking activities . . . MODELS, MOCKUPS, and PROTOTYPES. These are used for engineering, educational, and planning purposes. Models have proven to be very helpful in solving design problems and to check the workability of a design or idea before it is put into production.

MODEL

Industry's definition of a SCALED MODEL is a replica of a proposed, planned, or existing product. Even though computer graphics can be a great help in developing a new product design, a model may be constructed to see how the product will look in three-dimensions, to check out scientific theory, to demonstrate ideas, to aid in training, to use for advertising purposes. See Fig. 17-1.

MOCKUP

A MOCKUP is a full-size, three-dimensional copy of a product, Fig. 17-2. The mockup is usually made of plywood, plaster, clay, fiber glass, or a combination of materials.

PROTOTYPE

A PROTOTYPE is a full-size operating model of the production item, Fig. 17-3. It usually is handcrafted to check out and eliminate possible design and production "bugs."

HOW MODELS, MOCKUPS, AND PROTOTYPES ARE USED

Many industries employ models, mockups, and prototypes for design tools. See Fig. 17-4. A few of the more important applications are: automotive, aerospace, architecture, ship building, city planning, and construction engineering.

Fig. 17-1. Aerospace engineers study the feasibility of a new super sonic transport idea by using a scale wind tunnel model. (NASA)

Fig. 17-2. The space shuttle was first built as a full size mockup. (Rockwell International)

Fig. 17-3. When designing this 200 mph (320 km/h) train, it was important to know how it would stand up under actual operating conditions. Before it was put into production, a prototype was constructed and run more than a million miles. Only a few changes had to be made. (French National Railroad)

Fig. 17-4. Full size mockup of a radical design business aircraft. Potential customers can see how it will look. The mockup will also be used for advertising purposes. (Gates Learjet)

AUTOMOTIVE INDUSTRY

The automotive industry places great importance on the use of models, mockups, and prototypes. Mistakes in marketing and production can be very costly.

An automobile starts "life" as a series of sketches, Fig. 17-5. These are usually developed around specifications supplied by management. Promising sketches are usually drawn full size for additional evaluation.

Clay models are used for three-dimensional studies. Upon completion of further design development, a full size mockup is constructed, Fig. 17-6. If it is decided that the design has potential, a fiber glass prototype is generally fabricated, Fig. 17-7.

Fig. 17-5. The automotive industry uses sketches to develop ideas. Scale models are made of promising designs for future study and development. (General Motors Corp.)

Fig. 17-6. Final smoothing of the vinyl sheets coating the full size clay model of the Pontiac Trans Sport takes place before review by company officials. Only the tires are actual material. Modeled exactly to scale from the full-size line drawing on the wall, this clay body eventually will be destroyed after a two part epoxy/fiber glass mold is made. The resulting mold will be used to shape the body for a prototype vehicle.
(Pontiac Motor Div., GMC)

Fig. 17-7. Fiber glass prototype of the Pontiac Trans Sport. While only a few ''idea'' vehicles go into production, they are used for design studies. Many proposed innovations are utilized on production vehicles. (Pontiac Motor Div., GMC)

Fig. 17-8. Model of advanced USAF aircraft used for design studies. It was constructed employing information developed by CAD. (Grumman Aerospace Corp.)

Production fixtures (devices to hold body panels and other parts while they are welded together) and other tools needed for production are developed from the mockup.

After many months of development and production planning, the manufacturer is ready to make the new model vehicle available to the public.

AEROSPACE APPLICATIONS

The tremendous cost of aerospace vehicles makes it mandatory that they first be developed in model and mockup form, Fig. 17-8. Flight characteristics can be determined with considerable accuracy, without endangering human life, by testing complicated models in a wind tunnel, Fig. 17-9.

Prototype aircraft are the first two or three preproduction planes (usually handcrafted) that are flown to check the data obtained from wind tunnel and computer research, Fig. 17-10.

Fig. 17-9. Wind tunnel model of what could be the aircraft of the future. Wind tunnel models are EXACT scale models of the full size aircraft. While very expensive to construct, their cost is low when compared with the cost of the real aircraft. Flight characteristics of the full size aircraft can be determined without endangering the life of a pilot. (Grumman Aerospace Corp.)

Fig. 17-10. Prototype of the aircraft shown as wind tunnel model in Fig. 17-9. Flight tests were almost exactly as predicted by the model. (Grumman Aerospace Corp.)

Fig. 17-11. Architectural model of a proposed high school.

ARCHITECTURE

You have seen photos of proposed buildings in the real estate section of your newspaper. Some of these illustrations were of models that were very accurate miniature replicas of the proposed buildings, Fig. 17-11.

Many people use models when planning a new home, Fig. 17-12. This helps them to visualize how the house will look when completed. The model enables the owner to see the completed design in three-dimensions and also how paint colors and shrubbery plantings will look.

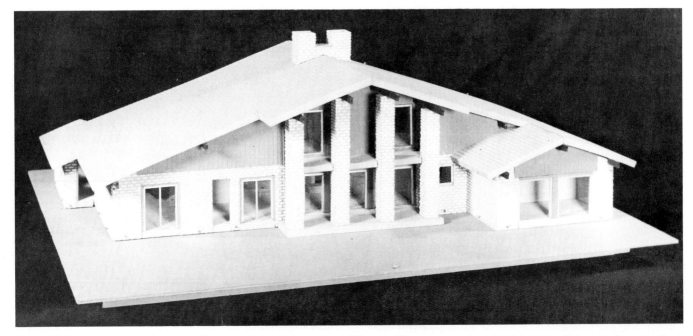

Fig. 17-12. Many potential home builders make models such as this one. They can see how their home will look before starting to build. (Taylor Made Models)

Fig. 17-13. Ship models at the Naval Ship Research and Development Center used to determine a proposed ship's characteristics. The model is tested by towing the model hull with instrumentation in the 140 ft. basin at the Center.

SHIPBUILDING

Ship hulls are tested in model form before designs are finalized and construction begins, Fig. 17-13. Specially designed equipment tows the model hull through the water test basin. The model hull behaves like the full size ship so design faults can be located and corrected. The models test a proposed ship's seaworthiness, power requirements, resistance to water, and stability characteristics.

CITY PLANNING

Most large cities use scale models to show city officials and planners how proposed changes and future developments will look. See Fig. 17-14. Models, while costly, permit intelligent decisions to be made before large sums of money are spent acquiring land and before existing buildings are torn down. The models can be used to plan for the movement of vehicles and people to avoid congestion after the plan is built. The models can be used to "help sell" the future development so financing can be obtained.

Fig. 17-14. Model showing the details of the tip of Manhattan and a portion of Brooklyn. The model is used for planning purposes. Many drafters will be needed to plan the future expansion of the city. (Lester Associates, Inc.)

CONSTRUCTION ENGINEERING

Many construction projects are designed from carefully constructed models, Fig. 17-15. By working from models, engineers can see how space can best be utilized. In some instances, they can determine how the proposed project will affect surrounding communities, Fig. 17-16. This helps minimize field problems and changes during construction.

Models may also be used to train personnel in plant operation.

CONSTRUCTING MODELS

Model making materials are readily available commercially, Fig. 17-17. Many products made for the model railroader are ideally suited for making architectural models. Kits are available for the small home builder who wants to design his or her own home.

A professional touch can be added to models by using accurately scaled furniture, automobiles, and figures that can be purchased at toy and hobby shops.

Preprinted sheets of brick and stone can be glued to suitable thicknesses of balsa wood for walls and partitions. Various types of abrasive paper are suitable for roofing, driveways, and walkways. Simulated window glass can be made from transparent plastic sheet. Several different scale sizes of windows and door frames are available molded in plastic. Most model shops can supply bushes and trees in various types and sizes.

Other types of models—autos, planes, and boats are made from bass wood, mahogany, balsa wood, metal, plaster, and various kinds of plastics.

Regular model making paint is produced in hundreds of colors and is ideal for painting all types of models. However, care must be exercised when painting models that have plastic in their construction. Be sure the paints are designed for plastics. If you are not sure, paint a scrap piece of similar plastic or a small portion of the plastic that is hidden from view. This will help you determine whether the paint is compatible with the material.

Fig. 17-15. Highly detailed model of a power generating plant. Note the clear plastic walls. The model is also used to train operating personnel.

Fig. 17-16. Model of a portion of the Susquehanna River in Maryland used to test how a proposed power generating plant will affect the river and its ecology. (Philadelphia Electric Co./Alden Research Lab)

Fig. 17-17. A few of the model making materials available at hobby shops.

SAFETY NOTE: Carefully read the instructions on the paint and adhesive containers. Use them only as specified.

Models may be assembled with model airplane cement, Elmer's, Titebond, cyanoacrylates, or epoxies. Use care and follow all safety directions according to the manufacturer's directions.

TECHNICAL VOCABULARY

Fabricated, Fiber glass, Fixture, Handcrafted, Mandatory, Mockup, Model, Preproduction, Prototype, Scaled Model, Simulated, Wind Tunnel.

UNIT 17—TEST YOUR KNOWLEDGE

Please do not write in the text. Place your answers on a sheet of notebook paper.
1. Industry uses models, mockups, and prototypes for what purposes?
2. A model is _____.
3. A mockup is a _____.
4. A prototype is _____.
5. List four (4) industries that employ models, mockups, and prototypes as design tools. Briefly describe how each industry listed used them.
6. Walls and partitions in a model home are usually constructed of _____ wood.
7. Roofs, driveways, and walkways can be made from different kinds and grades of _____ _____.
8. Models of cars, planes, and boats may be made from what kinds of materials?

OUTSIDE ACTIVITIES

1. Visit a model or hobby shop, then prepare a list of available materials which may be used for building architectural models.
2. Review technical magazines and clip illustrations that show models, mockups, and prototypes being used for engineering, educational, planning or other purposes. (Do not cut up library copies.)
3. Make a collection of materials suitable for building model homes.
4. Visit a professional model maker in your community. With permission, make a series of slides showing examples of various models and how these models are made. Then, give a talk to your class on professional model making.
5. Make a scale model of your drafting room. Discuss alternate layouts.
6. Construct a model helicopter similar to the one shown in Fig. 17-18. Design and construct from balsa wood a minimum building that would protect the helicopter from the elements.

Fig. 17-18. Plastic model helicopter to be used as a basis for constructing a hanger.

Unit 18

MAPS

After studying this unit, you will be able to define the uses of a map. You will be able to explain various types of maps and recognize their features. You will discover the types of maps drafters are able to produce.

A MAP is a graphic representation, usually on a flat plan, of a portion of the earth's surface. Maps are used to communicate ideas and concepts, as well as provide direction. Fig. 18-1 shows a typical map of a community.

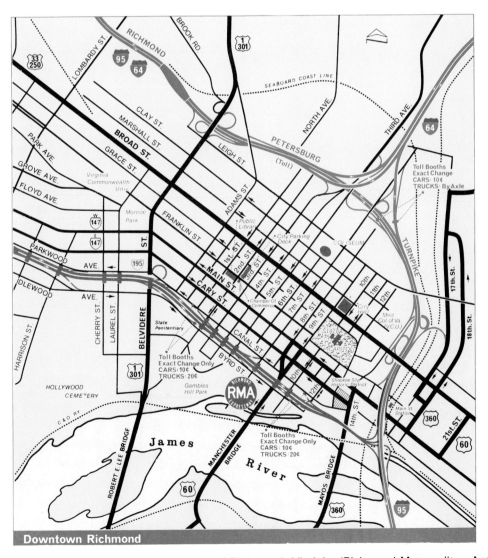

Fig. 18-1. Well-designed map shows a section of Richmond, Virginia. (Richmond Metropolitan Authority)

General reference maps show elevation and political features (cities and towns) of an area. United States Geological Survey maps show the exact location of features of the earth. Surveyors rely on these precision maps whenever they need data or facts.

Aeronautical and marine maps (often called CHARTS) provide navigational information for those using the air and sea lanes. Weather maps indicate meterological conditions for a given period of time, Fig. 18-2.

Road maps show the roads and highways in a region and aid in planning business and vacation trips. Other maps show or emphasize some particular features or surface characteristics of the earth (or planets).

A professional map maker is called a CARTOGRAPHER. He or she has the specialized skills to combine data from many sources to design and prepare accurate maps.

There are times when drafters are called upon to draw maps. This unit will describe the maps they would be expected to prepare.

PREPARING MAPS

Aerial photography, space satellites, and the computer have reduced the time needed to map large areas of land and sea. Accuracy has also been greatly improved. However, a great deal of "leg work" is still necessary to map small areas. A SURVEYOR, with an instrument called a TRANSIT, establishes the boundaries of the area being mapped, Fig. 18-3. Distances are measured with a TAPE. The BEARINGS (angles the lines make with due north and south) are recorded in the form of FIELD NOTES. With this information, maps such as the one shown in Fig. 18-4 are developed and drawn. North is towards the top of the sheet and is so indicated.

In metricated surveying, the areas of the surveyed land is given in HECTARES (ha) instead

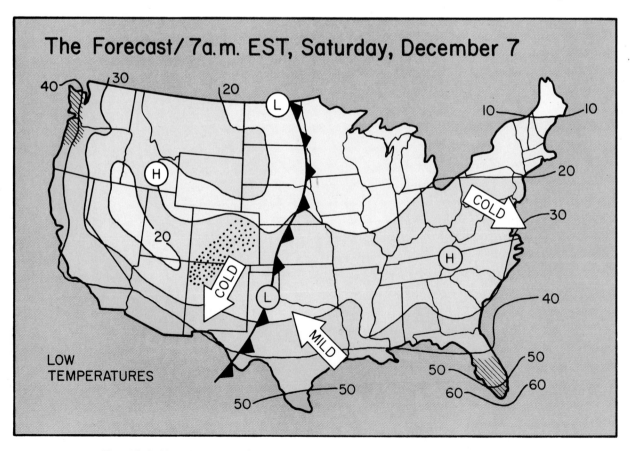

Fig. 18-2. Weather map. Most weather maps are now generated by computer.

Fig. 18-3. Surveyor using a transit to establish boundaries.

of acres, Fig. 18-5. Distances are given in meters and kilometers.

Maps are always drawn to scale because it is not possible to draw them full size. The size of the land area being mapped determines the scale.

In most towns, cities, and counties/parishes, a drawing of a building lot (called a PLAT) or farm may be drawn 1 in. = 30 ft. In places where metric units have been adopted, they will be drawn to a scale of 1:300. That is, 1.0 mm on the drawing equals 300.0 mm of the lot.

The UNITED STATES GEOLOGICAL SURVEY has started mapping the United States in metric. Each map will cover 4157 square kilometers. They will be printed to a scale of 1:25 000 (4.0 cm = 1.0 km). Such maps not yet metricated are drawn to a scale of 1:62,500 (approximately 1 in. = 1 mile).

Many maps use a graphic scale to indicate map scale, Fig. 18-6.

MAPS DRAWN BY DRAFTERS

The following maps are examples of a few of the many types a drafter is often requested to prepare. An important point to remember is to include the proper scale.

PLOT PLAN

The map of a lot on which a house is to be built is called a PLOT or PLAT PLAN, Fig. 18-7. The exact location of the lot and its size is given, as well as how the house or structure is situated on the lot.

The elevation or height of the land above sea level may also be shown as a series of irregular lines (called CONTOUR LINES) drawn on the map at predetermined differences in elevation.

CITY MAP

Street and lot layouts are shown on a CITY MAP, Fig. 18-8.

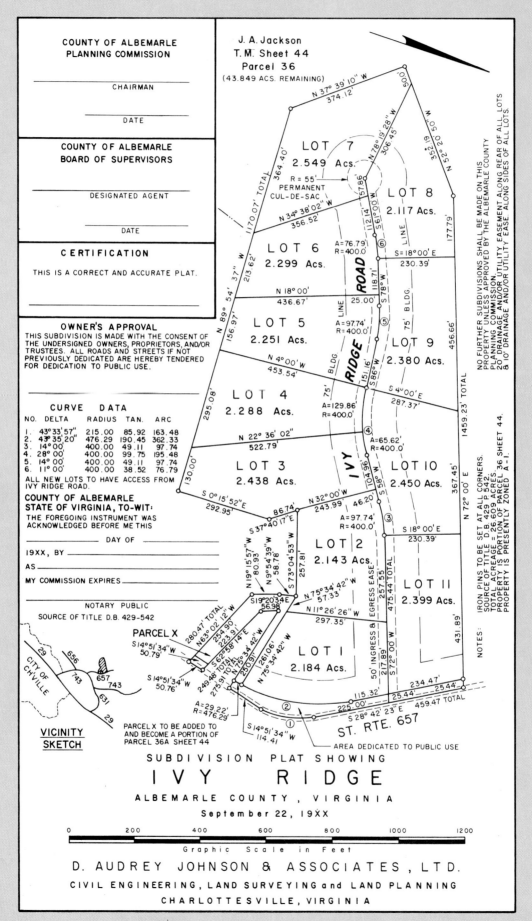

Fig. 18-4. Map typical of those prepared from information gathered by a surveyor. This is a plat of a housing subdivision.

Maps

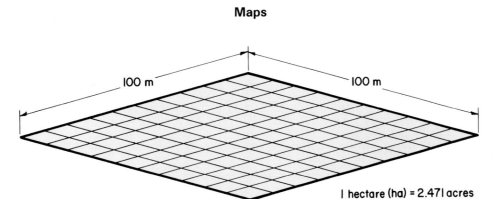

Fig. 18-5. A hectare (ha) is an area that is 100 meters long
and 100 meters wide. (It equals about 2.5 acres.)

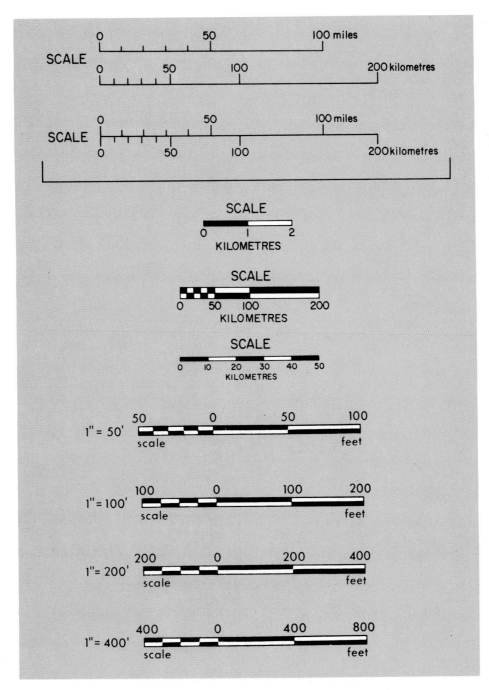

Fig. 18-6. Graphic scales found on many types of maps.

247

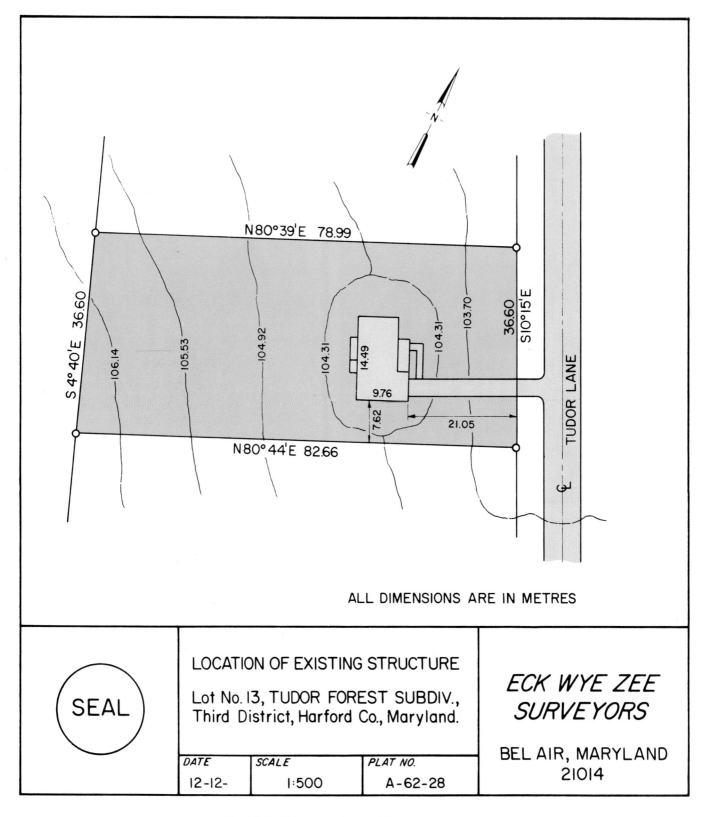

Fig. 18-7. Plat plan using metric measurements.

SPECIAL MAPS

Special maps such as LOCATION or VICINITY MAPS help you find your way around a specific area, Figs. 18-9 and 18-10. Other special maps aid in determining travel time and calculating distances. These maps are used to plan trips when traveling on vacations or for business, Fig. 18-11.

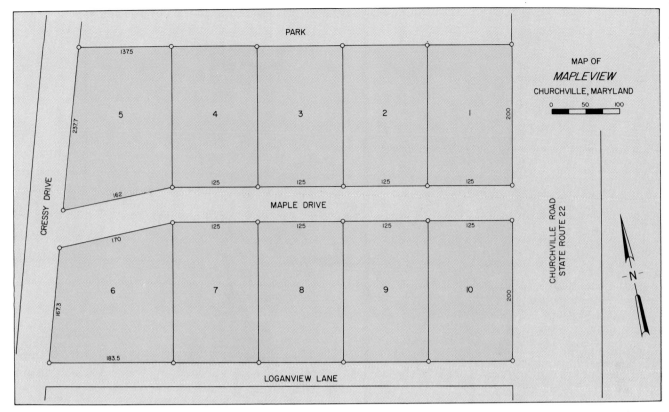

Fig. 18-8. Map of a section of a city showing the layout of streets and lots.

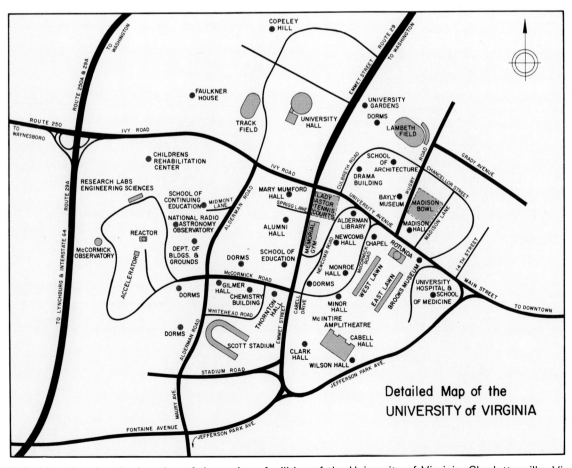

Fig. 18-9. Map showing the location of the various facilities of the University of Virginia, Charlottesville, Virginia.

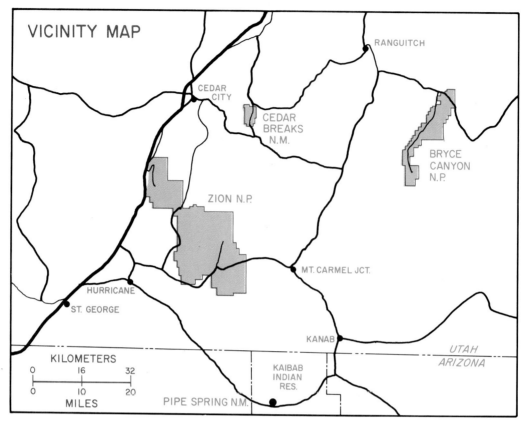

Fig. 18-10. Vicinity map of Zion National Park in Utah.

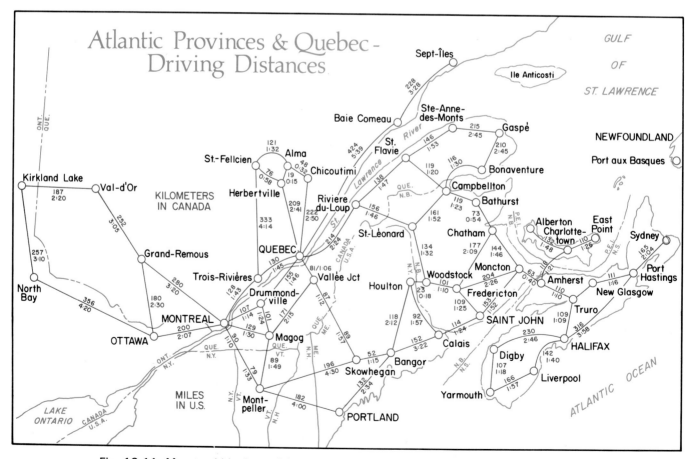

Fig. 18-11. Map to aid in determining travel time and calculating distances when traveling.

TOPOGRAPHIC MAPS

Information about physical features is included on a TOPOGRAPHIC MAP, Fig. 18-12. Crops, streams, airports, roads, bridges, buildings, and other land characteristics are shown. Contour lines are often included on topographic maps.

MAP SYMBOLS

By using SYMBOLS such as shown in Fig. 18-13, it is possible to include a large amount of information on a map. The symbol is often a stylized drawing (has a resemblance) of the feature it represents.

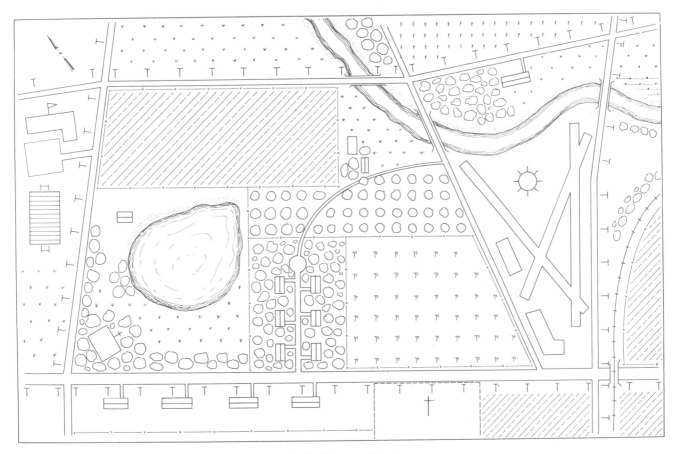

Fig. 18-12. Topographical map.

Fig. 18-13. A few of the hundreds of symbols used on maps. How many can you identify?

DRAFTING VOCABULARY

Acres, Aeronautical, Bearings, Boundaries, Cartographer, Concepts, Contour lines, Elevation, Emphasize, Field notes, Flat plane, Hectares, Marine, Meteorological, Metricate, Navigational, Plat, Political features, Specialized, Surveyor, Tape, Transit.

UNIT 18—TEST YOUR KNOWLEDGE

Please do not write in the book. Place your answers on a sheet of notepaper.
1. What is a map?
2. List some uses you have made of maps.
3. A professional map maker is called a _____.
4. A surveyor uses an instrument called a _____ to gather data to prepare a map.
5. Measurements made by a surveyor are recorded in the form of _____ _____.
6. As surveying is metricated, the area of the surveyed land is given in _____ instead of acres.
7. Why must maps be drawn to scale?
8. A plot or plat plan is a map of _____.
9. The elevation or height of land above sea level is often shown on a map with _____ lines.

OUTSIDE ACTIVITIES

1. Secure samples of special purpose maps and make a bulletin board display.
2. Prepare a map showing the school grounds. Label all athletic fields.
3. Make a plat plan of the property on which your home is located. Show the street and make sure North is up. Label all businesses, churches, and schools.
4. Prepare a map of your neighborhood.
5. Draw a map showing the route you take in coming to school.
6. CLASS PROJECT: Draw a map of the surrounding community and have each student draw in the route traveled in coming to school.
7. Secure a road map of your state. Plot the shortest route between your home town and the capitol of your state. If you live in the capitol city, plot the shortest route between it and the next largest city in the state.
8. Prepare a special map that will show the locations of the various schools in your school district.
9. MULTI-GROUP PROJECT: If survey equipment is available, have each group survey the school grounds and compare their results. Compare the results with an actual survey of the grounds.

Unit 19

CHARTS AND GRAPHS

After studying this unit, you will be able to create various charts and graphs used by industry and education. You will be able to apply charts and graphs to show trends, make comparisons, and measure progress. You will be able to analyze data or statistics and then create graphs or charts.

Industry and education have many uses for charts and graphs. A few of the more widely employed types are described and shown in this unit.

Charts and graphs make it possible to show trends, make comparisons, and measure progress quickly without studying a mass of statistics (numbers and information).

A drafter is often asked to prepare charts and graphs. Drafters take the statistical information and decides upon the most effective way to present this material in an interesting and easily understood manner.

GRAPHS

Graphs most frequently used are:
1. Line Graph.
2. Bar Graph.
3. Circle or Pie Graph (sometimes called an Area Graph).

LINE GRAPH

The LINE GRAPH may be used to make comparisons, Fig. 19-1. Lines that present the information are called CURVES. When only one curve appears on the graph, it should be drawn as a solid line.

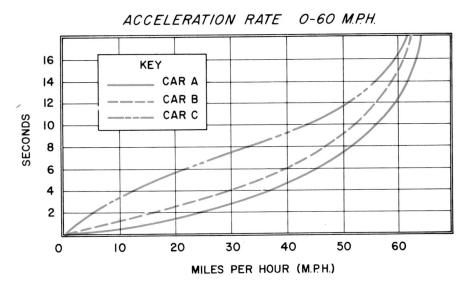

Fig. 19-1. LINE GRAPH that compares the times needed by three different cars to reach 55 mph.

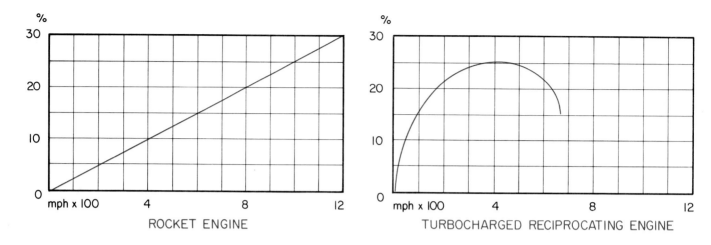

ENGINE EFFICIENCY-%

Fig. 19-2. LINE GRAPH showing that rocket engine efficiency increases the faster it travels while a turbocharged internal combustion engine's efficiency drops off at high speed.

When more than one curve is employed, each line should be clearly labeled and a KEY included with the graph to show what each curve represents.

A line graph may be used to show trends, that is, what has happened or what may happen. A line graph is illustrated in Fig. 19-2. In designing line graphs, be sure to use appropriate scales for each of the axes.

BAR GRAPH

Comparisons between quantities or conditions can also be made with BAR GRAPHS. Several different forms of the bar graph are available to the graph maker.

The HORIZONTAL BAR GRAPH presents information on a horizontal plane, Fig. 19-3.

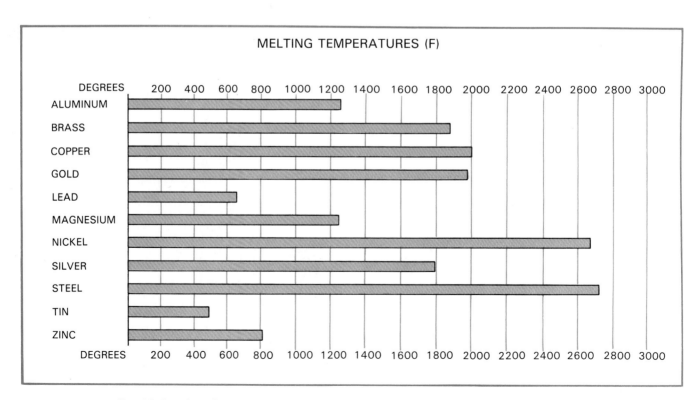

Fig. 19-3. HORIZONTAL BAR GRAPH used to indicate melting temperatures of metal.

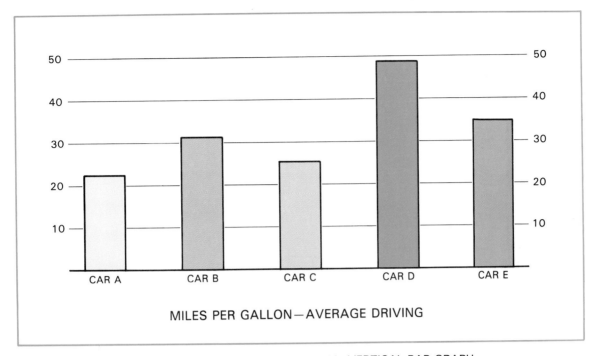

Fig. 19-4. Car economy is shown in this VERTICAL BAR GRAPH.

With the VERTICAL BAR GRAPH information is given in a vertical or upright position, Fig. 19-4, for easy comparisons.

The COMPOSITE BAR GRAPH can be drawn in either a vertical or horizontal position, Fig. 19-5.

This type of graph compares several items of information in the same graph.

A 100 PERCENT BAR GRAPH consists of a singular rectangular bar, Fig. 19-6. Information is presented on a percentage basis.

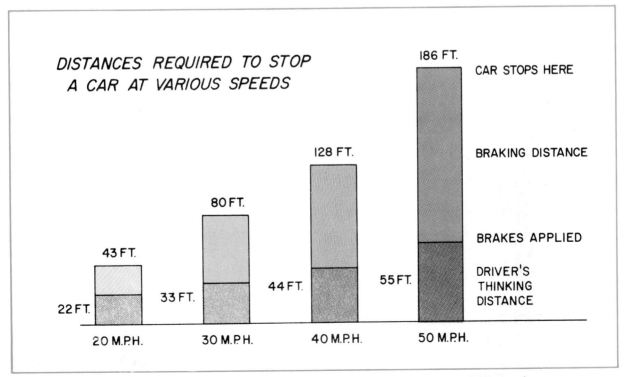

Fig. 19-5. COMPOSITE BAR GRAPH shows stopping distances at various speeds.

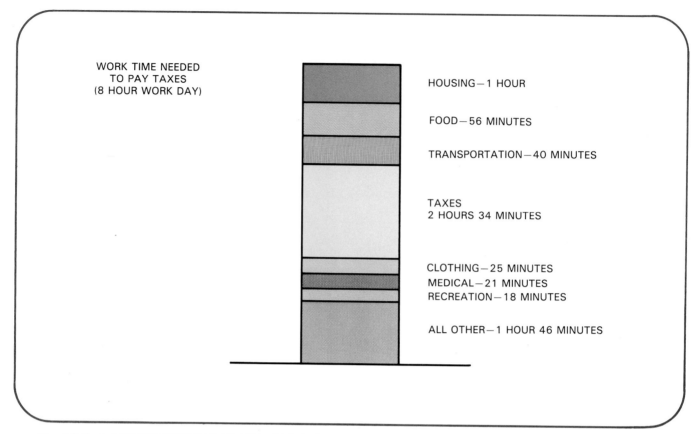

WORK TIME NEEDED
TO PAY TAXES
(8 HOUR WORK DAY)

HOUSING—1 HOUR

FOOD—56 MINUTES

TRANSPORTATION—40 MINUTES

TAXES
2 HOURS 34 MINUTES

CLOTHING—25 MINUTES
MEDICAL—21 MINUTES
RECREATION—18 MINUTES

ALL OTHER—1 HOUR 46 MINUTES

Fig. 19-6. A 100 PERCENT BAR GRAPH which indicates the amount of time one person works to pay their taxes.

A PICTORIAL BAR GRAPH is a variation of the bar graph. It employs pictures to represent the information instead of bars, Fig. 19-7. Pictures tend to make a graph more interesting.

CIRCLE OR PIE GRAPH

The CIRCLE or PIE GRAPH is shown in Fig. 19-8. It is composed of a segmented circle and shows the entire unit divided into comparable parts. It is also known as an AREA GRAPH.

CHARTS

CHARTS are another means employed to convey information rapidly. The most familiar of these is the ORGANIZATION CHART, Fig. 19-9. It shows the order of responsibility and the relation of persons and/or positions in an organization.

The PICTORIAL CHART is an ideal way of presenting comparisons in an interesting and easily understood manner, Fig. 19-10.

FLOW CHARTS may be used to show the sequence, or order of operations on how a product is manufactured and/or distributed, Fig. 19-11. A flow chart is often employed to show a cycle of events in the order they occur.

PREPARING CHARTS AND GRAPHS

There are many variations of the graphs presented in this unit. It is up to the ingenuity and creativity of the person preparing the material to decide which type will be best for the job.

DO NOT start a graph or chart until ALL of the information has been gathered. A rough layout should be prepared first to best determine what size and proportion presentation will be best.

Use color when ever possible to brighten your presentation and to make the information stand out. Hundreds of styles of transfer lettering, symbols, and figures are available, Fig. 19-12. They are useful in making the chart or graph more interesting.

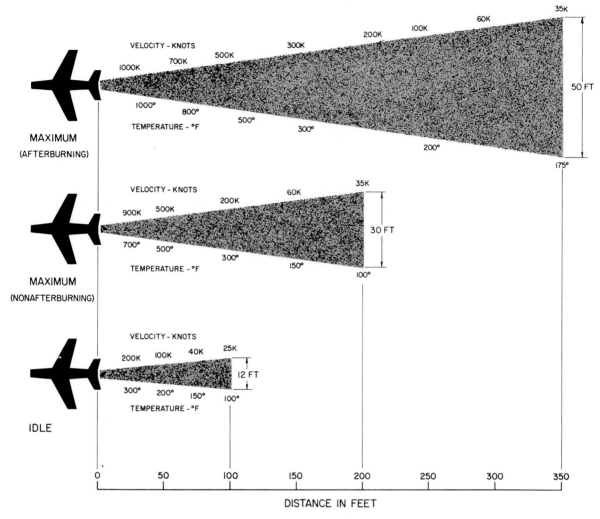

APPROXIMATE JET WAKE DANGER AREAS

Fig. 19-7. PICTORIAL BAR GRAPH showing the jet wake danger areas and the temperature and speed of the exhaust gases.

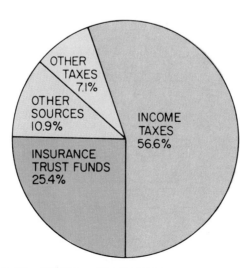

Fig. 19-8. The CIRCLE or PIE GRAPH is sometimes called an AREA GRAPH. The graph may be as simple as the above which shows where the federal government gets the money to operate, or it can be made more complex.

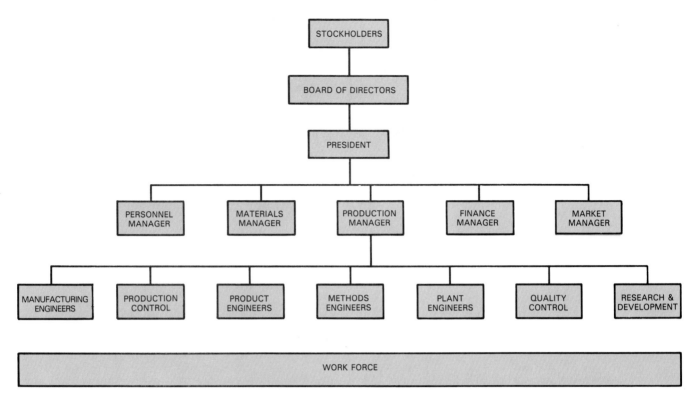

Fig. 19-9. An ORGANIZATION CHART.

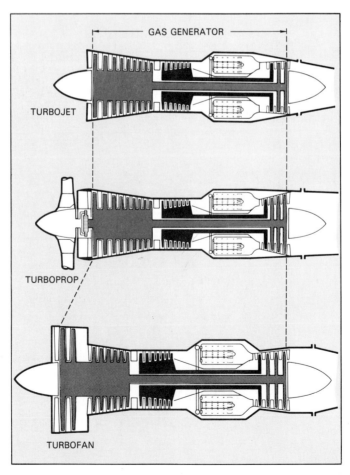

Fig. 19-10. PICTORIAL CHART comparing three types of jet engines. (Pratt & Whitney Aircraft)

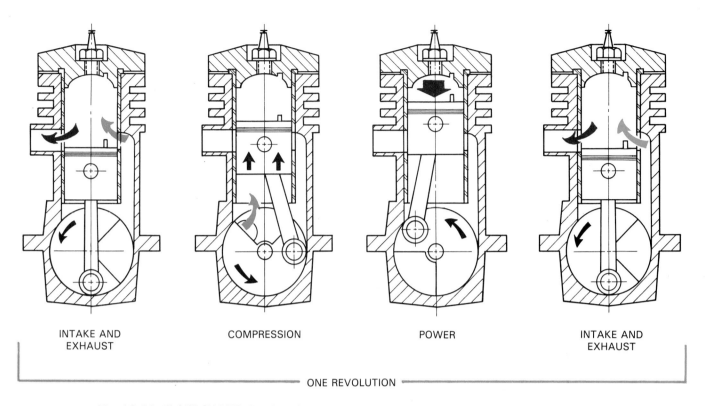

| INTAKE AND EXHAUST | COMPRESSION | POWER | INTAKE AND EXHAUST |

ONE REVOLUTION

Fig. 19-11. FLOW CHART showing the operation of a two-stroke internal combustion engine.

Fig. 19-12. A few of the commercially available transfer lettering and symbols that can be employed to make professional looking charts and graphs.

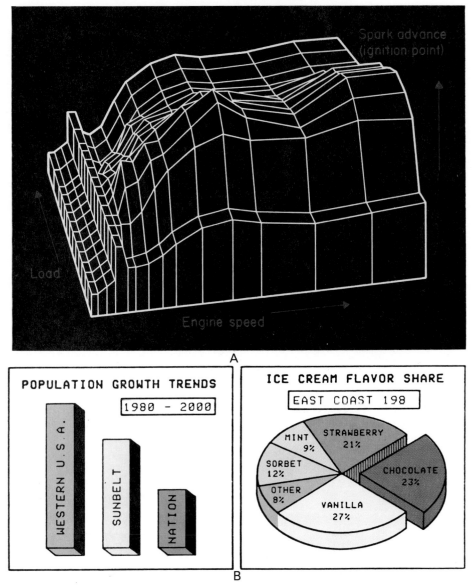

Fig. 19-13. Computer generated charts and graphs are finding wide acceptance in commerce and industry. A—Three-dimensional graph showing electronic ignition characteristics of an auto engine and how they vary with engine speed, load, and temperature. The graph enables engineers to tailor the ignition system to engine needs. If made manually from instrument readings, the graph would have required many hours to plot manually. The computer plotted the information as the test progressed. The computer generated information was available immediately after the test was completed. (The graph was redrawn from the original computer printout for approved legibility in reduced form.) B—Two examples of simple two-dimensional graphs that were developed by computer graphics.

COMPUTER GENERATED GRAPHS

Computer generated graphics has greatly speeded up the preparation of graphs. With the proper program, any type graph can be developed in two- and three-dimensional forms. Multicolor presentations are also possible. See Fig. 19-13.

DRAFTING VOCABULARY

Area graph, Bar graph, Chart, Circle graph, Comparisons, Curve, Flow chart, Graph, Ingenuity, Key, Line graph, Multicolor, Organization chart, Pictorial bar graph, Pictorial chart, Pie graph, Statistics, Transfer lettering, Trends, Variations.

TEST YOUR KNOWLEDGE—UNIT 19

Please do not write in the book. Place your answers on a sheet of notebook paper.
1. Why are charts and graphs used?

2. List the three most commonly used kinds of graphs.
   a. _____.
   b. _____.
   c. _____.
3. The line that presents the information on a _____ graph is called a _____.
4. Of what use is the KEY that should be included when more than one line is used on a graph?
5. Briefly describe each of the following graphs:
   a. Horizontal Bar Graph.
   b. Vertical Bar Graph.
   c. Composite Bar Graph.
   d. 100 Percent Bar Graph.
   e. Pictorial Bar Graph.
6. The _____ or _____ graph is also known as an Area Graph.
7. Briefly describe each of the following charts:
   a. Flow Chart.
   b. Organization Chart.
   c. Pictorial Chart.
8. Why should color be used on a chart or graph?

OUTSIDE ACTIVITIES

1. Develop a pie graph of how you spend your allowance.

2. Prepare a bar graph to show the approximate increase in horsepower in a particular make of automobile from 1940 to the present time. Use five year steps.
3. Prepare a line graph showing how the price of the automobile used in activity 2 has risen in the same periods.
4. Make a 100 percent bar graph showing a breakdown of the cost of a gallon of gasoline — including cost of the gasoline, state, federal, and local taxes.
5. Draw a pictorial bar graph showing how far an automobile will travel after the brakes are applied at 25 mph, 45 mph, 55 mph, and 65 mph.
6. Make an organization chart of the pupil personnel system used in your drafting room or Industrial Technology shops.
7. Develop a picture chart showing how the size of a particular make of auto has increased or decreased in the past 20 years.
8. Make a picture graph showing the enrollment in each grade of your school. Let each symbol represent 25 students.
9. Design a flow chart showing how a simple product could be manufactured in the Industrial Technology shop.

Unit 20

WELDING DRAWINGS

After studying this unit, you will be able to define the welding operation. You will be able to explain why weld symbols are used on drawings. You will be able to apply welding symbols to various fabrication situations. You will construct proper welding symbols showing exact welding specifications.

WELDING is a widely used industrial technique for fabricating metal pieces. It is a method of joining by heating metals to a high temperature which causes them to melt and fuse together. The high temperatures are generated by electricity (arc welding, for example) or by burning gases (oxygen and acetylene). See Fig. 20-1.

Fig. 20-1. Spot welding is one of the many types of welding employed by industry. Here robot welders are joining body sections of a sport car. (Pontiac Motor Division, GMC)

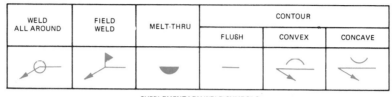

FILLET	PLUG OR SLOT	SPOT PROJEC- TION	SEAM	GROOVE							BACK OR BACKING	SURFACING	FLANGE	
				SQUARE	"V"	BEVEL	"U"	"J"	FLARE "V"	FLARE BEVEL			EDGE	CORNER

BASIC ARC AND GAS WELD SYMBOLS

WELD ALL AROUND	FIELD WELD	MELT-THRU	CONTOUR		
			FLUSH	CONVEX	CONCAVE

SUPPLEMENTARY WELD SYMBOLS

Fig. 20-2. Basic weld symbols.

Some method had to be devised to tell the welder what type of weld the engineer wanted on the job. The American Welding Society (AWS) developed and standardized the basic weld symbols shown in Fig. 20-2. These symbols provide a means of giving complete and specific welding information on the drawing of the part to be welded.

The welding symbol placed on the drawing is made up of basic and supplementary weld symbols that have been selected to describe the required weld, Fig. 20-3.

USING WELDING SYMBOLS

Before welding symbols can be employed effectively, the drafter must be familiar with the various types of joints used in welding, Fig. 20-4.

The location of the arrow with respect to the joint is important when using the welding symbol to specify the required weld. The side of the joint indicated by the arrow is considered the ARROW SIDE. The side opposite the arrow side of the joint is considered the OTHER SIDE of the joint.

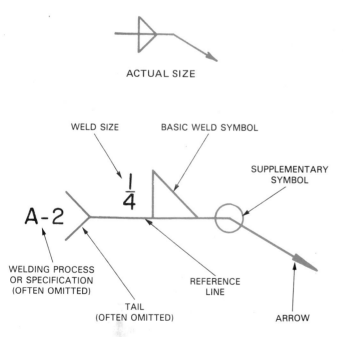

ACTUAL SIZE

WELD SIZE

BASIC WELD SYMBOL

SUPPLEMENTARY SYMBOL

A-2

REFERENCE LINE

WELDING PROCESS OR SPECIFICATION (OFTEN OMITTED)

TAIL (OFTEN OMITTED)

ARROW

Fig. 20-3. A welding symbol is made up of basic and supplementary weld symbols. It gives complete and specific welding information to the welder.

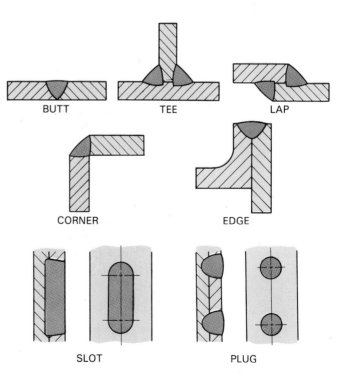

BUTT   TEE   LAP

CORNER   EDGE

SLOT   PLUG

Fig. 20-4. Basic joints used in welding.

When the weld is to be made on the ARROW SIDE of the joint, the weld symbol is placed on the reference line so it is TOWARDS THE READER, Fig. 20-5.

To make a weld on the OTHER SIDE of the joint the weld symbol is placed on the reference line AWAY FROM THE READER, Fig. 20-6.

Welds on both sides of the joint are specified by placing the symbol on BOTH SIDES of the reference line, Fig. 20-7.

A weld that is made all of the way around the joint is specified by drawing a circle at the point where the arrow is bent, Fig. 20-8. Field welds (welds not made in the shop) are specified as shown in Fig. 20-9.

Weld size is placed to the left of the symbol. The figure indicating the length of the weld is placed to the right of the symbol, Fig. 20-10.

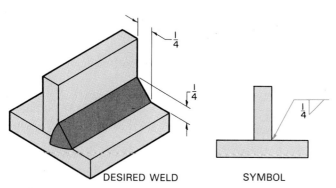

DESIRED WELD   SYMBOL

Fig. 20-5. Welding symbol that indicates the weld is to be made on the ARROW SIDE of the joint. Note the weld symbol is TOWARDS the reader.

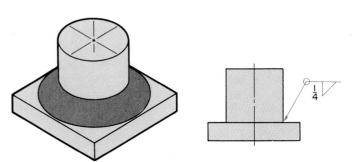

Fig. 20-8. Welding symbol that indicates the weld is to be made ALL AROUND the joint.

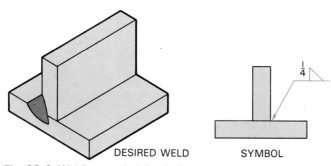

DESIRED WELD   SYMBOL

Fig. 20-6. Welding symbol that indicates the weld is to be made on the OTHER SIDE OF THE JOINT.

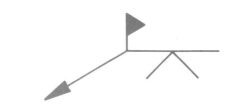

Fig. 20-9. Field welding symbol. This indicates the welding is to be done on the job and not in the shop.

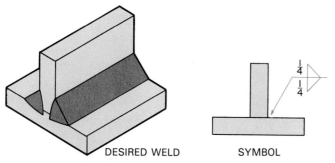

DESIRED WELD   SYMBOL

Fig. 20-7. Welding symbol that indicates the weld is to be made on BOTH SIDES of the joint.

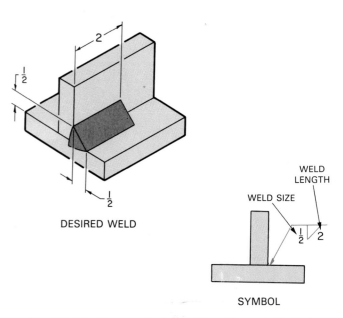

Fig. 20-10. How weld size and length are indicated.

DRAWINGS FOR ASSEMBLIES TO BE FABRICATED BY WELDING

Items that are to be assembled by welding are usually composed of several pieces, Fig. 20-11. Because the individual pieces are seldom cut to size, welded, and machined in the same general shop area, several drawings are often required. Each drawing will provide information on a specific operation. For simple jobs, all of the needed information can be included on a single drawing, Fig. 20-12.

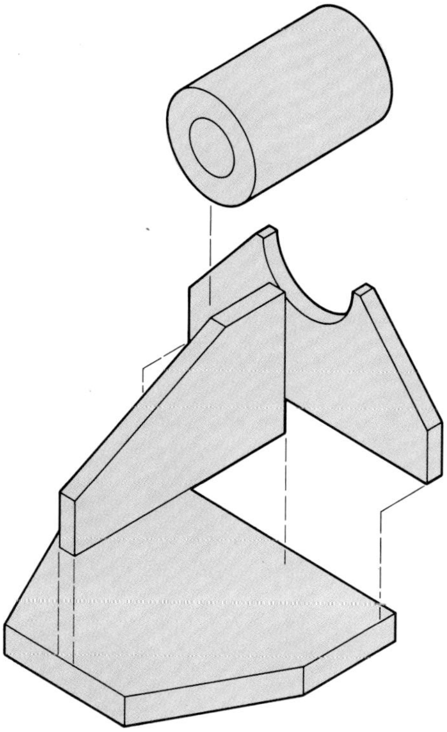

TECHNICAL VOCABULARY

American Welding Society, Arrow side, Fabricating, Field weld, Fuse, Generate, High temperature, Joining, Other side, Welding symbol.

TEST YOUR KNOWLEDGE—UNIT 20

1. How are the high temperatures needed for welding generated?
2. Welding symbols were developed to:
   a. Lessen the possibilities of the wrong type of weld being made.
   b. Tell the welder exactly what to do.
   c. Eliminate "hit or miss" welds.
   d. All of the above.
3. Sketch the symbol for a fillet weld on the arrow side of the joint.
4. Sketch the symbol for a fillet weld on the other side of the joint.
5. Sketch the symbol for a fillet weld on both sides of the joint.
6. Sketch the symbol for a fillet weld all of the

Fig. 20-11. A welded assembly is composed of different pieces.

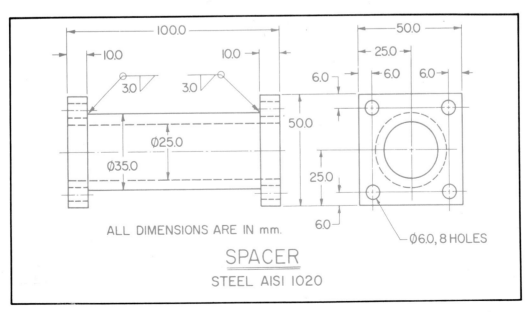

ALL DIMENSIONS ARE IN mm.

SPACER
STEEL AISI 1020

Fig. 20-12. Cutting, welding, and machining information is included on this drawing. Note the dimensions are in millimeters.

way around the joint.

7. What is a field weld?

8. Why are several different drawings needed for a job that is made up of several pieces and assembled by welding?

OUTSIDE ACTIVITIES

1. Secure samples of welded joints and mount them on a display board. Label the samples and add the proper drafting symbol to the display.

2. Visit a local industry that makes extensive use of welding and get samples of the work they produce.

3. Invite a professional welder to the school shop to demonstrate the safe and proper way to gas weld and to electric arc weld.

4. Report for the class on the welding techniques known as Gas Tungsten Arc Welding and Gas Metal Arc Welding.

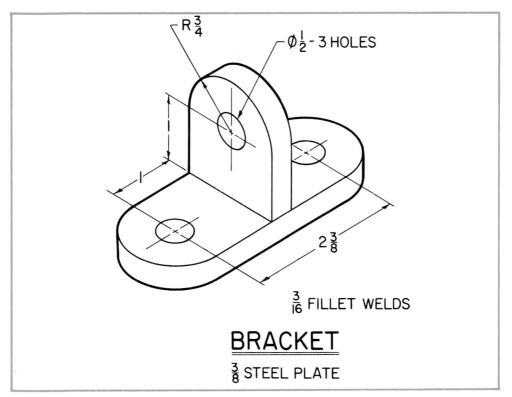

R$\frac{3}{4}$

$\emptyset\frac{1}{2}$ - 3 HOLES

1

2$\frac{3}{8}$

$\frac{3}{16}$ FILLET WELDS

BRACKET

$\frac{3}{8}$ STEEL PLATE

PROBLEM 20-1. BRACKET. Prepare a drawing showing the necessary views to fabricate it. Use the appropriate welding symbol.

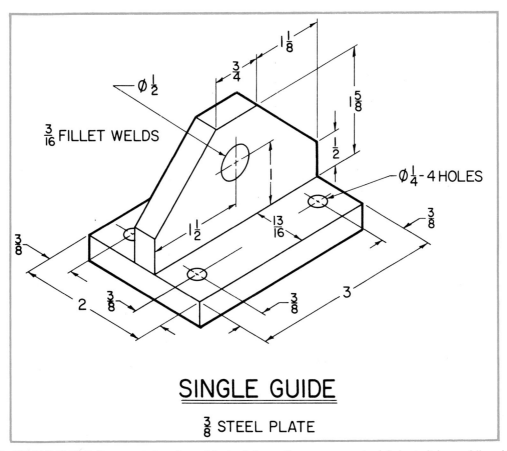

$\emptyset\frac{1}{2}$

$\frac{3}{16}$ FILLET WELDS

1$\frac{1}{8}$

$\frac{3}{4}$

1$\frac{5}{8}$

$\frac{1}{2}$

$\emptyset\frac{1}{4}$ - 4 HOLES

$\frac{3}{8}$

1$\frac{1}{2}$

$\frac{13}{16}$

$\frac{3}{8}$

2$\frac{3}{8}$

$\frac{3}{8}$

3

SINGLE GUIDE

$\frac{3}{8}$ STEEL PLATE

PROBLEM 20-2. SINGLE GUIDE. Prepare a drawing with the information necessary to fabricate it by welding. Welds are to be made on both sides of the joint.

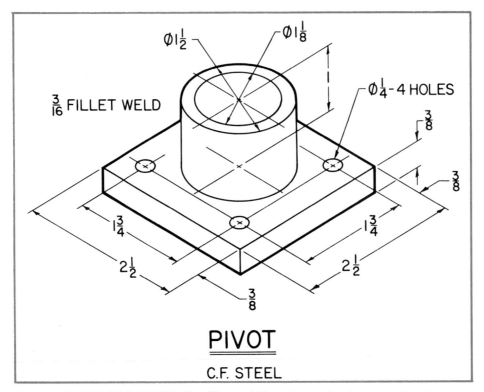

$\varnothing 1\frac{1}{2}$ $\varnothing 1\frac{1}{8}$

$\frac{3}{16}$ FILLET WELD

$\varnothing \frac{1}{4}$-4 HOLES

$\frac{3}{8}$

$\frac{3}{8}$

$1\frac{3}{4}$ $1\frac{3}{4}$

$2\frac{1}{2}$ $2\frac{1}{2}$

$\frac{3}{8}$

PIVOT

C.F. STEEL

PROBLEM 20-3. PIVOT. The fillet weld is to be made all around the joint.

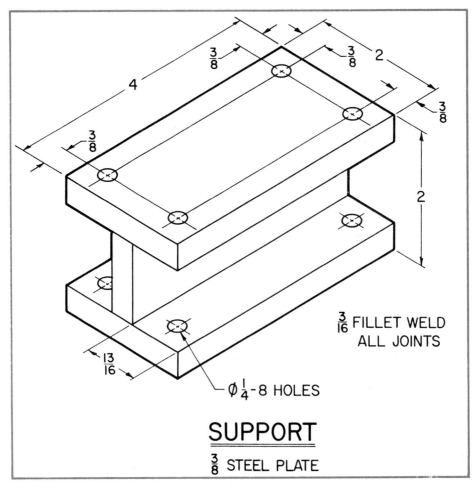

$\frac{3}{8}$ $\frac{3}{8}$ 2

4

$\frac{3}{8}$

$\frac{3}{8}$

2

$\frac{3}{16}$ FILLET WELD
ALL JOINTS

$\frac{13}{16}$

$\varnothing \frac{1}{4}$-8 HOLES

SUPPORT

$\frac{3}{8}$ STEEL PLATE

PROBLEM 20-4. SUPPORT. Prepare a drawing showing the views necessary to fabricate the object. Welds are to be made on both sides of each joint.

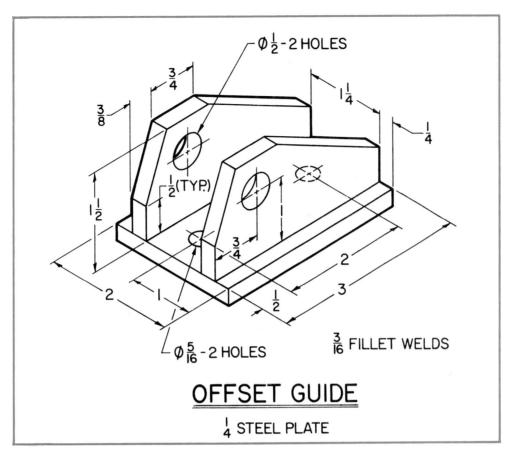

PROBLEM 20-5. OFFSET GUIDE. The weld is to be made on both sides of each vertical piece.

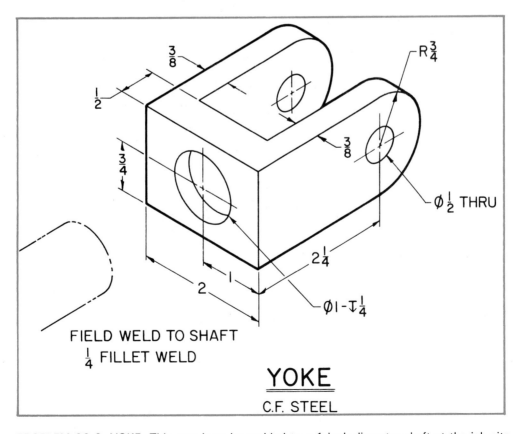

PROBLEM 20-6. YOKE. This part is to be welded to a 1 inch diameter shaft at the job site.

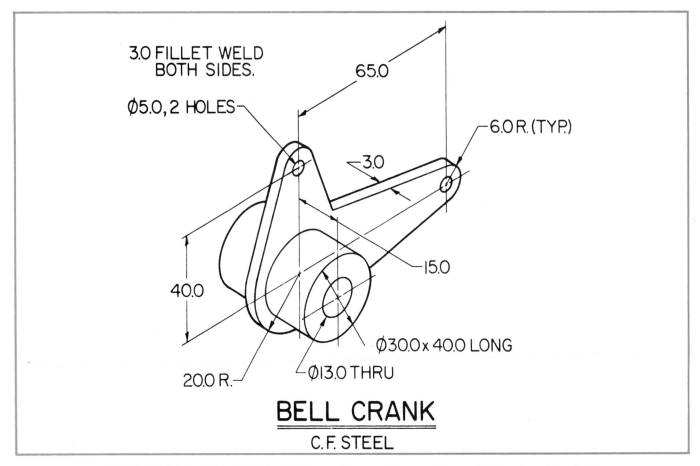

3.0 FILLET WELD
BOTH SIDES.

Ø5.0, 2 HOLES

65.0

6.0 R. (TYP.)

3.0

40.0

15.0

Ø30.0 x 40.0 LONG

20.0 R.

Ø13.0 THRU

BELL CRANK
C.F. STEEL

PROBLEM 20-7. BELL CRANK. A 3.0 mm fillet weld is specified on both sides of the joint.

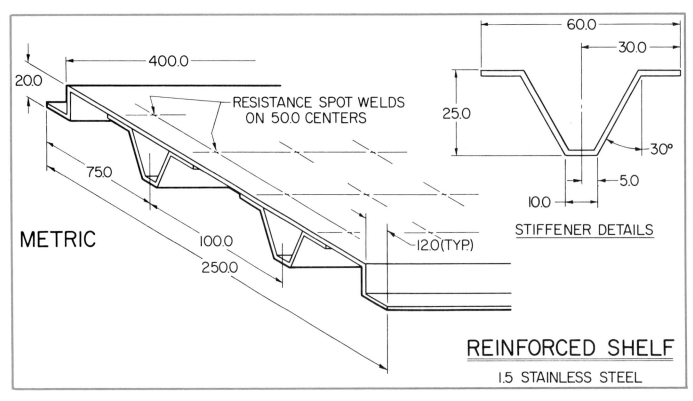

400.0

20.0

RESISTANCE SPOT WELDS
ON 50.0 CENTERS

75.0

100.0

250.0

12.0 (TYP.)

METRIC

60.0

30.0

25.0

30°

5.0

10.0

STIFFENER DETAILS

REINFORCED SHELF
1.5 STAINLESS STEEL

PROBLEM 20-8. REINFORCED SHELF. Stiffeners are spot welded to the stainless steel shelf.

Unit 21

FASTENERS

After studying this chapter, you will be able to define and identify a variety of fasteners. You will be able to list applications that make use of screw threads. You will show how inch based and metric based threaded fasteners are indicated on drawings. You will demonstrate three approved methods of drawing threads for bolts and nuts.

Manufactured products are assembled by many different kinds of fasteners, Fig. 21-1. A fastener is a device used to hold two or more parts together. They include screws, nuts and bolts, rivets, nails, etc. Because fasteners are so important to industry,

drafters, engineers, and designers must be familiar with them, know what types to use for a particular application, and how to draw them.

THREADED FASTENERS

Threads have many applications. They are employed to:
1. Make adjustments.
2. Transmit motion.
3. Assemble parts.
4. Apply pressure.
5. Make measurements.

How many examples of each can you name?

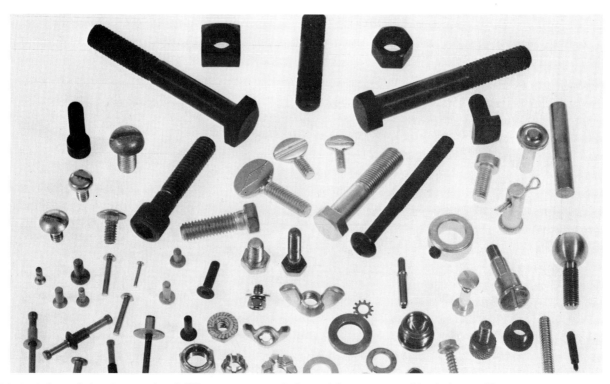

Fig. 21-1. A few of the thousands of different types and sizes of fasteners used by industry. How many can you identify?

271

Threaded fasteners employ the wedging action of the screw thread to hold items together. They are found extensively in manufactured products. See Fig. 21-2. Threaded fasteners vary in cost from several thousand dollars each for special bolts that attach the wings to the fuselage of large aircraft, to the small machine screw that costs a fraction of a penny. (One auto manufacturer uses more than 11,000 fasteners of various types and sizes.)

Until recently, industry usually worked only with the inch based UNIFIED THREAD SERIES. These fasteners are made in both coarse (identified by UNC) and fine (identified by UNF) thread series, Fig. 21-3. Now, industry must be familiar with metric based threaded fasteners.

While inch based UNC and UNF threads and metric based threads have the same basic profile (shape), Fig. 21-4, and some inch based and metric based threaded fasteners may appear to be the same size, THEY ARE NOT INTERCHANGEABLE. A metric based bolt CANNOT be used with an inch based nut that appears to have the same thread size as the bolt.

More than two million different kinds, shapes and sizes of inch based threaded fasteners are made. Add to this number, the metric based threaded fasteners. Both inch based and metric based fasteners will have to be kept in inventory

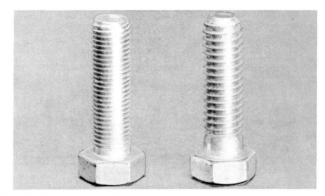

Fig. 21-3. Inch based fasteners and metric based fasteners are made in both fine and coarse thread series. Bolts shown are the same diameter and length. How do they differ?

for many, many years. Both inch based and metric based wrenches will also be needed.

A comparison of metric coarse pitch and UNC inch based thread sizes is shown in Fig. 21-5.

The user may have trouble telling a metric bolt from a similar inch based fastener. Some simple way will have to be devised to identify the metric fastener. At present in the United States, the imprinted hex head and a unique 12-element spline head are being considered. See Fig. 21-6.

DRAWING THREADS

It is very time consuming to show threads on a drawing as they would actually appear. For this

Fig. 21-2. These VTOL (Vertical Take Off and Landing) jets are being manufactured for the U.S. Marines. Metric standards are used. More than one million fasteners are used in each aircraft. (Hawker Siddeley Aviation Limited)

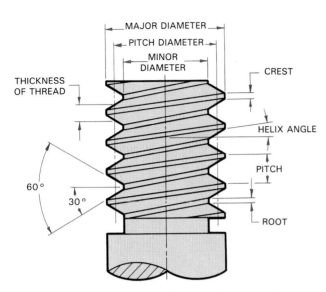

Fig. 21-4. Metric and inch based threads have exactly the same shape. The parts of a screw thread are shown.

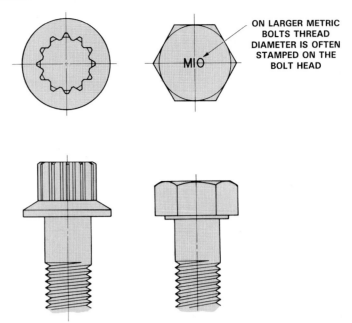

ON LARGER METRIC BOLTS THREAD DIAMETER IS OFTEN STAMPED ON THE BOLT HEAD

Fig. 21-6. Much study is being done to devise an easy way to identify metric based fasteners from inch based fasteners. The 12-element spline head and imprinted hex head (thread diameter is stamped on the head) are two methods being considered.

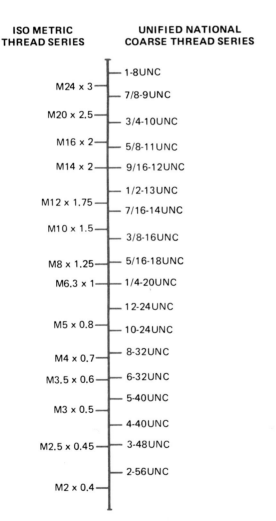

ISO METRIC THREAD SERIES	UNIFIED NATIONAL COARSE THREAD SERIES
	1-8UNC
M24 x 3	7/8-9UNC
M20 x 2.5	3/4-10UNC
M16 x 2	5/8-11UNC
M14 x 2	9/16-12UNC
	1/2-13UNC
M12 x 1.75	7/16-14UNC
M10 x 1.5	3/8-16UNC
M8 x 1.25	5/16-18UNC
M6.3 x 1	1/4-20UNC
	12-24UNC
M5 x 0.8	10-24UNC
M4 x 0.7	8-32UNC
M3.5 x 0.6	6-32UNC
	5-40UNC
M3 x 0.5	4-40UNC
M2.5 x 0.45	3-48UNC
	2-56UNC
M2 x 0.4	

ISO and Unified National Thread Series ARE NOT INTERCHANGEABLE

Fig. 21-5. A comparison of ISO metric coarse and Unified Coarse (UNC) inch based thread sizes. Even though several of them SEEM TO BE THE SAME SIZE, THEY ARE NOT INTERCHANGEABLE (one cannot be substituted for the other).

reason, either the SCHEMATIC or SIMPLIFIED representation is used. See the approved methods of thread representation in Fig. 21-7.

The DETAILED REPRESENTATION, Fig. 21-7 (top), looks like the actual screw thread. It is sometimes employed where confusion might result if the simplified representation were used.

The SCHEMATIC REPRESENTATION of a screw thread, Fig. 21-7 (center), is easier to draw. It should not be used for hidden threads or sections of external threads.

The SIMPLIFIED REPRESENTATION of a screw thread, Fig. 21-7 (bottom), is a fast and easy method used to draw threads. For this reason, it is widely used in drafting. It should be avoided where there is a possibility of this representation being confused with other details on a drawing.

Avoid mixing the various methods on the same drawing.

Regardless of which thread representation the drafter decides to draw, the thread size must be shown on the drawing. See Fig. 21-8 for the accepted way to present thread information.

273

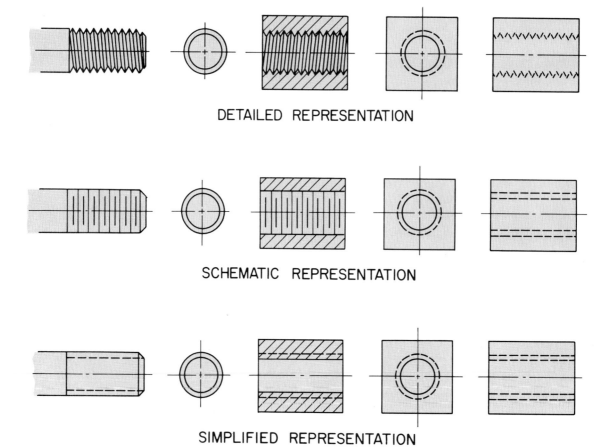

DETAILED REPRESENTATION

SCHEMATIC REPRESENTATION

SIMPLIFIED REPRESENTATION

Fig. 21-7. Approved ways of representing threads on drawings.

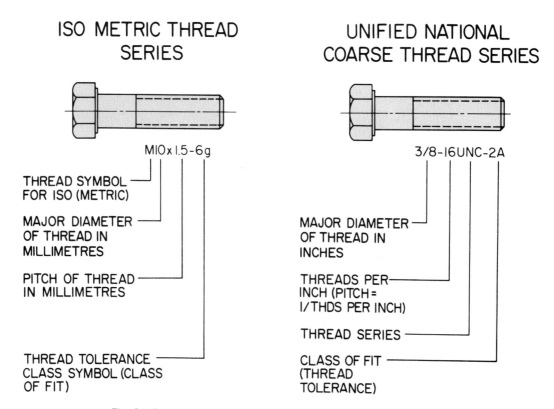

ISO METRIC THREAD SERIES

MlOx l.5-6g

THREAD SYMBOL
FOR ISO (METRIC)

MAJOR DIAMETER
OF THREAD IN
MILLIMETRES

PITCH OF THREAD
IN MILLIMETRES

THREAD TOLERANCE
CLASS SYMBOL (CLASS
OF FIT)

UNIFIED NATIONAL COARSE THREAD SERIES

3/8-l6UNC-2A

MAJOR DIAMETER
OF THREAD IN
INCHES

THREADS PER
INCH (PITCH =
1/THDS PER INCH)

THREAD SERIES

CLASS OF FIT
(THREAD
TOLERANCE)

Fig. 21-8. How thread size is noted and what each term means.

Threads are understood to be right-hand threads. Left-hand threads are represented by the letters LH following the class of fit (inch based thread) or thread tolerance (metric based thread).

TYPES OF THREADED FASTENERS

The fasteners described in this unit can usually be found in the school laboratory, typical hardware store, or automotive parts supply store. The majority of them are made of steel, brass, or aluminum. Special applications may require them to be made of other materials.

MACHINE SCREWS

MACHINE SCREWS are available with single slotted, cross slotted (Phillips), or hex style heads, and round, flat, fillister, pan and oval head shapes, Fig. 21-9. Nuts (square or hexagonal) are not furnished with machine screws and must be purchased separately. They have many applications in general assembly work where screws less than 1/4 in. (6.0 mm) in diameter are needed. Machine screws are available in many sizes starting as small as #0-80UNC (0.06 in. dia.) inch based and 1.4 x 0.3 metric based.

MACHINE BOLTS

MACHINE BOLTS are manufactured with square and hexagonal heads, Fig. 21-10. They are used to assemble machinery and other items that do not require close tolerance fasteners which are more expensive. Machine bolts are secured by tightening the matching nut.

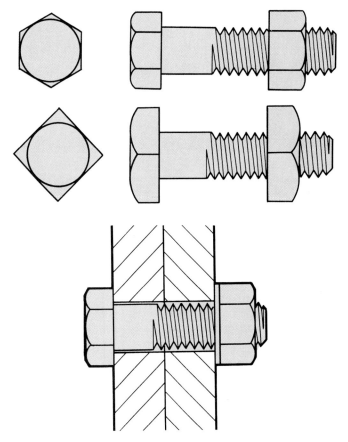

Fig. 21-10. Machine bolts.

CAP SCREWS

CAP SCREWS are used when the assembly requires a stronger, more precise, and better appearing fastener, Fig. 21-11. They are primarily employed to bolt two pieces or sections together. The screw passes through a CLEARANCE HOLE in one part and screws into a THREADED HOLE in the other part, Fig. 21-12. Machine tools are usually assembled with cap screws.

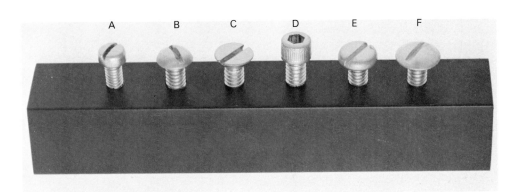

Fig. 21-9. A few of the many styles of machine screw head types available. A—Fillister. B—Round. C—Flat. D—Socket. E—Pan. F—Truss.

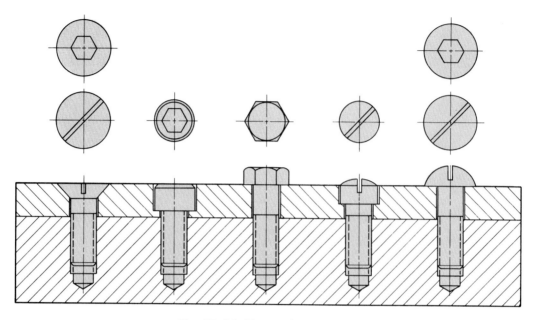

Fig. 21-11. Types of cap screws.

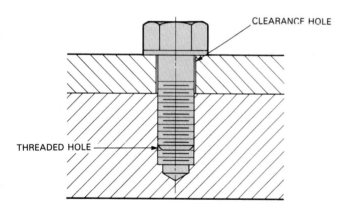

CLEARANCE HOLE

THREADED HOLE

Fig. 21-12. How a cap screw works. What type thread representation is used on this drawing?

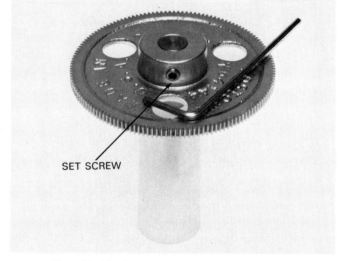

SET SCREW

Fig. 21-13. A typical setscrew application.

SET SCREWS

SET SCREWS prevent slippage of pulleys and gears on shafts, Fig. 21-13. They are available in many different head and point styles. Set screws are made of heat treated steel to make them stronger and less likely to fail.

STUD BOLTS

A STUD BOLT is threaded on both ends. One end is threaded into a tapped hole. The piece to be clamped is fitted over the stud bolt and the nut screwed on to clamp the two sections tightly together. Refer to Fig. 21-14.

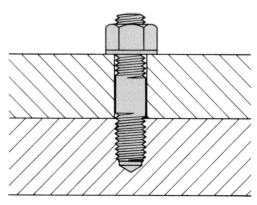

Fig. 21-14. Stud bolt. What type thread representation is used on this drawing?

NUTS

NUTS are screwed down on bolts or screws to tighten or hold together the sections through which the bolt passes. A few of the many types of nuts available are shown in Fig. 21-15.

WASHERS

WASHERS are used with nuts and bolts to distribute the clamping pressure over a larger area. They also prevent the fastener from marring the work surface when the nut is tightened. Many types are manufactured, Fig. 21-16.

NONTHREADED FASTENERS

Nonthreaded fasteners comprise a large group of holding devices.

RIVETS

Permanent assemblies are made with RIVETS, Fig. 21-17. Two or more sections of material are held together by these headed pins. Holes are drilled in the material to be riveted. The rivet shank passes through the hole. After aligning the sections, the plain end of the rivet is upset or headed by hammering to form a second head. The parts are drawn together by the heading operation.

BLIND RIVETS are mechanical fasteners that have been developed for applications where the joint is accessible from one side only. They require special tools to put them in place, Fig. 21-18. Blind rivet types are shown in Fig. 21-19.

COTTER PINS

The COTTER PIN is fitted into a hole drilled crosswise in a shaft, Fig. 21-20. The ends are bent over after assembly to prevent parts from slipping or turning off the shaft.

KEYS, KEYWAYS, AND KEYSEATS

A KEY is a small piece of metal partially fitted into the shaft and partially into the hub to prevent rotation of a gear, wheel, or pulley on the

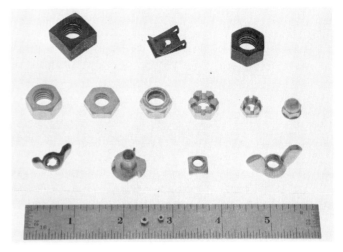

Fig. 21-15. Common nut styles. How many can you identify?

Fig. 21-16. Washer types. A—Plain washer. B—Split type lock washer. C—External type lock washer. D—Internal type lock washer. E—Internal-external type washer; used when the mounting holes are oversize. F—Countersunk type washer.

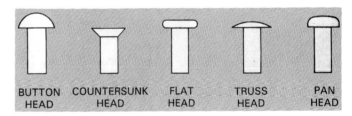

BUTTON HEAD COUNTERSUNK HEAD FLAT HEAD TRUSS HEAD PAN HEAD

Fig. 21-17. Rivet head styles.

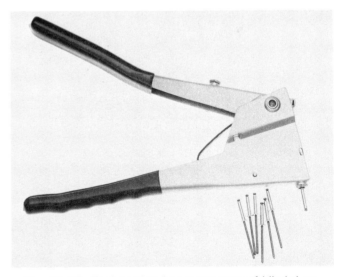

Fig. 21-18. Tool used to insert one type of blind rivet.

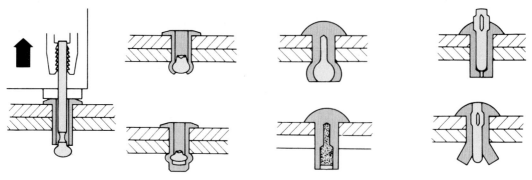

Fig. 21-19. Types of blind rivets.

Fig. 21-20. Cotter pin.

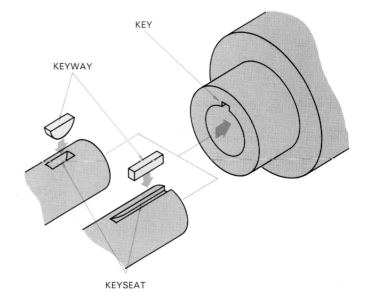

Fig. 21-21. Two types of keys are shown. The half-round key is called a Woodruff key. The other is a rectangular key.

shaft, Fig. 21-21. The KEYSEAT is machined in the shaft. The KEYWAY is cut in the hub of the mating part. Different types of keys have been devised for special applications.

FASTENERS FOR WOOD

The most common fasteners used in wood are nails and screws.

NAILS

NAILS are an easy way to fasten wood pieces together, Fig. 21-22. They are usually made of mild steel. For exterior work, nails made from mild steel are given a galvanized (zinc) coating so they will not rust. Nail size is given as "penny" and is abbreviated with the lower case "d."

WOOD SCREWS

WOOD SCREWS are manufactured from many kinds of metal. Screw size is indicated by the shank diameter and length. Head styles and how each is measured is shown in Fig. 21-23. Wood screws are available in lengths from 1/4 in. to 6 in. (6.5 mm to 150.0 mm).

HOW TO DRAW BOLTS AND NUTS

The dimensions given in Fig. 21-24 are approximations but are acceptable for most drafting ap-

plications. Information needed to complete the drawing includes: 1. Bolt diameter. 2. Bolt length. 3. Type of head and/or nut.

To draw square and hexagonal headed bolts and nuts follow the procedure in Fig. 21-25.
1. Draw center lines and lines representing the diameter (D).
2. On center line, draw a circle (diameter = 1 1/2D).
3. Using 30-60 degree triangles, circumscribe a hexagon (or with a 45 degree angle, a square) about the circle.
4. Develop the side view of the bolt.
5. Draw arcs in the bolt head and nut using radii given in Fig. 21-24. (Templates for drawing bolt heads and nuts are available and make the job faster and easier.)

Fig. 21-22. The home building industry uses vast quantities of nails. More than a thousand different kinds are available.

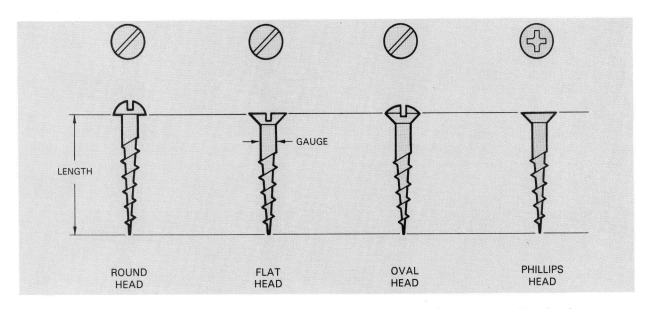

GAUGE

LENGTH

ROUND HEAD   FLAT HEAD   OVAL HEAD   PHILLIPS HEAD

Fig. 21-23. How the various types of wood screws are measured. Screw sizes and head styles do not change in metrics, but their sizes will be expressed in millimeters.

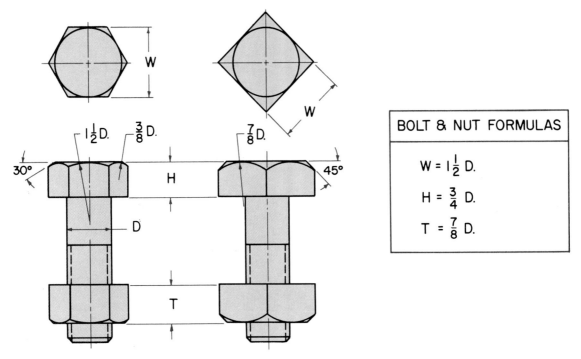

BOLT & NUT FORMULAS

$$W = 1\frac{1}{2} D.$$

$$H = \frac{3}{4} D.$$

$$T = \frac{7}{8} D.$$

Fig. 21-24. Information needed to draw a bolt and nut.

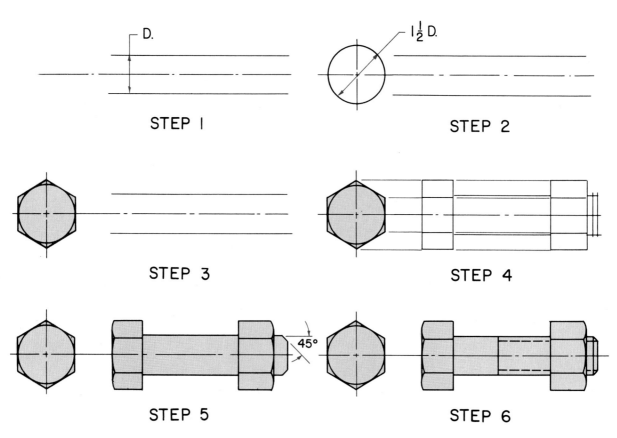

STEP 1

STEP 2

STEP 3

STEP 4

STEP 5

STEP 6

Fig. 21-25. Steps in drawing a bolt and nut. The same procedure is followed whether the bolt or nut is to have a square or hexagonal head.

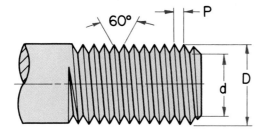

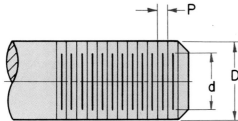

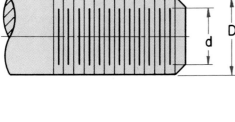

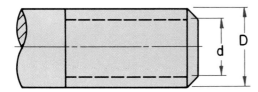

$P = \text{PITCH} = \dfrac{1}{N}$

$N = \text{NUMBER OF THREADS PER INCH}$

$D = \text{MAJOR DIAMETER OF THREADS}$

$d = \text{MINOR DIAMETER OF THREADS}$

$d = D - \dfrac{1.300}{N}$

Fig. 21-26. Dimensions used when drawing detailed, schematic, and simplified thread representations.

6. Complete by drawing the chamfers on the bolt head and nut (not necessary if a template is used). Draw threads—either simplified or schematic—on the bolt. The dimensions for drawing schematic and simplified threads are shown in Fig. 21-26.

TECHNICAL VOCABULARY

Abbreviated, Aligning, Accessible, Adjustment, Application, Clearance, Comparison, Crosswise, Distribute, Element, Extensively, Fillister head, Galvanized, Heat treated, Hex head, Interchangeable, Pan head, Pitch, Profile, Representation, Schematic, Series, Simplified, Spline, Tapped hole, Tolerance, Transmit motion, Wedging.

UNIT 21—TEST YOUR KNOWLEDGE

Please do not write in the text. Place your answer on a sheet of note paper.
1. Identify 10 different fasteners. Underline the threaded fasteners.
2. What are fasteners?
3. Give five (5) applications that make use of screw threads.
4. List three (3) uses of threaded fasteners.
   a.
   b.
   c.
5. What is the difference between a coarse thread and a fine thread on a given size and length bolt?
6. Make sketches of a detailed representation of

a screw thread, a schematic representation and a simplified representation.

7. Explain the meaning of the following screw thread size: 1/2-2OUNC-3LH.

8. What fasteners are most commonly used to join wood?

Place the letter of the correct definition in the blank to the left of the fastener.

9. ____ Machine screw
10. ____ Cap screw
11. ____ Machine bolt
12. ____ Set screw
13. ____ Nut
14. ____ Washer
15. ____ Cotter pin
16. ____ Rivets
17. ____ Key
18. ____ Keyseat

a. Threaded on both ends.
b. Fitted on a bolt.
c. Distributes clamping pressure of nut or bolt.
d. Prevents slippage of a pulley on a shaft.
e. Used for any general assembly work where screws less than 1/4 in. dia. are needed.
f. Used when close tolerance fasteners are not required.
g. Used when assembly requires a stronger, more precise and better appearing fastener.
h. Small piece of metal partially fitted into a shaft and pulley, gear, or wheel to prevent turning on a shaft.
i. Makes a permanent assembly.
j. Fits in a hole drilled near the end of a shaft.
k. Machined on a shaft for a key.

DRAWING PROBLEMS

1. Draw 3/4-1OUNC-2 x 4 in. long hexagonal and square head bolts and nuts on the same sheet. Allow 3 inches between the drawings. Use a simplified thread representation.

2. Draw 1-8UNC-2 x 3 in. long hexagonal and square head bolts and nuts on the same sheet. Allow 3 inches between the drawings. Use a schematic thread representation.

UNIT 21—OUTSIDE ACTIVITIES

1. Secure samples of machine screws, cap screws, machine bolts, set screws, stud bolts, nuts, washers, rivets, cotter pins, nails, and wood screws. Make a display for use in the drafting room. Label all samples.

2. Get a sample showing a key, keyseat and keyway in use. Describe how the components are assembled.

3. Secure an example of a setscrew application. How does this example differ from the use of a key?

Unit 22

ELECTRICAL AND ELECTRONICS DRAFTING

After studying this unit, you will gain a basic knowledge of the electrical and electronics drafting area. You will be able to describe why electrical and electronics drafting is diagrammatic using standard symbols. You will be able to demonstrate four types of diagrams used in this field.

Our world, as we know it today, could not exist without electricity and the electronic devices it powers. You will realize how true this is by just listing the electrical and electronic devices you may depend upon in your home—TV set, radio, stereo, VCR, personal computer, Fig. 22-1, washer, refrigerator, range, and the lighting to name but a few.

Fig. 22-1. Typical personal computer.

283

Fig. 22-2. Drafters preparing drawings for this robot welder (whether manually or with CAD) must have an extensive knowledge of electrical/electronic drafting techniques and standards. Electronic sensors on the robot "read" a code on the fixture holding the auto body components. The robot's computer adapts welding sequence for the body design moving into position. This means that hatchbacks, station wagons, four-door sedans, and convertibles can be welded on the same assembly line. (Pontiac Motor Div., GMC)

There are many other electrical devices that may not be as familiar as those found at home or at school. Examples include the computers that are important to business and medicine. Also the programmable devices which control the machines that make, inspect, assemble, and test so many of the products we use, Fig. 22-2. Many hobbies are electronically oriented, Fig. 22-3.

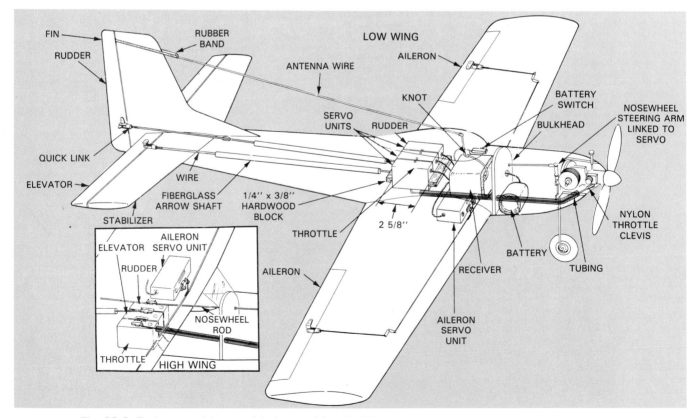

Fig. 22-3. Radio control in a model airplane. Many hobbies are electronically oriented. (Heath Company)

COMPUTERIZED ELECTRICAL AND ELECTRONICS DRAFTING

Computer aided design, Fig. 22-4, has found extensive use in the design and layout of printed circuits, Fig. 22-5, and integrated circuits (IC's), Fig. 22-6, for electronic devices. CAD does the work in a fraction of the time required to do it by hand. Many complex integrated circuit designs would be obsolete before they could be put into production if the design and layout work were done manually.

As in other areas of computer aided design, the designer must have a thorough knowledge of electrical and electronic drafting techniques and standards before they can become skilled in using CAD to its fullest potential.

Fig. 22-4. Computer aided design has found extensive use in the design and layout of electronic circuits. They do the work in a fraction of the time needed to do it manually and changes can be made easily. (CALCOMP, A Sanders Graphics Company)

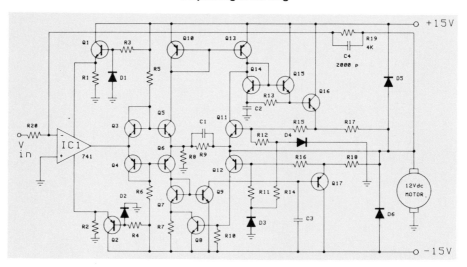

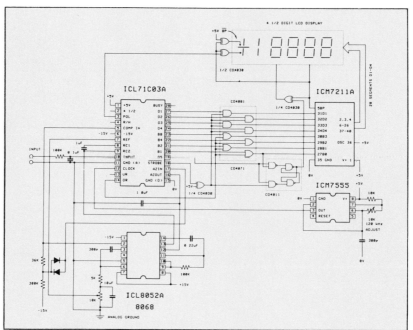

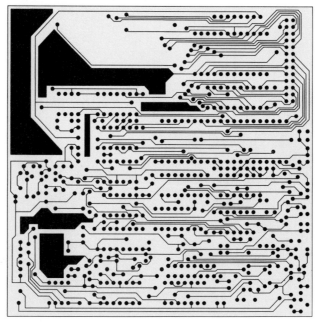

Fig. 22-5. Samples of computer designed circuits and circuit board. (ROBO Systems)

Fig. 22-6. Emission control microcomputer is slightly larger than a paperback book. It receives data from engine-mounted sensors. At a rate of 300,000 calculations per second, the computer analyzes this sensor information and commands the carburetor to send the optimum air/fuel mixture to the engine's cylinders. It was computer designed. (Delco Electronics, GMC)

ELECTRICAL AND ELECTRONIC DRAFTING

Electrical and electronic drafting is done manually in much the same manner as conventional drafting. The same equipment is used.

However, instead of using regular multiview drawings, a large portion of electrical and electronic drafting is diagrammatic in character. That is, considerable use is made of symbols. The symbols represent the various components and wires that make up the electrical/electronic circuit, Fig. 22-7. They are easier and quicker to draw than the actual part. Symbols are combined on a diagram that shows the function and relation of each component in the circuit.

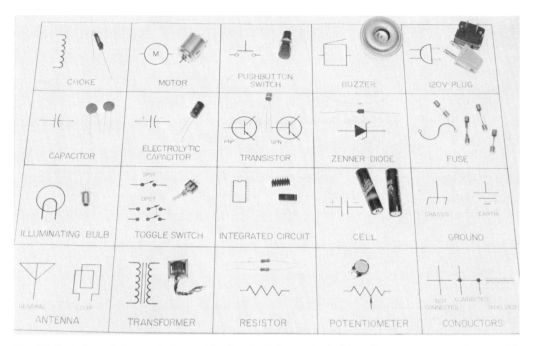

Fig. 22-7. A few of the symbols used in electrical/electronic drafting. Components are shown with their symbols. These are few of the hundreds of symbols used.

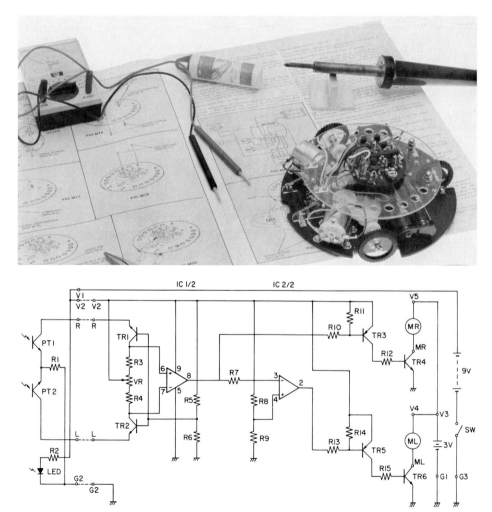

Fig. 22-8. LINE TRACER ROBOT with schematic diagram. This miniature robot operates very similar to the large industrial robot carts that bring materials to machines in automated factories. The industrial robot carts have no human operators but follow wires embedded in the floor or lines drawn on the floor. The miniature device follows a line drawn on the floor. (Graymark International, Inc.)

TYPES OF DIAGRAMS USED IN ELECTRICAL AND ELECTRONIC DRAFTING

The drafter working in this area of drafting must be familiar with the different types of diagrams.

SCHEMATIC DIAGRAM, Fig. 22-8. A drawing with symbols and single lines to show electrical connections and functions of a specific circuit. The various components that make up the circuit are drawn without regard to their actual physical size, shape, or location.

CONNECTION OR WIRING DIAGRAM, Fig. 22-9. While commonly used to show the distribution of electricity on architectural drawings, this type of diagram may be drawn to show the general physical arrangement of the IC's, transistors, diodes, resistors, switches, etc., that make up an electronic circuit, Fig. 22-10.

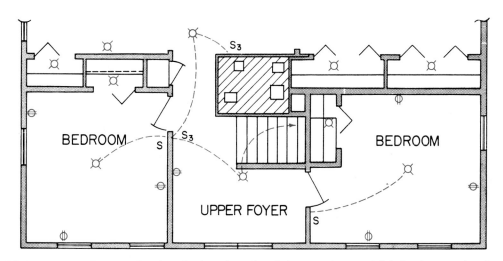

Fig. 22-9. A wiring diagram showing the location of switches, outlets, and lighting in a modern home.

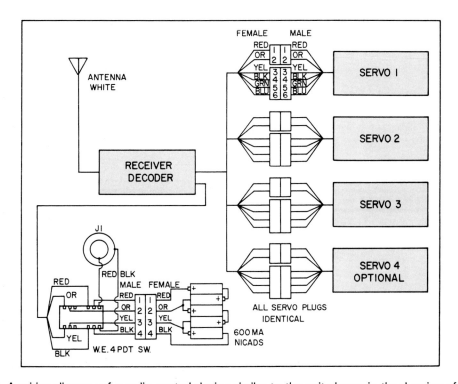

Fig. 22-10. A wiring diagram of a radio control device similar to the unit shown in the drawing of Fig. 22-3.

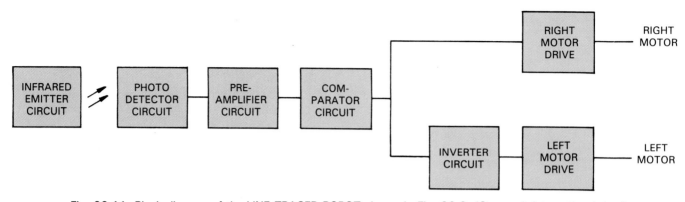

Fig. 22-11. Block diagram of the LINE TRACER ROBOT shown in Fig. 22-8. (Graymark International, Inc.)

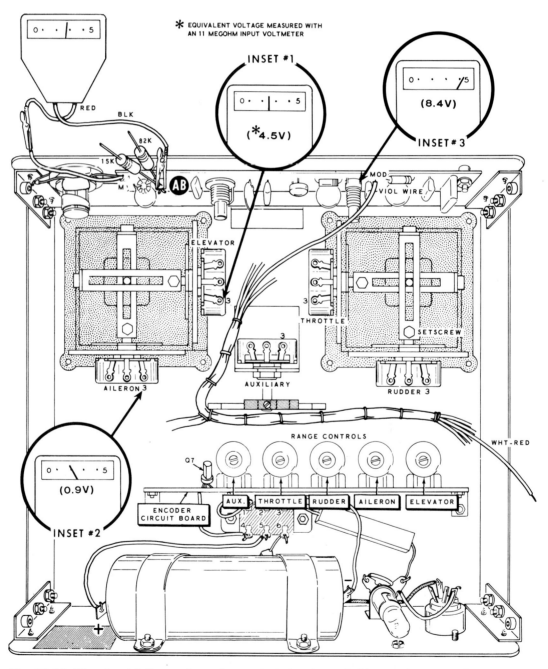

Fig. 22-12. The pictorial diagram is used extensively by electronic kit manufacturers. You do not have to be an expert to construct the kits.

BLOCK DIAGRAM, Fig. 22-11. A simplified way to show the operation of an electronic device. This utilizes "blocks" (squares, rectangles, triangles, etc.) joined by a single line. It reads from left to right.

PICTORIAL DIAGRAM, Fig. 22-12. In this diagram, the components are drawn in pictorial form and in their proper location. The pictorial diagram is used extensively by electronic kit manufacturers because it is so easy to understand.

DRAWING ELECTRICAL AND ELECTRONIC SYMBOLS

Symbols in circuit diagrams need not be drawn to any particular scale. However, they should be shaped correctly, large enough to be seen clearly, and in proportion, Fig. 22-13. The easiest way to do this is to draw them with the aid of an electrical and electronic template, Fig. 22-14.

Symbols and lines are drawn the same weight as a visible object line. A darker line may be employed when a portion of the diagram must be emphasized.

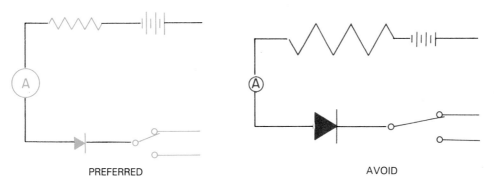

PREFERRED   AVOID

Fig. 22-13. Draw symbols in proportion.

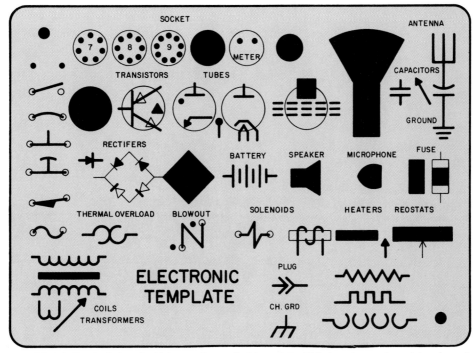

Fig. 22-14. Typical electrical/electronic symbol template. This is only one template of a set. Templates are the easiest way to draw the symbols in proper proportion to each other.

DRAFTING VOCABULARY

Block diagram, Diagram, Diagrammatic, Electrical, Electronic, Emphasized, Extensive, Integrated circuit, Obsolete, Oriented, Potential, Printed circuit, Proportion, Schematic, Symbols, Transistor.

UNIT 22—TEST YOUR KNOWLEDGE

Please do not write in the text. Place your answers on a sheet of notebook paper.

1. How does electrical and electronic drafting differ from conventional drafting?
2. Where has computer aided design been used extensively by the electronics industry?
3. A SCHEMATIC DIAGRAM is a drawing _____.
4. How does a BLOCK DIAGRAM differ from a WIRING DIAGRAM?
5. What is a PICTORIAL DIAGRAM?
6. Symbols and lines used in electrical/electronic drafting are drawn the same weight as a _____ _____ line.
7. Electrical diagrams used on house plans are called _____ diagrams.

UNIT 22—OUTSIDE ACTIVITIES

1. Make a schematic diagram of a desk lamp or light used on a drafting table.
2. Prepare a schematic diagram of a two cell flashlight.
3. Make a complete wiring diagram of your bedroom.
4. Secure a small battery powered toy. Examine how it operates. Make a suitable electronics diagram showing your findings.
5. Prepare a pictorial diagram of a crystal radio.
6. Make a suitable diagram showing four batteries, switch, and lamp wired in series.
7. Make a suitable diagram showing four batteries, switch, and lamp wired in parallel.
8. Prepare the wiring diagram showing how to connect a receiver, amplifier, CD, turntable, and two speakers.

Unit 23

ARCHITECTURAL DRAFTING

After studying this unit, you will be able to explain what architectural plans are and the importance of being able to understand them. You will be able to produce a typical set of house plans including the plot plan, elevations, floor plans, sectional views, and details. You will be able to read an architect's scale. You will learn how to plan a house based on the knowledge obtained in this unit.

Plans which provide craftworkers with the information needed to construct buildings in which people live, work, and play are called ARCHITECTURAL DRAWINGS. See Fig. 23-1. Architectural drawings or plans are created, designed, and produced by architects.

Fig. 23-1. Artist's sketch of a home to be built. Since this house was to be built on speculation (there was no buyer when construction started), this sketch was needed for advertising purposes.

Fig. 23-2. The speculation house after it was constructed. The optional garage seen on the sketch was not built.

Buying a home is probably one of the largest investments you will make in your lifetime. An understanding of the basic principles of architectural drafting will be a great help if you plan to design or construct a new home, modernize an existing home, or judge the soundness and value of a home offered for sale.

The ability to read and interpret architectural drawings is essential to those in the construction industry, such as carpenters, masons, plumbers, electricians, roofers, etc., Fig. 23-2. It is also useful to workers in lumber yards or hardware and building supply stores.

BUILDING A HOME

To make sure your home is built to your specific requirements, it is important that you have a good plan and a well defined contract with your builder. Plans provide the vast amount of information needed to construct a modern home.

BUILDING CODES

Plans should include all applicable aspects of the local and state building codes. Building codes are laws which provide for the health, safety, and general welfare of the people in the community. Building codes are based on standards developed by government and private agencies.

BUILDING PERMIT

In most communities, the building contractor or owner must file a formal application for a building permit. Plans and specifications are submitted for the proposed structure. The plans are reviewed by building officials to determine whether they meet local building code requirements.

An inspection card is usually posted on the building site and the work is inspected by local building officials as construction progresses. The card is signed as each phase of construction is approved.

PLANS

Because it is not possible to include all construction details on a single sheet, a set of typical house plans will include a plot plan, foundation and/or basement plan, floor plans, elevations (front, rear, and side views of the home), wall sections, built-in cabinet, and fireplace details.

SCALE

The plans are generally drawn to a one-fourth inch scale (1/4" = 1'-0"). This means that 1/4 inch on the drawing equals 1 foot on the building being constructed.

A larger scale, 1" = 1'-0", is used when greater detail is needed on a structural part. Framing planes are often drawn to 1/8" = 1'-0" scale.

PLOT PLAN

The PLOT PLAN shows the location of the structure on the building site, Fig. 23-3. Plot plans also show walks, driveways, and patios. Overall building and lot dimensions are included. Contour lines (lines indicating the slope of the site surface) are sometimes shown.

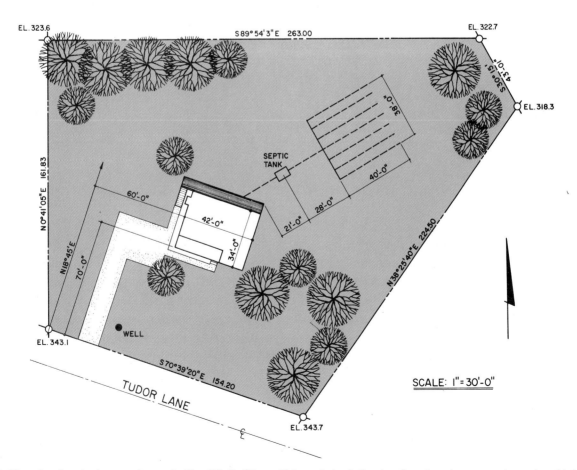

Fig. 23-3. Plot plan for the house shown in Fig. 23-2. (Note: This and the following five drawings were traced and inked from the architect's original drawings for better reproduction in the book.)

240" FIBERGLASS SHINGLES WITH
ADHESIVE TABS ON 15" FELT

ALUM. FLASHING

4"(TYP)

FIN. FLOOR

1"x 8" CEDAR SIDING
W/6½" EXPOSED

FIN. FLOOR

FOUNDATION:
I COAT THOROSEAL
2 COATS PORTLAND CEMENT
I COAT HOT TAR

FRONT ELEVATION
¼"=1'-0"

Fig. 23-4. Elevation drawing.

ELEVATIONS

ELEVATIONS are the front, rear, and side views of the house, Fig. 23-4. They are made of lines which are visible when the building is viewed from various positions. Included on the elevations are the floor levels, grade lines, window and door heights, roof slope, and the types of materials to be used on the walls and roof. Foundation and footing lines below grade level are indicated with hidden lines.

FLOOR PLANS

The size and shape of the building, as well as the interior arrangement of the rooms, are shown on the FLOOR PLANS, Fig. 23-5. Additional information such as location and sizes of the interior partitions, doors, windows, stairs, and utility installations (plumbing, electrical, etc.) is also included.

Foundation and basement plans are frequently combined on a single sheet, Fig. 23-6.

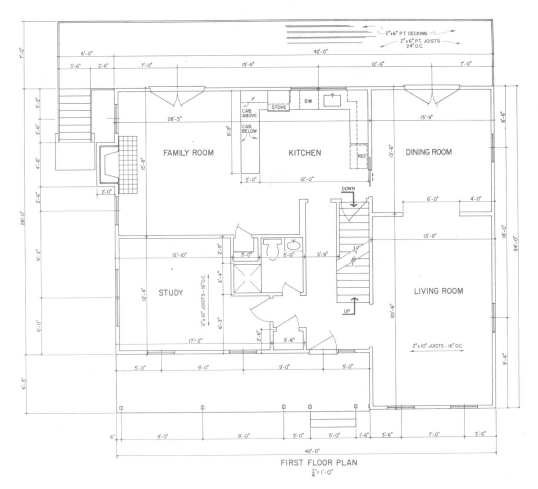

FIRST FLOOR PLAN
¼" = 1'-0"

Fig. 23-5. Floor plan. Electricals and climate control units are shown on other drawings.

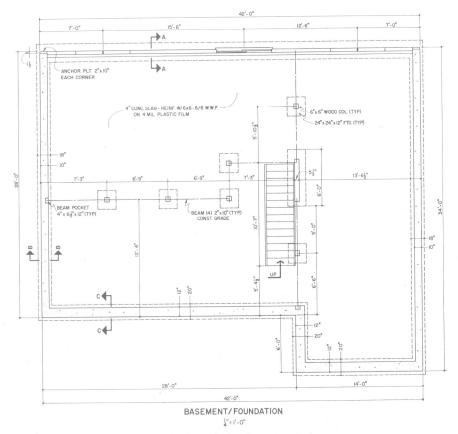

BASEMENT/FOUNDATION
¼" = 1'-0"

Fig. 23-6. Foundation/basement plan.

297

SECTIONS

SECTIONAL VIEWS are used to give construction details of the structure from basement or foundation footing, to the ridge of the roof, Fig. 23-7. Break lines are often incorporated into the section to reduce the drawing size and save time in drawing the plan.

Along with the sizes of the framing materials, the types and kinds of materials to be used for sheathing, insulation, interior and exterior wall surfaces, along with other miscellaneous information, are also indicated.

DETAILS

Information to assist the craftworker in constructing such things as built-in cabinets, fireplaces, and other pertinent information are given on DETAIL SHEETS, Fig. 23-8.

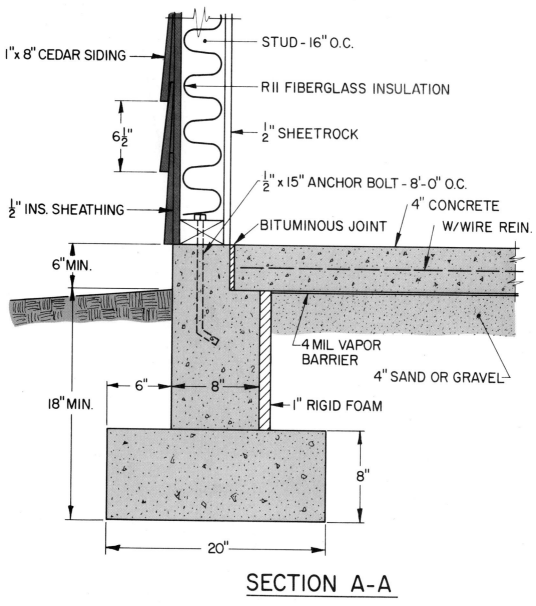

SECTION A-A

Fig. 23-7. Section showing foundation details noted on Fig. 23-6.

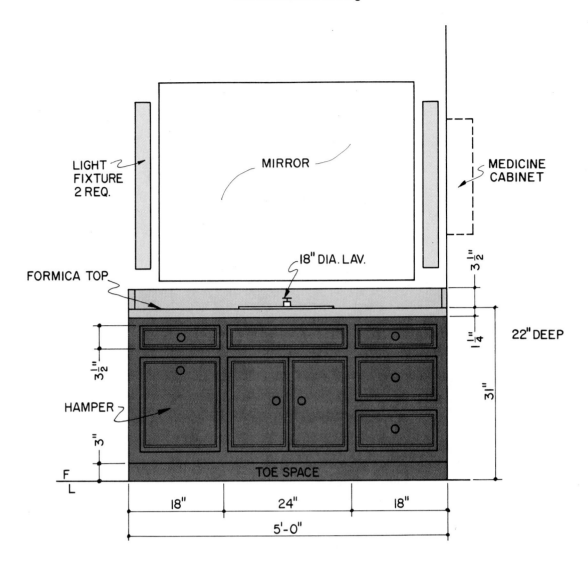

GUEST ROOM LAVATORY CABINET DETAILS
UNIT I - UNIT 2 IS REVERSED
SCALE: 1/2"= 1'-0"

Fig. 23-8. Details found on a typical architectural drawing.

ARCHITECTURAL DRAFTING TECHNIQUES

In general, most conventional drafting techniques will apply to architectural drafting.

SYMBOLS, ''SPEC'' SHEETS

Extensive use of SYMBOLS will be noted in architectural drafting, Fig. 23-9. Symbols are employed because it is not practical to show on plans items such as doors, windows, plumbing fixtures, etc., as they would actually appear in the structure.

Sheets of specifications usually accompany house plans. These "SPEC" SHEETS describe in writing things that cannot be easily indicated on the drawings such as quality of materials, how installation of specific items are to be made, etc.

Exploring Drafting

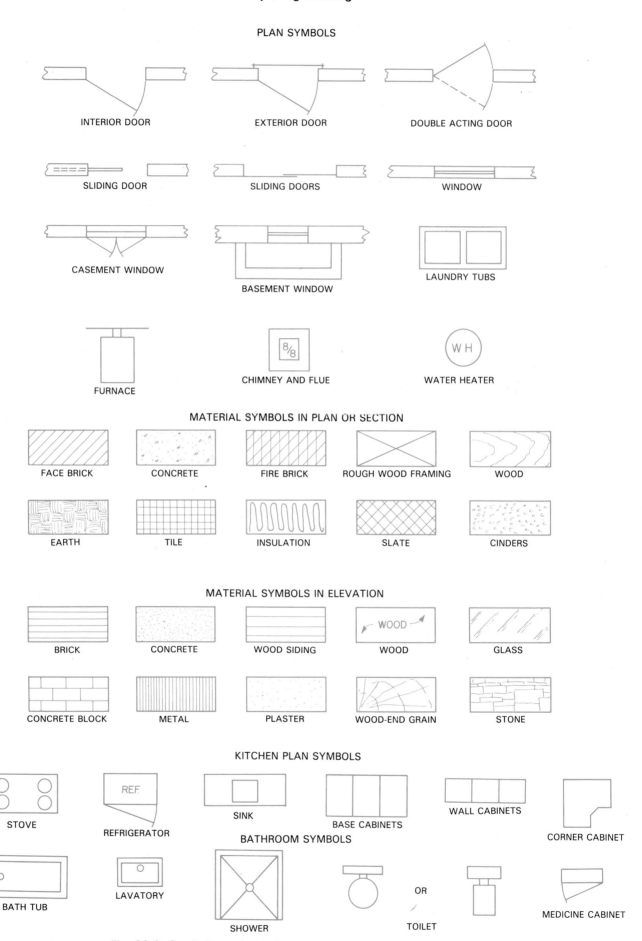

PLAN SYMBOLS

INTERIOR DOOR   EXTERIOR DOOR   DOUBLE ACTING DOOR

SLIDING DOOR   SLIDING DOORS   WINDOW

CASEMENT WINDOW   BASEMENT WINDOW   LAUNDRY TUBS

FURNACE   CHIMNEY AND FLUE   WATER HEATER

MATERIAL SYMBOLS IN PLAN OR SECTION

FACE BRICK   CONCRETE   FIRE BRICK   ROUGH WOOD FRAMING   WOOD

EARTH   TILE   INSULATION   SLATE   CINDERS

MATERIAL SYMBOLS IN ELEVATION

BRICK   CONCRETE   WOOD SIDING   WOOD   GLASS

CONCRETE BLOCK   METAL   PLASTER   WOOD-END GRAIN   STONE

KITCHEN PLAN SYMBOLS

STOVE   REFRIGERATOR   SINK   BASE CABINETS   WALL CABINETS   CORNER CABINET

BATHROOM SYMBOLS

BATH TUB   LAVATORY   SHOWER   TOILET   OR   MEDICINE CABINET

Fig. 23-9. Symbols used to indicate materials and fixtures on drawings.

ELECTRICAL SYMBOLS

OUTLET   DUPLEX OUTLET   SINGLE OUTLET   WEATHERPROOF OUTLET   DROP CORD

CIRCUIT BREAKER   RANGE OUTLET   FLOOR OUTLET   TELEPHONE   BELL

SWITCH SINGLE POLE   SWITCH THREE-WAY   SWITCH FOUR-WAY   PUSH BUTTON   CHIME

Fig. 23-9 continued.

SCALE

Architectural plans are drawn to scale, with few exceptions. When drawn to scale, the views presented on the drawing are shown in an accurate proportion of the full-sized structure. A 1/4" = 1'-0" scale would mean that 1/4 inch on the drawing equals 1 foot in the structure to be built.

Scale drawings are relatively easy to make if an ARCHITECT'S SCALE is used. This scale is divided into various major units, each unit representing a distance of one foot. Shown in Fig. 23-10 is a 1/4" = 1'-0" unit of the scale.

This unit and the other units on the scale are further subdivided into equal parts or multiples of twelve to represent inches, Fig. 23-11.

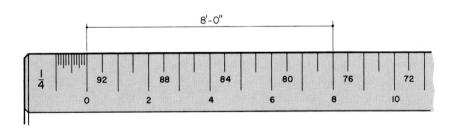

Fig. 23-10. Measuring feet on an architect's scale. The 1/4" = 1'-0" is being used. Count to the right of zero.

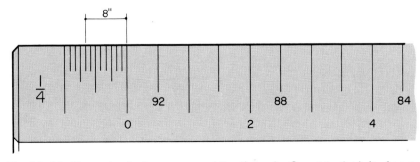

Fig. 23-11. Measuring inches on an architect's scale. Count to the left of zero.

When using the architect's scale to make a measurement, for example, 8'-8" (eight feet eight inches), start at the 0 and go to the right to locate 8' (eight feet). Then, moving in the opposite direction from 0 locate the 8" (eight inches). The combined measurement would be 8'-8". See Fig. 23-12.

Fig. 23-13 will give you an opportunity to practice reading the architect's scale. List your answers on a sheet of notebook paper.

TITLE BLOCK

A TITLE BLOCK, Fig. 23-14, for architectural drawings should include the following information:
1. Type of structure (house, garage, shed, etc.).
2. Where it is to be located.
3. Architect's name (and/or drafter).
4. Date.
5. Drawing scale or scales used.
6. Sheet number and number of sheets making up the full set of drawings (Sheet 1 of 10, etc.).

The title block may be made any convenient size; however, 2 in. by 4 in. is a suitable size for most home plans. It is usually located in the lower right corner of the drawing sheet.

LETTERING ARCHITECTURAL PLANS

Most architectural lettering follows the Roman alphabet of letters and numbers, Fig. 23-15. Conventional single stroke Gothic lettering is sometimes used on architectural drawings. Many architects and drafters develop their own distinctive style of lettering. Whether you decide to use the single stroke Gothic letter form or develop a style of your own, remember that your lettering must be legible.

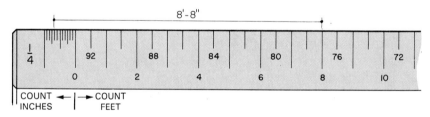

Fig. 23-12. Measuring both feet and inches on an architect's scale.

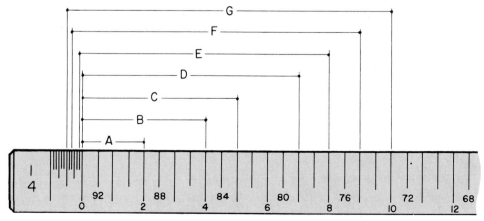

Fig. 23-13. How many of these dimensions can you read correctly? Do not write in the book. Place your answers on a piece of notebook paper.

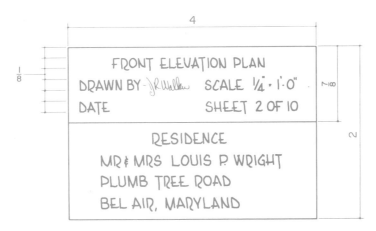

Fig. 23-14. Plan title block.

A B C D E F G H I J K L M N O P Q R S T
U V W X Y Z & 1 2 3 4 5 6 7 8 9 0
a b c d e f g h i j k l m n o p q r s t u v w x y z-

Fig. 23-15. One style of architectural lettering.

DIMENSIONING ARCHITECTURAL PLANS

There is only a slight difference between the dimensioning techniques used on architectural drawings and on other forms of drafting. The dimensions are read from the bottom and right sides of the drawing sheet. Measurements over twelve inches (12") are expressed in feet and inches. Foot and/or inch marks (' and ") are used on all dimensions.

Dimensions may be placed on all sides of the drawing and through the drawing itself. Dimension figures, usually 1/8 in. high, are written above the dimension line or in a break in the dimension line. The dimension line may be capped with arrowheads, small dark circles, or a short 45 deg. slash mark, as shown in Fig. 23-16.

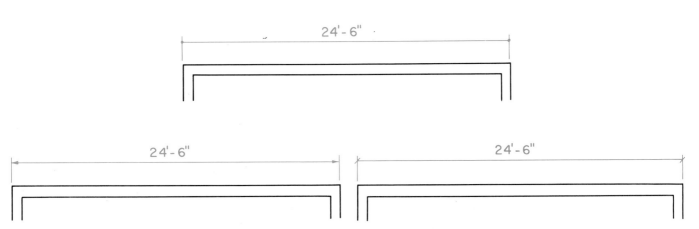

Fig. 23-16. Ends of dimension lines may be capped by arrowheads, small darkened circles, or short 45 deg. slash marks.

ARCHITECTURAL DIMENSIONING RULES

1. Make sure that the dimensioning is complete. A builder should never have to assume or measure a distance on the drawing for a dimension.
2. ALWAYS letter dimensions full-size regardless of the drawing's scale.
3. Dimensions and notes should be at least 1/4 in. away from the view.
4. Align dimensions across the view whenever possible.
5. Place overall dimensions outside the view.
6. Overall dimensions are determined by adding detail dimensions. Do not scale the drawing.
7. Dimension partitions center to center or from center to outside wall.
8. Room size may be indicated by stating the width and length of the room.
9. Windows, doors, beams, etc., are dimensioned to their centers.
10. Window and door sizes and types are given in a SCHEDULE, Fig. 23-17.

ARCHITECTURAL ABBREVIATIONS

Time and space can be saved when lettering architectural drawings by using abbreviations. Fig. 23-18 shows some commonly used abbreviations.

OBTAINING INFORMATION

A great deal of research into catalogs and reference books must be made on the part of the student to find the many standard sizes that are essential in completing an accurate set of plans.

Additional information may be found in mail order catalogs, lumber and building supply company literature, from the Federal Housing Administration, and copies of your local building ordinances. Specialized textbooks in architecture, carpentry, plumbing, and other building areas are excellent sources of information when planning a home or other structure.

MARK	REQ'D	OVERALL SASH SIZE WIDTH	HEIGHT	DESCRIPTION
W-1	4	5'-4" x	4'-0"	ALUMINUM SLIDING WINDOW WITH INSULATED SASH-FIBERGLASS SCREEN
W-2	3	5'-4" x	2'-0"	SAME AS ABOVE
W-3	1	4'-0" x	3'-0"	"
W-4	1	2'-7" x	5'-8"	1/4" PLATE GLASS
W-5	1	9'-6" x	2'-4"	FIXED, SAME AS ABOVE

WINDOW SCHEDULE

Fig. 23-17. Window schedule of the type found on a typical house plan.

asphalt tile—AT	concrete-CONC	footing—FTG
beam—BM	drawing—DWG	grade line—GL
bedroom—BR	door—DR	lavatory—LAV
brick—BRK	elevation—EL	living room—LR
ceiling—CLG	exterior—EXT	plaster—PL
center line—CL or ₵	floor—FLR	room—RM

Fig. 23-18. Common abbreviations used on architectural plans.

PLANNING A HOME

When planning a home, you must determine how much money will be available for building purposes. Most people building homes must borrow a major portion of the construction money. Lending institutions use formulas based on the cash available for a down payment and the owners income (salary, etc.) to establish how much money they will lend to build or purchase a home.

One way to ascertain the size of a home you plan to build is by the square foot method. A local contractor can give you the approximate cost per square foot of construction in your community.

Divide the cost of construction per square foot into the money available to build the house, and you will find the floor area of a home you can afford to build.

After determining the approximate size of the proposed house, you should try to develop a suitable floor plan. Say, for example, calculations show that you can afford a home with an area of 1200 square feet. Outlines of homes having 1200 square feet of floor space are shown in Fig. 23-19. Room use and space available must be carefully considered and balanced.

Be sure to include closet and storage facilities. Using scale cutouts of furniture and appliances you plan to use in the various rooms will aid in your planning. The preliminary floor plan can be drawn on graph paper. Each square can equal one foot. Changes can be made easily on the grid.

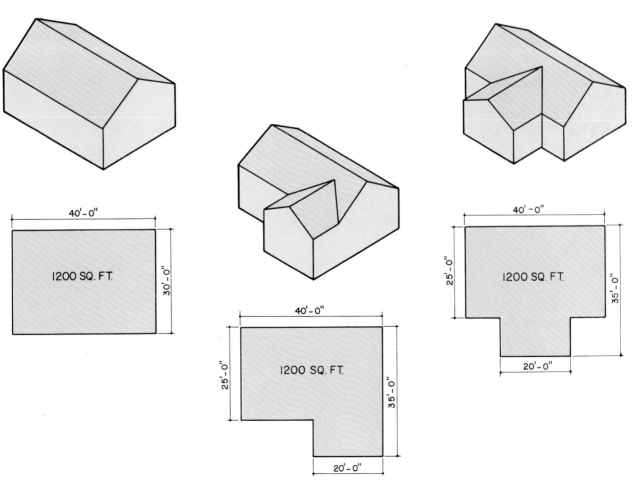

Fig. 23-19. Examples of homes with floor areas of 1200 square feet.

The cutouts MUST be made to the same scale as the floor plan. Your drawing should also show all openings and the space doors will take up when opened.

By moving the cutouts, it is easy to find out whether the room is large enough for the intended use.

Sizes of some furniture and fixtures are given in Fig. 23-20. Dimensions of items not shown may be found in manufacturers' catalogs, home magazines, and mail order catalogs.

Cutout planning for a typical room is shown in Fig. 23-21.

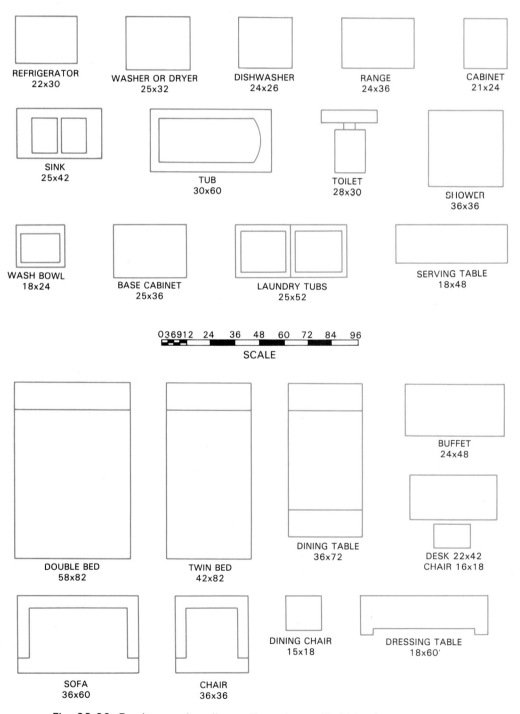

Fig. 23-20. Furniture and appliance dimensions will aid in planning a home.

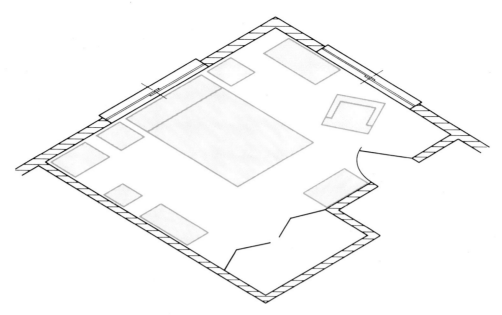

Fig. 23-21. How cutouts may be used to plan a bedroom.

METRICATION IN ARCHITECTURE

The building construction industry probably will be one of the last business enterprises to go metric in the United States. However, lumber producers have decided upon a "soft" conversion of existing lumber sizes. This will not require a basic redimensioning of lumber sizes as would the SI or "hard" conversion. A "soft" conversion simply means that inches have been changed to millimeters, pounds to kilograms, etc. A "hard" conversion with metric engineering standards is one in which all of the designing is done in preferred metric sizes.

Some plumbing fixtures made to SI standards are being imported. Problems have been encountered trying to join these fixtures to inch-based pipes and fittings.

In the United States, most buildings are designed to a 4-inch base module. See Fig. 23-22. All lumber, blocks, bricks, panel stock (plywood, hardboard, etc.) and components such as windows and doors are specified in multiples of the 4-inch module. Windows, for example, usually are 2'-8" or 3'-0" wide.

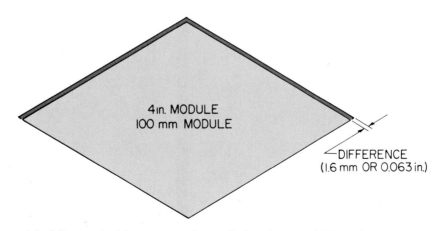

Fig. 23-22. Most buildings and building materials now used are designed to multiples of the basic 4 in. module. A 100 mm × 100 mm module is the metric based unit being proposed.

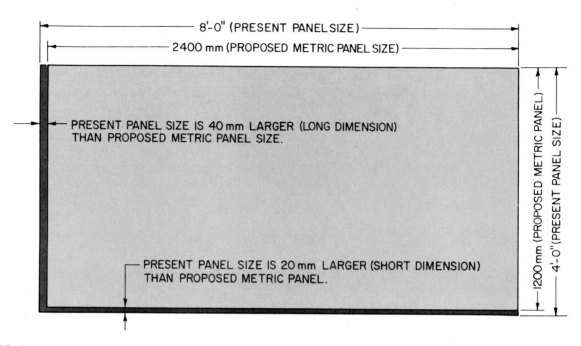

Fig. 23-23. A comparison of the conventional size building panel and the proposed metric size panel. The proposed metric size panel would be too small to use for a replacement of the conventional 4 ft. × 8 ft. panel.

A 100 mm by 100 mm module is being recommended as the basic metric module. All metric based building materials and components will be based on multiples of this module.

A 100 mm by 100 mm module is almost, BUT NOT QUITE, the same size as the customary 4 in. by 4 in. module. The difference is small, but is enough to cause major problems if ALL materials used in constructing a building are not designed to the same basic module. See Fig. 23-23.

METRIC ARCHITECTURAL DRAFTING

There is little difference between metric and customary measuring systems in architectural drafting. Dimensions will be given in millimeters and meters instead of inches and feet.

Scales for use in metric based architectural drafting are shown in Fig. 23-24. All structures will be designed and constructed on the 100 mm module.

METRIC SCALE	USE	CUSTOMARY EQUIVALENT
1:10	Construction Details	1″ = 1′−0″ (1:12)
1:20	Construction Details	3/4″ = 1′−0″ (1:16)
1:25	Construction Details	1/2″ = 1′−0″ (1:24)
1:50	Plans, Elevations	1/4″ = 1′−0″ (1:48)
1:100	Plot Plan	1/8″ = 1′−0″ (1:96)
1:200	Plot Plan	1/16″ = 1′−0″ (1:192)
1:500	Site Plan	1/32″ = 1′−0″ (1:384)

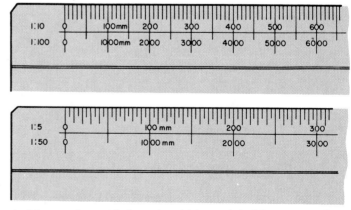

Fig. 23-24. A few of the metric scales recommended for use in metric based architectural drafting.

Fig. 23-25. As with other areas of industry, computer aided design (CAD) and computer aided engineering (CAE) are changing the field or architecture. (Heath/Zenith Educational Systems)

COMPUTER AIDED DESIGN AND DRAFTING

As with other areas of industry, computer aided design (CAD) and computer aided engineering (CAE) are changing the field of architecture, Fig. 23-25. Talented people working with microcomputers are able to produce quality work, faster and more profitably than ever before. Time is greatly reduced between concept, design, working drawings, and construction scheduling of a building project.

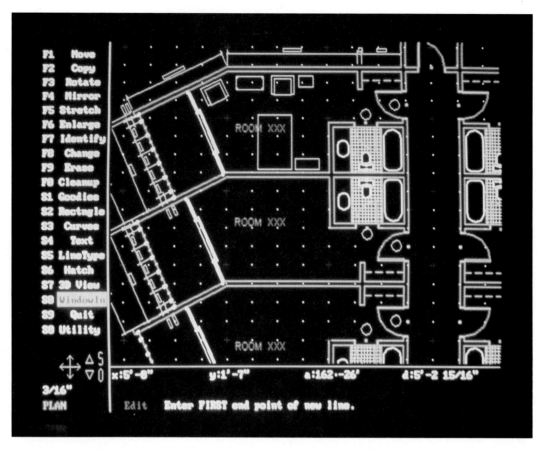

Fig. 23-26. CAD can help plan the most efficient floor plan layout. Note the menu down the side of the CRT. (Microtecture)

An important advantage of CAD in architecture allows two- and three-dimensional details to be developed and combined to create new structures and layouts, Fig. 23-26. Ideas can be viewed in three-dimensional space and may be rotated and seen from any angle. Design changes can be viewed immediately.

Architectural CAD programs are designed to reduce repetitive work to a minimum. See Fig. 23-27. Other particulars such as the outline of a structure can be developed with the windows and doors located. Structural details, meeting engineering standards, and window and door details, will be inserted automatically. A bill of material and estimated cost of materials can be computed using the same information. New architectural CAD programs under development will offer architects advanced design and management capabilities.

DRAFTING VOCABULARY

Architect, Architecture, Basement, Building code, Building permit, Built-in, Contract, Construction, Detail Sheets, Elevation, Essential, Floor plan, Formula, Foundation, Framing, Grade line, Inspection card, Insulation, Modernize, Module, Ordinances, Partition, Pertinent, Plot plan, Proposed, Ridge, Scale, Site, Specification, Standards, Structure, Wall sections.

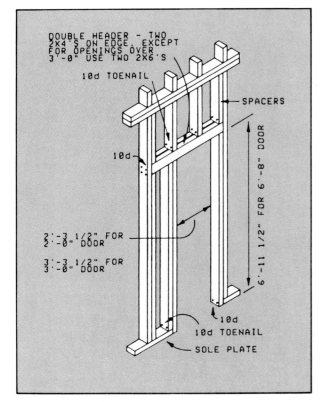

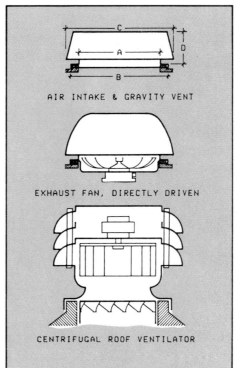

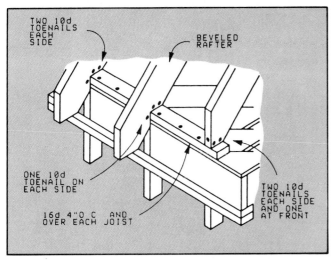

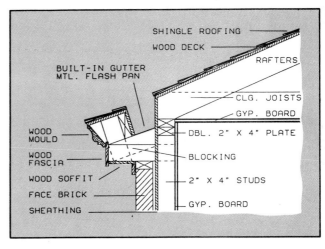

Fig. 23-27. Examples of architectural CAD developed structural details. (Richards Community College)

UNIT 23—TEST YOUR KNOWLEDGE

Please do not write in the text. Place your answer on a sheet of notebook paper.

1. Architectural drawings are _____.

2. The ability to read and interpret architectural drawings is essential to workers such as:

    a. _____

    b. _____

    c. _____

    d. _____

    e. _____

3. What are building codes?
4. Architectural drawings are usually drawn to a scale of _____.
5. The location of the building(s) on the construction site is shown on the _____ _____.
   This drawing also shows the walks, driveways, and patios.
6. Drawings that show the various views of the structure (front, rear, and side views) are called
   _____.
7. The floor plan shows the: (Check the correct answer.)
   a. Size and shape of the rooms.
   b. Location and sizes of the windows and doors.
   c. Location of utilities (plumbing, heating, and electrical fixtures).
   d. All of the above.
   e. None of the above.
8. Make a sketch showing three recommended methods for dimensioning architectural drawings. Refer to Fig. 23-16.

UNIT 23—OUTSIDE ACTIVITIES

1. Make a scale drawing (1/4" = 1'-0") of the school drafting room. Use cutouts to determine an efficient layout for the furniture in the room.
2. Prepare a drawing of your bedroom showing the location of the furniture in the room. Use a scale of 1/2" = 1'-0".
3. Draw the floor plan of a two-car garage with a workshop in one end.
4. Design and draw a full set of plans for a two room summer cottage or hunting cabin.
5. Draw the plans necessary to convert your basement or other space into a workshop, a recreation room, or a dark room.
6. Design and draw the plans needed to construct a storage or tool house. See Fig. 23-28.
7. Design a small two bedroom house that can be constructed at minimum cost. Construct a model of it.
8. Plan a structure to house a sports car agency.
9. Secure samples of computer generated architectural drawings.

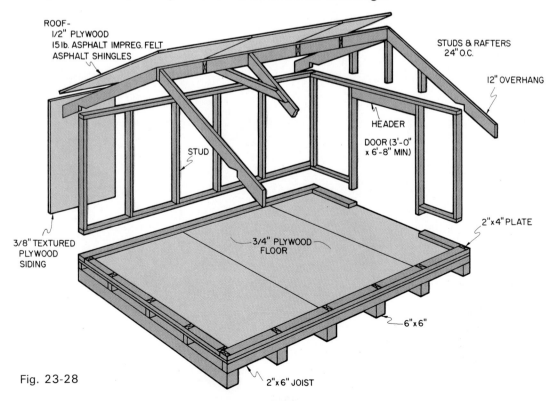

ROOF-
1/2" PLYWOOD
15 lb. ASPHALT IMPREG. FELT
ASPHALT SHINGLES

STUDS & RAFTERS
24" O.C.

12" OVERHANG

HEADER

DOOR (3'-0"
x 6'-8" MIN)

STUD

3/8" TEXTURED
PLYWOOD
SIDING

3/4" PLYWOOD
FLOOR

2"x4" PLATE

6"x6"

2"x6" JOIST

Fig. 23-28

Unit 24

COMPUTER GRAPHICS

After studying this unit, you will be able to explain how computer technology is revolutionizing drafting and engineering. You will be able to describe the basic equipment and applications of computer aided design and drafting. You will be able to define the fundamental terminology of computer graphics.

Computer graphics is revolutionizing drafting and engineering, Fig. 24-1.

A computer, in its simplest form, is a device that performs mathematical calculations. The first computer was the human hand. It can solve many mathematical problems. The ABACUS was the earliest MECHANICAL computer, Fig. 24-2. No one knows when it was invented. The abacus is still widely used in many parts of the world.

A modern computer is electronic. It can perform high speed computations, including mathematical calculations and logical decisions, under the

Fig. 24-1. Three dimensional view of bicycle hub created using computer graphics. (Applicon)

313

Fig. 24-2. The abacus was the first mechanical computer. It is still used in many parts of the world.

direction of a PROGRAM. A program is a series of instructions that "tells" a computer what needs to be done and how it is to be accomplished. The computer is a system that consists of a processor unit, monitor (cathode ray rube or CRT), operator console (keyboard), input devices, output devices, and auxiliary data storage devices, Fig. 24-3.

The MAINFRAME COMPUTER is a large computer. It is capable of processing large amounts of data at very fast speeds with access to billions of characters of data. Some super mainframe computers can process 200 million instructions a second.

A MICROCOMPUTER is a computer system commonly consisting of a display screen (CRT or cathode ray tube), keyboard, and limited data storage based upon a small silicon chip called a MICROPROCESSOR, Fig. 24-4. It is sometimes referred to as a personal computer (PC).

The MINICOMPUTER falls somewhere between a microcomputer and a mainframe computer. However, as microcomputers become more powerful, it is becoming difficult to distinguish them from minicomputers.

Today, computers operate factories, diagnose medical problems, control fuel and air mixtures in automobile engines, and have thousands of other applications. They also aid in designing mechanical products, buildings and other structures, solving engineering problems, etc., through the use of computer generated graphics, Fig. 24-5.

Fig. 24-3. A computer system used for computer aided design and drafting. Many schools and colleges use CAD equipment similar to this. (Heathkit/Zenith Educational Systems/Div. of the Heath Co.)

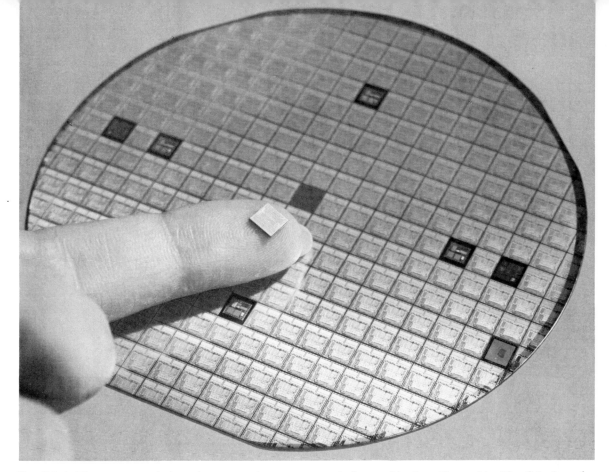

Fig. 24-4. The computer is based upon a microprocessor similar to this tiny silicon chip. The 3-inch wafer of silicon shown produces 190 chips. (Delco Electronics Div./General Motors Corp.)

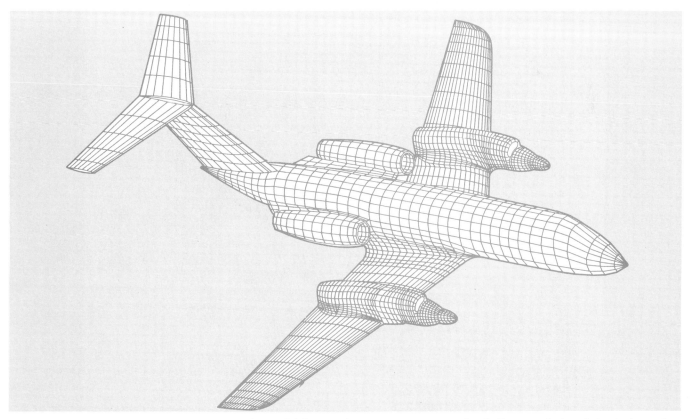

Fig. 24-5. A computer generated aerodynamic model for propfan test assessment of a Gulfstream 2 aircraft in dual propfan configuration. The model was generated by Lockheed Advanced Aeronautics Co. to analyze subsonic airflow where the engine nacelle joins the wing. The model has 5400 surface elements and was generated in 22.5 minutes on a special supercomputer. Using a conventional computer it would have required 17.5 hours to complete.

Fig. 24-6. The aerospace industry manufactures many complex and sophisticated products. Without the computer and computer graphics, the space shuttle would have required many additional years to design and manufacture. (NASA)

Since their products are usually the most sophisticated (complex) manufactured, the aerospace industry has been a pioneer in computer technology, Fig. 24-6.

COMPUTER GRAPHICS

Computer technology has revolutionized (completely changed) all areas of engineering and drafting. It was first employed for aircraft design in the 1950's. Computer graphics is now a required tool of industrial technology, Fig. 24-7.

Computer Aided Design (CAD) is a computer graphics technology that places designs, drawings, graphs, and pictures (in color) on a display screen, Fig. 24-8. The screen (CRT) is similar to a television screen but offers two-way communication between engineer/designer/drafter and computer. The technology is sometimes referred to as Computer Aided Design and Drafting (CADD).

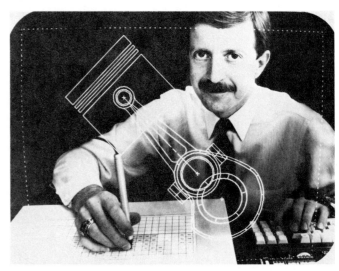

Fig. 24-7. An engineer checking a 2.5-liter four-cylinder engine piston/crankcase drawing on screen. The computer aided engineering (CAE) system permits an engineer to prepare a working drawing without lifting a pencil. After developing the basic drawing on the screen by operating the keyboard and using a wire-connected stylus, the engineer can rotate the drawing to view it from any angle, change its scale, or add depth to it to produce a three-dimensional version.
(Pontiac Motor Division/GMC)

316

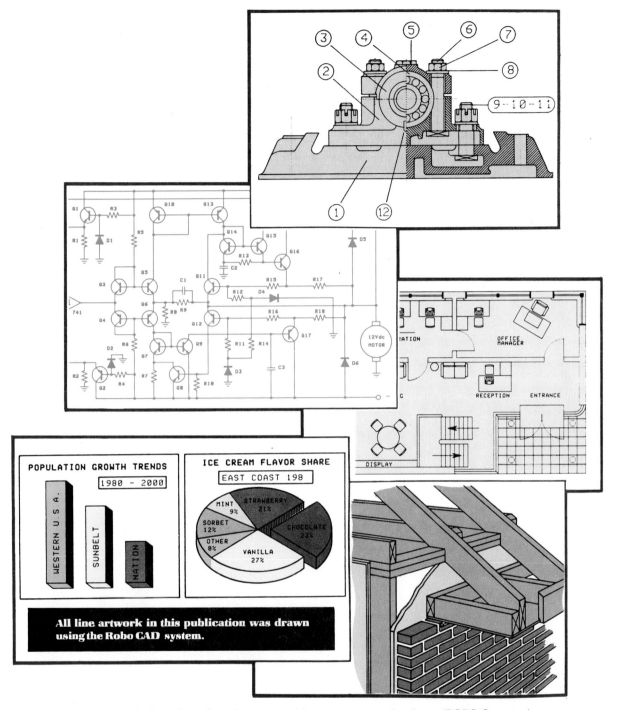

Fig. 24-8. Samples of work generated by computer technology. (ROBO Systems)

CAD does not require designers/engineers/drafters to learn computer programming. Specialized programs enable them to use computer graphics after a short training period. HOWEVER, THE PRINCIPLES OF DRAFTING ARE COMMON TO TRADITIONAL DRAFTING AND CAD. A working knowledge of basic drafting standards, techniques, and procedures is absolutely necessary before CAD is attempted.

With computer graphics two- and three-dimensional images of the object being designed can be generated, Fig. 24-9. The computer can be instructed to rotate the object through various positions for study and evaluation, Fig. 24-10. Design changes can be made quickly and reevaluated.

Design changes can be made (depending upon the CAD system employed) by touching a LIGHT

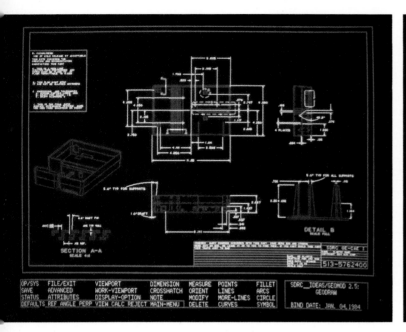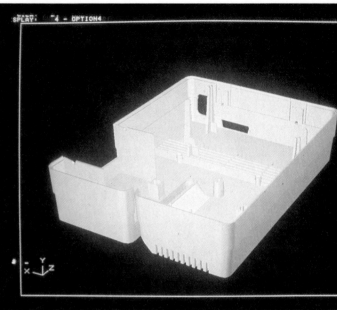

Fig. 24-9. Computer graphics can be used to generate images in two- and three-dimensional form and in full color or in black and white. (General Electric CAE International Inc.)

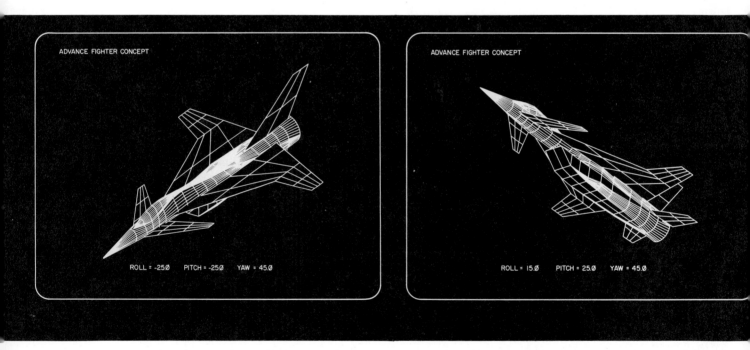

Fig. 24-10. The computer was instructed to rotate the object through various positions. Note in drawing B a stabilizer has been added because studies indicated it was needed. (NASA)

PEN, Fig. 24-11, or moving a CURSOR, Fig. 24-12, to the area being altered (symbol, line, point, etc.) and selecting and keying in the necessary commands to make the required changes through the INPUT TERMINAL. A LIGHT PEN is a stylus-shaped photosensitive pointing device. A CURSOR is a special character, Fig. 24-13, that indicates your present point of interest on the display screen. By moving the pen or cursor to a position on the screen the coordinates of that point are input as data to the computer memory. Coordinates are a series of points of measurement located along x, y, and z axes from a fixed origin, Fig. 24-14. The z-axis is needed for three-dimensional graphics, Fig. 24-15.

SQUARE   UNDERLINE   POINTER   CROSS-HAIR

Fig. 24-13. A cursor indicates your point of interest on the CRT image. Shown are examples of cursors used by various CAD systems.

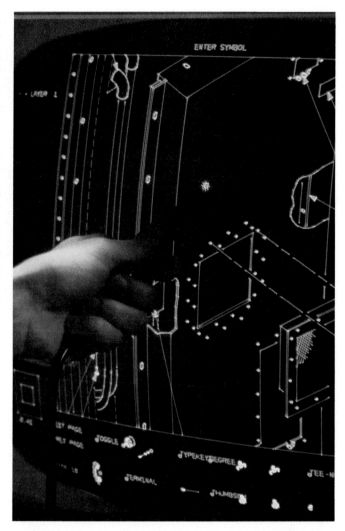

Fig. 24-11. A light pen being used on the CRT.

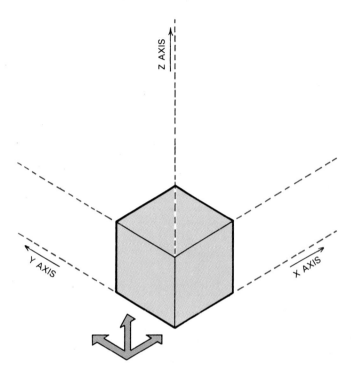

Fig. 24-14. Coordinates are a series of points of measurement located along x, y, and z axes from a fixed point of origin.

Fig. 24-12. The cursor on this CAD system is moved by means of a JOY STICK. (ROBO Systems)

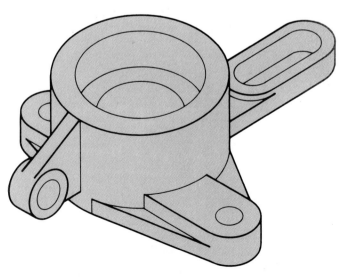

Fig. 24-15. A z axis is needed to generate three-dimensional images like the one shown above. Some CAD systems are not powerful enough to generate true three-dimensional graphics. (ROBO Systems)

It is also possible to "zoom in" and enlarge a section on the proposed design for study and/or modification. Other portions of a DATA BASE (organized information on a particular subject) describing the part's design can be called up instantaneously on the screen.

When a design has been finalized (completed), CAD has the capability of turning design data into working drawings (called HARD COPY) by means of AUTOMATED DRAFTING MACHINES or HIGH SPEED PLOTTERS, Fig. 24-16. The design data (mathematical shape and size description of the part) is retained in computer memory, or put on tape or disks, for later use in COMPUTER AIDED MANUFACTURING (CAM) systems that will cut the material and make the part, Fig. 24-17. Computer aided manufacturing or CAM refers to automated manufacturing operations. Computers control the manufacturing processes.

When computer aided design and computer aided manufacturing systems are used to gather the technology it is called CAD/CAM.

Computer graphics can also simulate how the part will "work" with other parts in an assembly, Fig. 24-18. This permits the engineer/designer/drafter to determine whether possible conflicts exist (for example, two parts in the same place at one time).

Fig. 24-16. Two sizes of plotters. Plotters produce hard copy (working drawings) from CAD developed information. (CALCOMP, A Sanders Graphics Company)

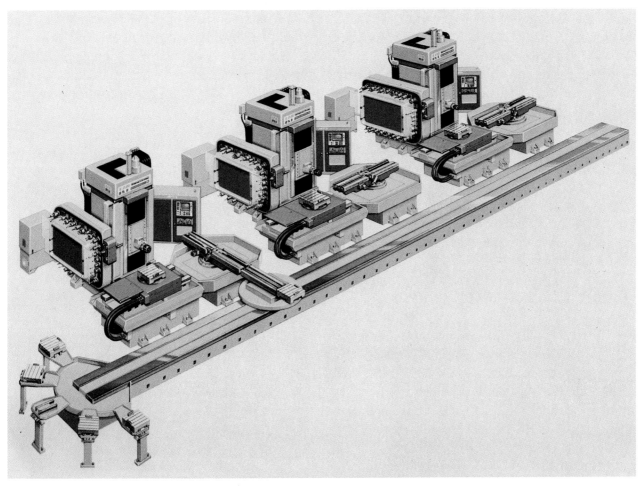

Fig. 24-17. A group of machine tools that are part of a CAD/CAM system. Each machine has a microcomputer control system. Note how the work moves from station to station. (Kearney & Trecker Corp.)

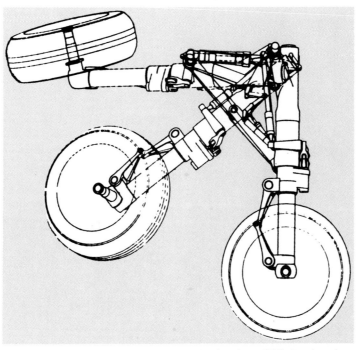

Fig. 24-18. Computer graphics can simulate how CAD developed parts will "work" together. This illustration shows a landing gear during retraction. Note how it pivots as it is raised.

The most advanced form of computer graphics is known as COMPUTER AIDED IMAGERY (CAI). CAI simulates the object in full color, three-dimensional form, and animates it so the parts that make up the design can be filmed as if they were in actual operation, Fig. 24-19.

While originally conceived for engineering purposes, CAI is used to create the galactic special effects in motion pictures.

Computer aided imagery requires a super computer because each frame can require up to 72 million calculations.

WHY USE COMPUTER GRAPHICS

The main function of an engineer/designer/drafter is to define (explain) the basic shape of a part, assembly, or product, Fig. 24-20. This is called PRELIMINARY DESIGN. The process involves many modifications (changes) and refinements (improvements) before a design is finalized.

Until the advent of CAD, designers/engineers/drafters had to imagine and then evaluate a three-dimensional object that was drawn in two-dimen-

sions on a flat sheet of paper. The only way a design could be verified in three-dimensions was to make a wood, clay, or plastic model. A new model usually had to be constructed when major design changes were made. This is a time consuming and expensive process.

The introduction of CAD provided the engineer/designer/drafter with a dynamic new tool. CAD technology permits more time for creative work. There is no need for the mathematical work and the repetitive drafting required before CAD.

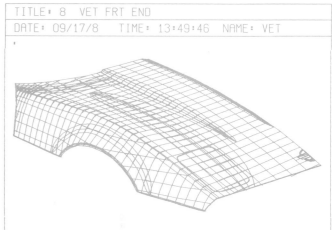

TITLE: 8 VET FRT END
DATE: 09/17/8 TIME: 13:49:46 NAME: VET

Fig. 24-19. Computer aided imagery (CAI) simulates how an Air Force tanker would behave during takeoff under various weight and atmospheric conditions. There was no need to endanger the aircraft or lives of the crew by using an actual tanker to conduct the tests during marginal weather conditions.
(Evans & Sutherland)

Fig. 24-20. Preliminary design of the front end of a Corvette developed from sketch. Final product shown below.
(General Motors Advanced Concept Center)

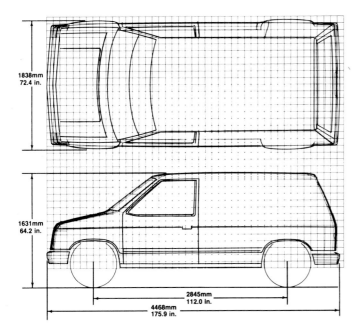

Fig. 24-21. CAD generated outline of the Dodge MINI VAN. This is the first step in developing working drawings for the manufacture of a new model. (Dodge Div., Chrysler Corp.)

HOW CAD WORKS

A normal sequence of CAD starts with the generation of a geometric model of the proposed design on the display screen, Fig. 24-21. Generally, only the outline of the design is created on the geometric model. Details are added later.

A CAD program is activated through a MENU, Fig. 24-22. A menu is a program generated list of options from which the drafter can select to execute desired procedures. A menu can appear as grids of commands and symbols on a DIGITIZER BOARD, Fig. 24-23, or as a list of options on the CRT display.

Digitizer menu selection is made using a light pen, stylus, puck, or joystick, Fig. 24-24. Pointing at a command is the same as keying it in on the keyboard. Pointing at a menu symbol selects it for

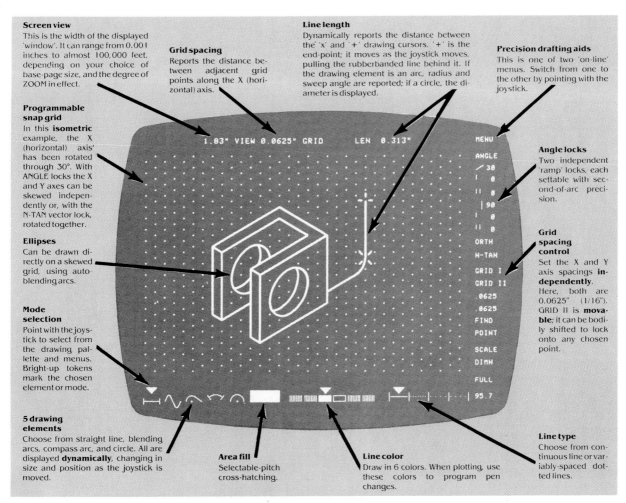

Screen view
This is the width of the displayed 'window'. It can range from 0.001 inches to almost 100,000 feet, depending on your choice of base-page size, and the degree of ZOOM in effect.

Programmable snap grid
In this **isometric** example, the X (horizontal) axis' has been rotated through 30°. With ANGLE locks the X and Y axes can be skewed independently or, with the N-TAN vector lock, rotated together.

Ellipses
Can be drawn directly on a skewed grid, using auto-blending arcs.

Mode selection
Point with the joystick to select from the drawing pallette and menus. Bright-up tokens mark the chosen element or mode.

5 drawing elements
Choose from straight line, blending arcs, compass arc, and circle. All are displayed **dynamically**, changing in size and position as the joystick is moved.

Grid spacing
Reports the distance between adjacent grid points along the X (horizontal) axis.

Line length
Dynamically reports the distance between the 'x' and '+' drawing cursors. '+' is the end-point; it moves as the joystick moves, pulling the rubberbanded line behind it. If the drawing element is an arc, radius and sweep angle are reported; if a circle, the diameter is displayed.

Precision drafting aids
This is one of two 'on-line' menus. Switch from one to the other by pointing with the joystick.

Angle locks
Two independent 'ramp' locks, each settable with second-of-arc precision.

Grid spacing control
Set the X and Y axis spacings **independently**. Here, both are 0.0625" (1/16"). GRID II is **movable**; it can be bodily shifted to lock onto any chosen point.

Area fill
Selectable-pitch cross-hatching.

Line color
Draw in 6 colors. When plotting, use these colors to program pen changes.

Line type
Choose from continuous line or variably-spaced dotted lines.

`1.03" VIEW 0.0625" GRID     LEN 0.313"`

```
MENU
ANGLE
/ 30
| 0
|| 0
| 90
| 0
|| 0
ORTH
N-TAN
GRID I
GRID II
.0625
.0625
FIND
POINT
SCALE
DIMN
FULL
95.7
```

Fig. 24-22. The menu shown to the right of the screen is a program generated list of options. The drafter can refer to the menu and can select desired procedures. (ROBO Systems)

Fig. 24-23. The menu of this CAD system can be seen on both the digitizer board and on the screen. This engineer is designing an IC (integrated circuit). Only a small part of the design is shown on the display screen.

Fig. 24-24. This computer system makes use of both a puck and a digitizer board and stylus. The puck is on the large digitizer board. (CALCOMP, A Sanders Graphics Company)

placement on your drawing. Menu selection is made on a CRT by moving the cursor to the appropriate block.

Once graphic features are in the system, they can be moved, deleted, or mirrored (reversing a graphic image around one, or both axes), Fig. 24-25.

Extension lines, dimension lines, arrow heads, text, and dimensions are added by selecting the appropriate menu function and keying in the required material. The drafter can select text size, style, and orientation (location on the graphics), Fig. 24-26.

HOW COMPUTER GRAPHICS ARE GENERATED

Numbers and letters (alphanumerics) are the language of computers, not pictures. Before a picture (drawing) can be developed or generated on the display screen, it must first be translated into computer language.

There are two ways to convert a picture into numbers and letters—RASTER and VECTOR, Fig. 24-27. With VECTORS, each picture is broken into lines. Each line (vector) has two end points. It can be located on a grid programmed into the computer. The grid is a drawing aid for determining distance. It may be shown on the screen as a network of uniformly spaced points (fine dots).

The technique is similar to drawing pictures by the graph method. See page 93 on enlarging or reducing by the graph method.

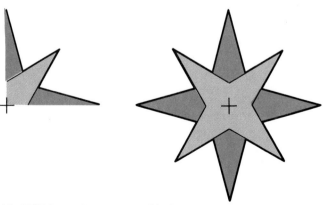

AS FIRST DEVELOPED   IMAGE MIRRORED ON TWO AXES

Fig. 24-25. A symmetrical drawing can be mirrored around the x and y axes for perfect symmetry.

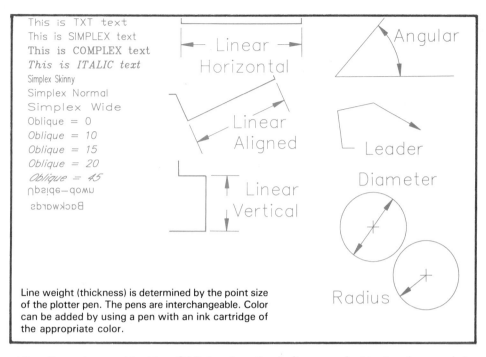

Fig. 24-26. When adding dimensions and text to a CAD drawing, the drafter can select text and numeral size, style, and orientation (position and/or location) on the graphics. The material shown above was prepared using the AUTOCAD™ computer design and drafting program. (AUTOCAD/Autodesk, Inc.)

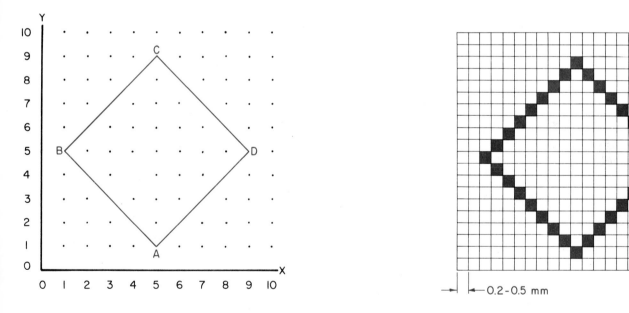

Fig. 24-27. A comparison of vector and raster displays.

Horizontal measurement of the end point is its x-coordinate. The vertical measurement is its y-coordinate. (Coordinates represent units of real measurement from a fixed point.) Most CAD programs permit grid size to be changed at will to determine sheet size—A, B, C, D and E, of the hard copy (drawing printout). Plotter capacity limits actual printout size.

Since alphanumerics are understood by the computer, a vector picture can be described by entering the x- and y-coordinates of the end points with the list of the connections between the points.

Circles and arcs are generated by specifying three points or a center and radius. Curves must be defined mathematically.

Vector picture insertion involves a digitizer board (also called a tablet) and an input device such as a stylus or puck. They convert object shapes into computer understandable language.

Existing drawings can be traced or new drawings created by tracing the lines or locating their end points with the input device. The boards surface senses the position of the stylus or puck and transmits these coordinates to the computer.

The image being created appears directly on the display screen. The tablet may have a menu with predefined graphic symbols that can be used in creating drawings.

A RASTER utilizes many tiny PICTURE ELEMENTS, called PIXELS. They are arranged in a fixed, precise manner. Each pixel is the same size and shape. They can be made to match a series of computer memory locations. Your TV is an example of a raster display. Raster pictures are inserted by scanning, with either a TV camera or a specially designed scanner.

Most CAD systems use a vector display because less computer memory is needed.

INDUSTRY EXAMPLE OF COMPUTER GRAPHICS

The series of illustrations in Fig. 24-28 through Fig. 24-38 follows a design-through-manufacture session. The system is employed to construct an aircraft wing, its supporting ribs, and the fuel and hydraulic lines which pass through the wing. After redefining the model's geometry, one of the ribs is used to demonstrate dimensioning, and numerical control tool path generation.

The series of illustrations in Fig. 24-28 through Fig. 24-38 is reproduced by permission of Tektronix, Inc. These copyrighted illustrations show the flow from design concept using computer graphics right into manufacturing using computer numerical control.

By studying these illustrations you will learn how computer graphics is used as a drafting tool.

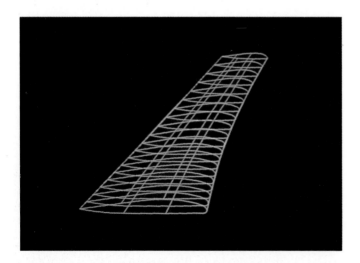

Fig. 24-28. A "curved mesh surface" defines the shape of an airfoil for a proposed aircraft wing.
(Copyright 1986 by permission of Tektronix, Inc.)

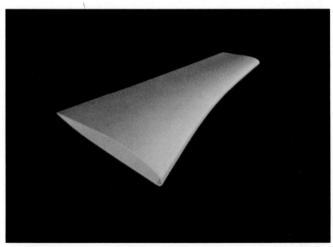

Fig. 24-29. The skin is formed. Skin thickness is specified.

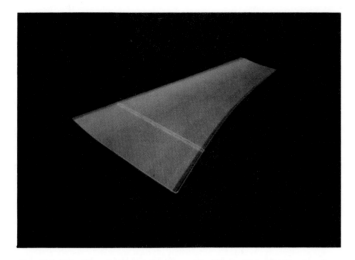

Fig. 24-30. The drafter selects a quadrilateral (solid having four sides and four angles) that represents a cross-section of the rib through the wing cord.

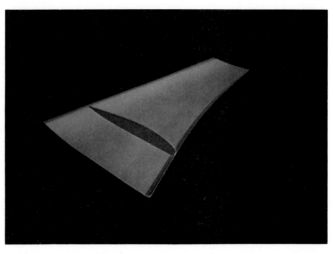

Fig. 24-31. Using the "slab projected to a surface" construction method, the drafter projects the cross-section to the airfoil surface to generate the solid rib.

Fig. 24-32. Two axial sweeps (linear projections of sets of curves) define the two non-circular shapes that are positioned on the rib, along with cylinders of the appropriate radius. The drafter subtracts the six shapes, producing holes in the rib.

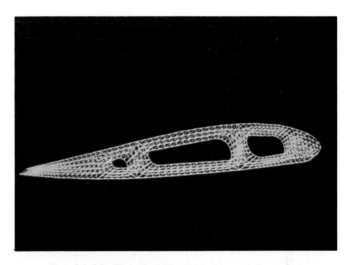

Fig. 24-33. The rib with the holes cut through.

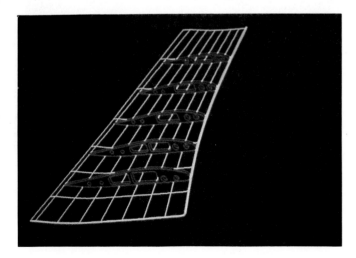

Fig. 24-34. Repeating the process shown in Figs. 24-30 through 24-32 results in the wing's supporting structure.

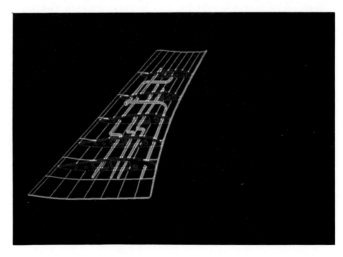

Fig. 24-35. To expand center lines to circles which represent piping or tubing, the drafter uses a construction method specifying a radius along the center line curve.

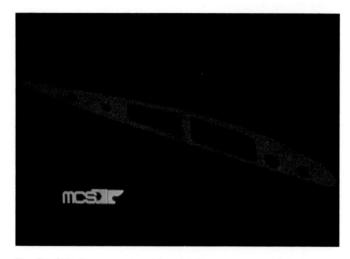

Fig. 24-36. Computer graphics quickly generates a 3-D view of the rib.

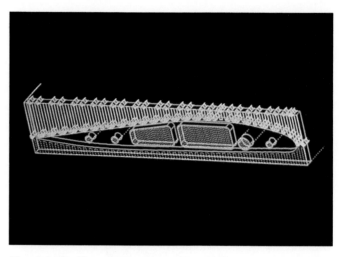

Fig. 24-37. To provide dimensional information, the drafter "blanks" all of the solid except one of the ribs. This allows the computer's drafting module to generate the appropriate dimensions. Notes and labels can also be added.

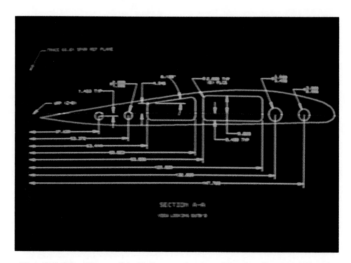

Fig. 24-38. The solid rib is cut from a block of material using the numerical control function called "area clearance." The numerical control "pocketing" function removes the circular and non-circular holes.

TECHNICAL VOCABULARY

Activate, Automated, Complex, Computations, Computer aided design, Computer aided imagery, Computer aided manufacturing, Console, Coordinates, CRT, Cursor, Data, Data base, Digitizer tablet, Grid, Input terminal, Instantaneously, Keyboard, Light pen, Logical decisions, Mainframe computer, Mathematical calculations, Menu, Microcomputer, Microprocessor, Minicomputer, Mirrored, Monitor, Plotter, Preliminary design, Processor unit, Program, Raster, Refinements, Sophisticated, Vector, Verify, Zoom in.

TEST YOUR KNOWLEDGE

Please do not write in the book. Place your answers on a sheet of notebook paper.
1. A computer, in its simplest form, is a device that performs _____ _____.
2. The oldest computer is the _____. An abacus is the oldest _____ computer.

3. A modern computer can perform high speed _____.

4. List the components that make up a computer system.
   a. _____
   b. _____
   c. _____
   d. _____
   e. _____
   f. _____

5. A program is a series of instructions that "tells" a computer _____.

6. The term CAD means _____ _____ _____.

7. CAD is a computer technology that _____ _____.

8. Preliminary design means: (Check the correct answer/s.)
   ____ a. The main function of an engineer/designer/drafter.
   ____ b. To define the basic shape of a part, assembly or product.
   ____ c. Final design of a product.
   ____ d. All of the above.
   ____ e. None of the above.

9. A CAD program is activated by a menu. A menu is a _____.

10. A cursor is _____.

11. Computer generated designs are converted into working drawings on a _____ _____ _____.

OUTSIDE ACTIVITIES

1. Secure samples of various types of drawings prepared using a computer. Prepare a bulletin board display around these drawings.

2. Visit an industrial drafting room that uses computer graphics. Interview an operator of the computer stations. How does the computer save time? How does the computer assist with repetitive work? How does the computer improve productivity? What did the computer installation cost the business? Report to the class what you observed and learned.

3. Visit a computer store that sells CAD equipment. Request literature on the equipment they sell. Prepare a bulletin board around the material.

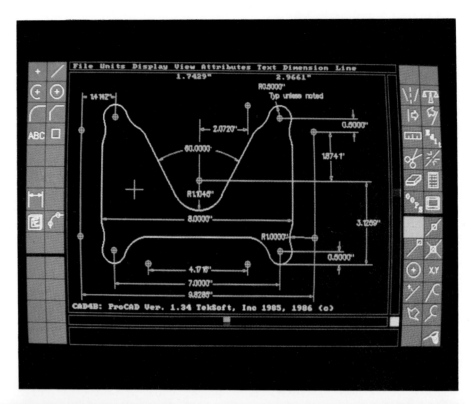

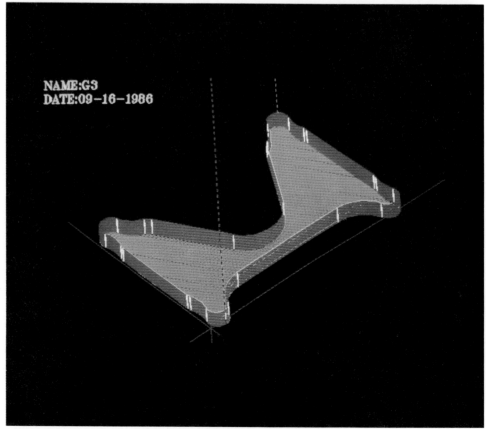

Industry often designs component parts on a screen using Computer Aided Design software. The computer is able to then display the component to the operator as a finished item. Coupled with various manufacturing processes, the computer can produce the components in a coupled process called CAD/CAM which stands for Computer Aided Design and Computer Aided Manufacturing. (TekSoft, CAD/CAM Systems)

Unit 25

MANUFACTURING PROCESSES

After studying this chapter, you will understand many of the basic manufacturing processes. The knowledge and recognition of these processes will aid in your drafting applications to various products you design and draw. You will be able to explain and illustrate methods of shaping, forming, and fabricating both metals and plastics.

The purpose of most drawings is to describe a part or a product that is to be manufactured. It is therefore important that the drafter have an understanding of how the materials can be cut, shaped, formed, and fabricated. Metals, plastics, and other materials are available in a variety of shapes and sizes. See Fig. 25-1. By understanding manufacturing processes, the drafter can better visualize the objects being drawn.

MACHINE TOOLS

The world of today could not exist without MACHINE TOOLS. They produce the accurate and uniform parts needed for the many products we use.

A definition, according to the National Machine Tool Builders' Association follows: "A machine tool is a power driven machine, not portable by hand, used to shape or form metal (or other materials) by cutting, impact, pressure, electrical techniques, or by a combination of these processes."

Machine tools are manufactured in a large range of styles and sizes. Only basic tools and manufacturing techniques will be covered in this unit.

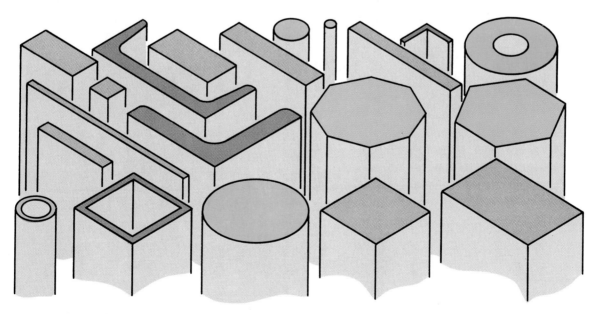

Fig. 25-1. Metals, plastics, and other materials used in manufacturing are made in many sizes and shapes.

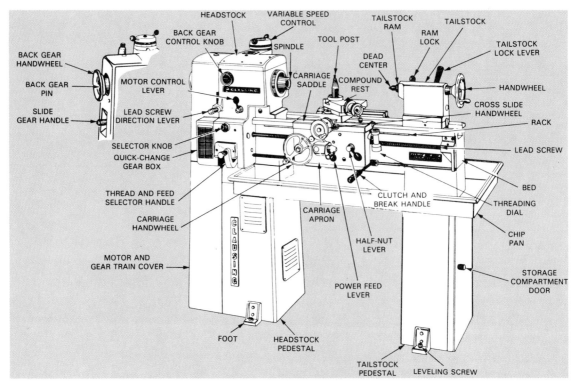

Fig. 25-2. The lathe and its major parts. (Clausing Machine Tools)

LATHE

The LATHE is one of the oldest and most important of the machine tools, Figs. 25-2 and 25-3. It operates on the principle of the work being rotated against the edge of a cutting tool, Fig. 25-4. The cutting tool can be controlled to move lengthwise and across the face of the material being machined (turned).

Operations other than turning can also be performed on the lathe. It is possible to bore, Fig. 25-5, ream (finishing a hole to exact size), and cut threads and tapers.

There are many variations of the basic lathe. The TURRET LATHE is used when a number of identical parts must be turned, Fig. 25-6. It is a conventional lathe fitted with a six-sided tool holder called a TURRET. Different cutting tools fitted in the turret rotate into position for machining operations.

Other lathes range in size from the small lathe needed by the instrument and watchmaker, Fig. 25-7, to the large lathes that machine the forming rolls for steel mills, Fig. 25-8.

Fig. 25-3. The lathe is a very important machine tool.

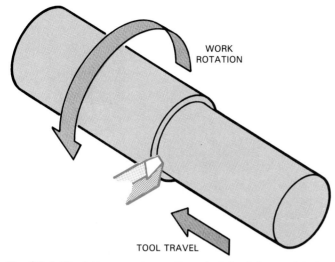

Fig. 25-4. The lathe operates on the principle of the work being rotated against the edge of a cutting tool.

Fig. 25-5. Boring (internal machining) on a lathe. (Clausing Machine Tools)

Fig. 25-6. Turret lathe. (Clausing Machine Tools)

DRILL PRESS

The DRILL PRESS is probably the best known of the machine tools, Fig. 25-9. A cutting tool called a TWIST DRILL is rotated against the work with sufficient pressure to cut its way through the material, Fig. 25-10. The spiral flutes on the twist drill do not pull the drill into the work. Pressure must be applied to the rotating drill to make it cut.

Other operations which can be performed on a drill press include: reaming (finishing a drilled hole to exact size), countersinking (cutting a chamfer on a hole so a flat head fastener can be inserted), and tapping (cutting internal threads in a drilled hole).

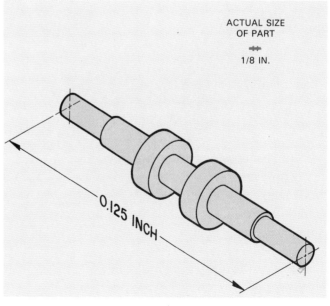

Fig. 25-7. Part made on a lathe used by instrument and watch makers.

Fig. 25-8. Lathe for machining forming rolls for a steel mill.

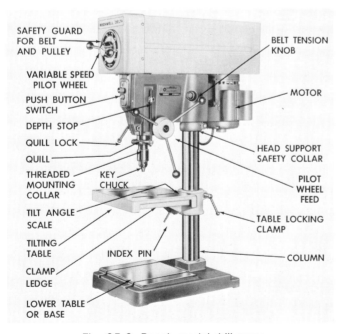

SAFETY GUARD
FOR BELT
AND PULLEY

BELT TENSION
KNOB

VARIABLE SPEED
PILOT WHEEL

PUSH BUTTON
SWITCH

MOTOR

DEPTH STOP

QUILL LOCK

QUILL

HEAD SUPPORT
SAFETY COLLAR

THREADED
MOUNTING
COLLAR

KEY
CHUCK

PILOT
WHEEL
FEED

TILT ANGLE
SCALE

TILTING
TABLE

TABLE LOCKING
CLAMP

INDEX PIN

CLAMP
LEDGE

COLUMN

LOWER TABLE
OR BASE

Fig. 25-9. Bench model drill press.

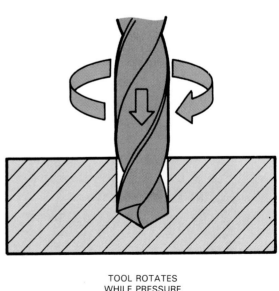

TOOL ROTATES
WHILE PRESSURE
IS APPLIED TO
FORCE DRILL INTO
MATERIAL

Fig. 25-10. The operating principle of a drill press.

MILLING MACHINE

A MILLING MACHINE is a very versatile machine tool. It can be used to machine flat and irregularly shaped surfaces, drill, bore, and cut gears.

The VERTICAL MILLING MACHINE uses a cutter mounted vertical to the worktable, Figs. 25-11 and 25-12. The HORIZONTAL MILLING MACHINE uses a cutter mounted horizontal to the worktable, Figs. 25-13 and 25-14. Metal is

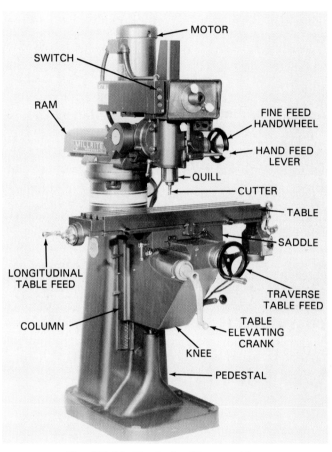

Fig. 25-11. Vertical milling machine.

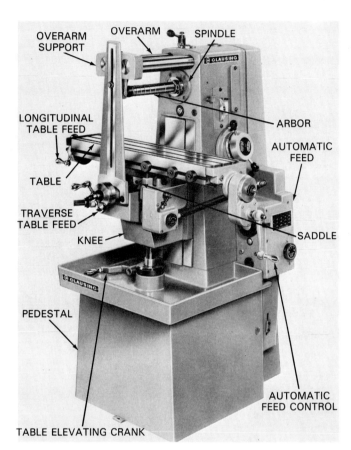

Fig. 25-13. Horizontal milling machine.

Fig. 25-12. Cutter (called an end mill) is mounted in a vertical position. It cuts on both the face and perpihery (side).

Fig. 25-14. Cutter on a horizontal milling machine.

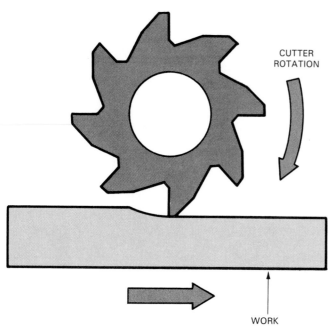

CUTTER
ROTATION

WORK

Fig. 25-15. The operating principle of a horizontal milling machine.

removed by means of a rotating cutter that is fed into the moving work, Fig. 25-15. There are many types of milling machines. They vary in size and by types of automatic controls.

PLANER AND BROACH

Large flat surfaces are machined on a PLANER, Fig. 25-16. It operates by moving the work against a cutting tool or tools, Fig. 25-17.

BROACHING employs a multi-tooth cutting tool. Each tooth has a cutting edge that is a few thousandths of an inch higher than the one before and increases in size to the exact finished size required, Fig. 25-18. The broach is pushed or pulled over the surface being machined. Many flat surfaces on automobile engines are broached. Broaching can also be utilized to do internal

Fig. 25-16. A 144 inch by 126 inch by 40 ft. (3.7 m by 3.2 m by 12.2 m) double housing (two cutting heads) planer. The two people add perspective to the machine's size. (G.A. Gray Co.)

machining like cutting keyways, splines, and irregular shaped openings, Fig. 25-19.

GRINDING

GRINDING is an operation that removes material by rotating an abrasive wheel against the work, Fig. 25-20. A BENCH GRINDER is the simplest and most widely used grinding machine, Fig. 25-21.

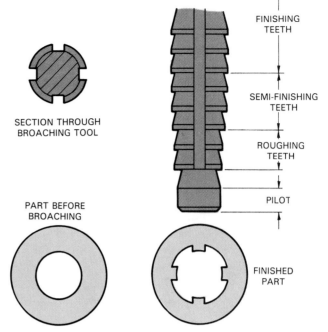

SECTION THROUGH BROACHING TOOL

FINISHING TEETH

SEMI-FINISHING TEETH

ROUGHING TEETH

PILOT

PART BEFORE BROACHING

FINISHED PART

Fig. 25-19. Typical broaching tool and the work it produces.

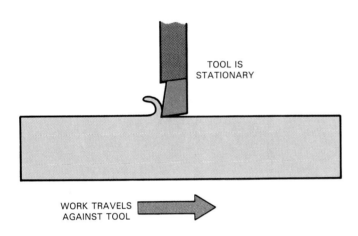

TOOL IS STATIONARY

WORK TRAVELS AGAINST TOOL

Fig. 25-17. Operating principle of a planer.

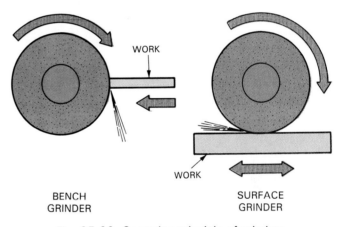

WORK

WORK

BENCH GRINDER

SURFACE GRINDER

Fig. 25-20. Operating principle of grinders.

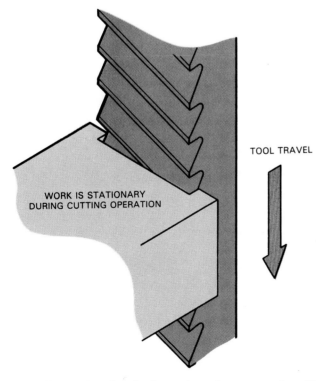

WORK IS STATIONARY DURING CUTTING OPERATION

TOOL TRAVEL

Fig. 25-18. Drawing showing how a broach operates. A multi-tooth cutter moves against the work. The operation may be on a vertical or horizontal plane.

Fig. 25-21. Bench grinder. (Stanley Tools)

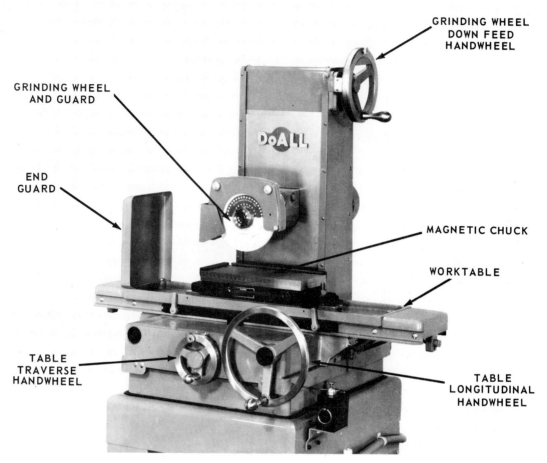

GRINDING WHEEL
DOWN FEED
HANDWHEEL

GRINDING WHEEL
AND GUARD

END
GUARD

MAGNETIC CHUCK

WORKTABLE

TABLE
TRAVERSE
HANDWHEEL

TABLE
LONGITUDINAL
HANDWHEEL

Fig. 25-22. Manually operated surface grinder. (The DoAll Co.)

ABRASIVE
WHEEL

Fig. 25-23. A cylindrical grinder. Closeup shows part being ground and the abrasive wheel doing the work. (Landis Tool Co.)

Flat surfaces are ground to very close tolerances on a SURFACE GRINDER, Fig. 25-22. It is possible to surface grind hardened steel parts to tolerances of 1/100,000 (0.00001 in.) (0.0002 mm) with very fine (almost mirrorlike) surface finishes.

Round work can be ground to the same tolerances and surface finishes on a CYLINDRICAL GRINDER, Fig. 25-23.

OTHER MACHINING TECHNIQUES

In addition to conventional machining, industry employs many unusual techniques to shape metal and other materials. There are electrical discharge machining (EDM), electro chemical machining (ECM), electron beam machining, laser machining, and ultrasonic machining to name but a few of the new techniques. They are usually used to machine materials that are difficult or impossible to machine by conventional methods.

If interested, you might want to research these newer machining techniques.

MACHINE TOOL OPERATION

Machine tools are controlled in a number of different manners.

1. MANUALLY. A machinist feeds the cutter into the work and manually guides it through the operations that will produce the part specified on the drawing, Fig. 25-24.

2. NUMERICAL CONTROL (N/C). All machine movements are controlled by electric motors called servos. They position the work and feed the cutter(s). Instructions from a punched tape, magnetic tape, or directly from a computer, control electronic impulses that tell the servos when to start, in what direction to move, and how far they are to move, Fig. 25-25.

3. COMPUTER-ASSISTED NUMERICAL CONTROL (CNC). CNC machine tools are designed around a microcomputer that is part of the machine's control unit (MCU). Using one of the many N/C languages, the machining instructions are entered directly into the system. Fig. 25-26.

Fig. 25-25. Numerically controlled (N/C) machine tool. Instructions from a punched tape or magnetic tape tell the servos when to start, in what direction to move, and how far they are to move the work being machined.

Fig. 25-24. The operations on this lathe are manually operated. (U.S. Army)

Fig. 25-26. CNC vertical milling/drilling machine designed for school shop/lab and industrial use. (Dyna Electronics Inc.)

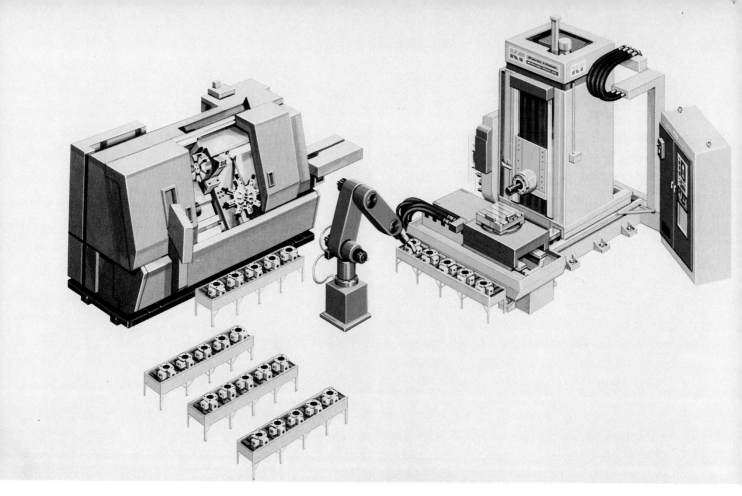

Fig. 25-27. This robot, or automated cell, consists of several machines with robotics material handling. The cell becomes a fully automatic process through the application of robotics, power clamping of parts, special tools, and other forms of automation. (Kearney & Trecker Corp.)

It is possible, using computer graphics (see Unit 24), for an engineer/designer/drafter to design a part and see how it will fit and work with the other parts that make up the product. After the design has been confirmed, the computer can be programmed to analyze the geometry and calculate the tool paths required to make the part. The program is verified by machining a sample part from some inexpensive material like plastic or special wax.

The system that makes all of this possible is called CAD/CAM (computer aided design/computer aided manufacturing).

Some CAD/CAM operations employ robotics in their operation, Fig. 25-27.

SHEARING AND FORMING

SHEARING is a process where the material (usually in sheet form) is cut to shape using actions similar to cutting paper with scissors.

STAMPING is divided into two separate classifications: CUTTING and FORMING. The cutting operation is also known as BLANKING, Fig. 25-28. It involves cutting flat sheets to the shape of the finished part. FORMING is a process where flat metal is given three-dimensional form, Fig. 25-29.

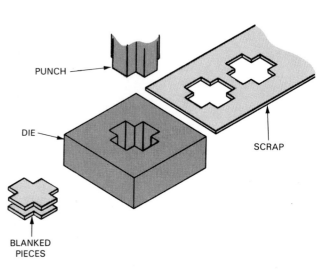

Fig. 25-28. Blanking operation.

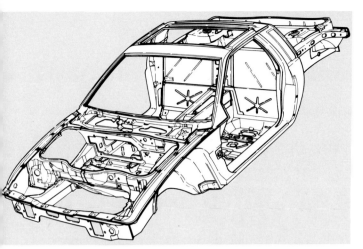

Fig. 25-29. Forming gave flat metal sheet three-dimensional form for use in this auto body. It also added strength and rigidity. (Pontiac Motor Division, GMC)

CASTING PROCESSES

Materials can also be given shape and form by reducing them to liquid in a molten state and pouring them into a mold of the desired shape. This is called CASTING.

SAND CASTING

In the SAND CASTING PROCESS the mold that gives the molten metal shape is made of sand, Fig. 25-30. This is one of the oldest metal forming techniques known.

A sand mold is made by packing sand in a box called a FLASK, around a PATTERN, Fig. 25-31, of the shape to be cast. Because metal contracts as it cools, patterns are made slightly oversize to allow for this shrinkage. After the pattern has been drawn (removed) from the sand, the mold halves, called the COPE and the DRAG, are reassembled. A cavity or mold of the required shape remains in the sand.

Before assembling the mold, openings called SPRUES, RISERS, and GATES are made in the mold. A SPRUE is an opening into which the molten metal is poured. RISERS allow the hot gases to escape. GATES are trenches that run from the sprues and risers to the mold cavity. This permits molten metal to reach and fill the mold. Sand molds must be destroyed to remove the casting.

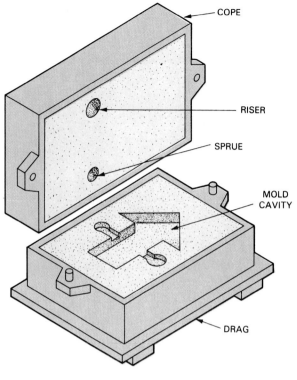

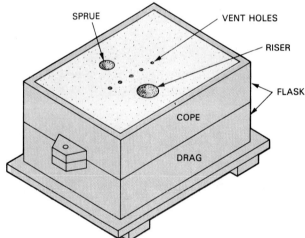

Fig. 25-30. Parts of a typical sand mold.

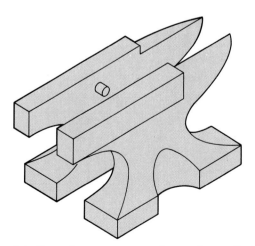

Fig. 25-31. Two piece pattern used to make cavity in sand mold.

341

Fig. 25-32. Familiar items made by the permanent mold process.

PERMANENT MOLD CASTING

Some molds used for casting metal are made from metal. These molds are called PERMA-NENT MOLDS because they do not have to be destroyed (like sand molds) to remove the casting. The process produces castings with a fine surface finish and a high degree of accuracy.

Auto pistons, fishing sinkers, and toy lead soldiers are familiar products made using the permanent mold process, Fig. 25-32.

DIE CASTING

DIE CASTING is a process where molten metal is forced into a DIE or MOLD under pressure, Fig. 25-33. The pressure is maintained until the metal solidifies. The mold is opened and the casting is ejected, Fig. 25-34. The mold or die is made of metal. Die castings are denser than sand castings. The castings are usually very accurate.

PLASTICS

Plastics are made into usuable products by many different processes. Some plastics can be shaped directly into final form. Other plastics require several operations to transform them into usuable products. Many of the processes for forming plastics are similar to those used to shape metal.

COMPRESSION MOLDING

In COMPRESSION MOLDING a measured amount of plastic is placed in a two-part mold. See Fig. 25-35. The mold is heated and closed. Pressure is applied. Plastic, melted by the heat and pressure, flows into all parts of the mold cavity.

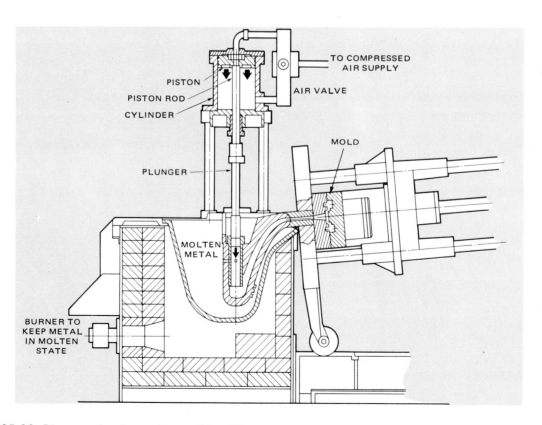

Fig. 25-33. Diagram of a die casting machine. The metal is forced into the die cavity (mold) under pressure.

Fig. 25-34. A die cast aluminum wheel emerging from the die. This wheel is more than a third lighter than a similar wheel produced by other casting techniques. (Kelsey-Hayes)

The mold is opened and the molded plastic is removed for trimming.

TRANSFER MOLDING

TRANSFER MOLDING is similar to compression molding. It differs in that the plastic is heated in a separate section of the mold before it is forced into the mold cavity, Fig. 25-36.

INJECTION MOLDING

Plastic model kits are one of the many plastic products made by INJECTION MOLDING, Fig. 25-37. Plastic granuals are placed in an injection molding machine and heated until they are soft enough to flow, Fig. 25-38. The softened plastic is forced into the mold cavity, Fig. 25-39. When cooled, the mold is opened, and the part is ejected.

REINFORCED PLASTICS

REINFORCED PLASTICS are resins that have been strengthened with some type of fiberglass cloth, boron, graphite, Fig. 25-40. Resin saturated fibers can be formed mechanically using heat and pressure. Body panels for the Chevrolet Corvette are made this way. They can also be laid-up by hand, Fig. 25-41. Reinforced plastic products are usually formed in polished metal molds.

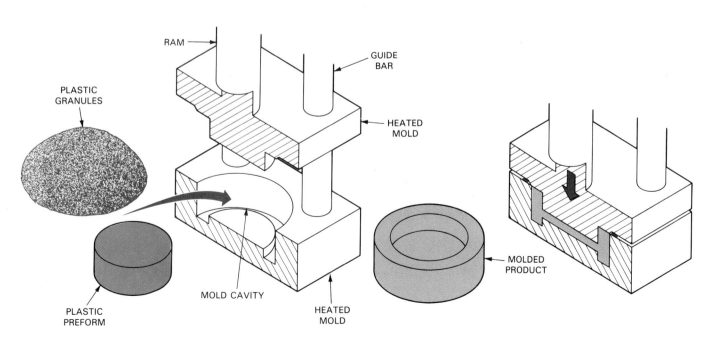

Fig. 25-35. Compression molding process.

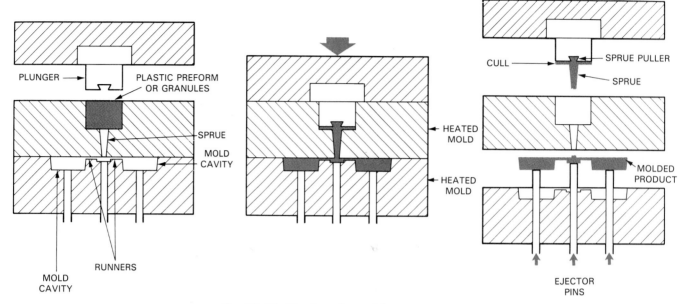

PLUNGER

PLASTIC PREFORM
OR GRANULES

SPRUE

MOLD
CAVITY

RUNNERS

MOLD
CAVITY

HEATED
MOLD

HEATED
MOLD

CULL

SPRUE PULLER

SPRUE

MOLDED
PRODUCT

EJECTOR
PINS

Fig. 25-36. The transfer molding process.

Fig. 25-37. Plastic model kits are typical products made by
injection molding.

Fig. 25-38. Plastic injection molding machine which produces
automotive products.

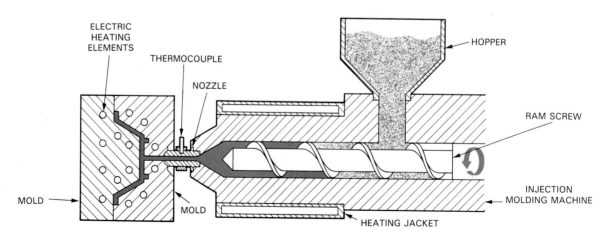

ELECTRIC
HEATING
ELEMENTS

THERMOCOUPLE

NOZZLE

HOPPER

RAM SCREW

MOLD

MOLD

INJECTION
MOLDING MACHINE

HEATING JACKET

Fig. 25-39. How injection molding is done.

Fig. 25-40. Aircraft made from reinforced plastics (also known as composites). One engine has been mounted on the Avtek 400 prior to installing elevators, flaps, and ailerons. The ''Windecker Eagle,'' in the background, is the first and only all-composite aircraft to be awarded a Type certificate by the FAA. (Avtek Corp.)

Fig. 25-41. The aft section of fuselage and vertical stabilizer of the Avtek 400 being removed from the mold by two technicians. The plane's strong, lightweight airframe is made primarily of honeycomb of ''Nomex'' aramid fiber sandwiched between skins of ''Kevlar'' aramid fiber. The plane's maximum weight is 5500 pounds, less than half that of most aluminum turboprop aircraft. (Avtek Corp.)

TECHNICAL VOCABULARY

Automatic controls, Blanking, Bore, Boron, Broach, CAD/CAM, Casting, Chamfer, CNC, Countersinking, Cylindrical grinding, Drill, Electrical discharge machining (EDM), Electron beam machining, Electro-chemical machining (ECM), Electronic impulses, Flask, Gate, Graphite, Keyway, Laser machining, Lathe, Machined, Machine tool, Milling machine, Mold, Mold cavity, Numerical control (N/C) Pattern, Plastic, Ream, Resin, Riser, Saturated, Shearing, Spline, Sprue, Stamping, Surface grinder, Tapping, Twist drill, Ultrasonic machining, Versatile.

TEST YOUR KNOWLEDGE—UNIT 25

Please do not write in the text. Place your answers on a sheet of notebook paper.

1. Define a machine tool.
2. Make a sketch showing how a lathe operates.
3. Make a sketch showing how a drill press operates.
4. List three operations that can be performed on a drill press.
   a. _____
   b. _____
   c. _____
5. Make a sketch showing how a milling machine operates.
6. Make a sketch showing how a planer operates.
7. List three machine tools used to machine flat surfaces.
   a. _____
   b. _____
   c. _____
8. Make a sketch showing how a surface grinder operates.

Match each of the following terms with the correct statement listed below by placing the appropriate letter in the blank next to each term.

9. ____ Pattern	16. ____ Forming
10. ____ Sand casting	17. ____ Permanent mold casting
11. ____ Manual machine operation	18. ____ Compression molding
12. ____ N/C	19. ____ Transfer molding
13. ____ CNC	20. ____ Injection molding
14. ____ CAD/CAM	
15. ____ Blanking	

a. Sheet metal is given three-dimensional shape.
b. The mold is not destroyed when the casting is removed.
c. The machinist guides the machine through machining operation by hand.
d. Mold must be destroyed to remove casting.
e. Softened plastic is forced into the mold cavity by a revolving screw. Plastic model kits are made by this process.
f. A measured amount of plastic is placed in a two-part mold. The mold is heated and closed. Pressure is applied. The heat and pressure melts the plastic and forces it into all parts of the mold.
g. Computer-assisted numerical control. Machine movement is controlled by a computer that is part of the machine's control unit.
h. Used to make the mold cavity in a sand mold.
i. Cutting sheet metal to shape.
j. Computer aided design/computer aided manufacturing.
k. Machine operations controlled by electric motors called servos.
l. Plastic is heated in a separate section of the mold before it is forced into mold cavity.

OUTSIDE ACTIVITIES

1. Prepare a bulletin board using clippings which illustrate basic machine tools.
2. Secure samples of work manufactured on various machines including a lathe, drill press, milling machine, and a surface grinder. Make a display and call out the advantages of using the different machine tools.
3. Bring into class examples of products made and cast in sand molds, die castings, and permanent molds. Describe the differences among the products.
4. Report to the class, using visual aids, the developments in Flexible Manufacturing Stations.

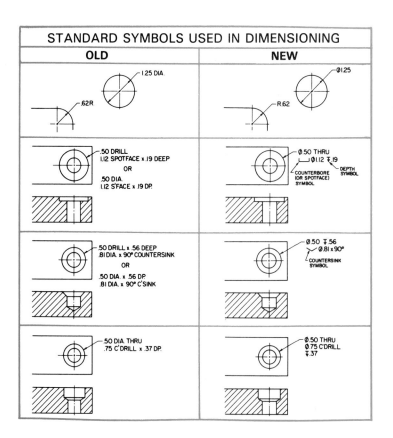

Recently ANSI (American National Standards Institute) recommended certain changes in specifying circles and holes. Study and compare the examples shown above.

USEFUL INFORMATION

ANSI DRAFTING STANDARDS

The American National Standards Institute (ANSI) revised the standards to be used on drawings. The new symbols and standards will replace the older abbreviations of certain English words. These new symbols will make drafting a more international language. Refer to the illustration on page 346.

Drafters will use the symbols to dimension their drawings. In addition, drafters will use the symbols in the notes added to drawings. As older drawings are revised, drafters will add them to the revisions. CAD operators will be able to select the symbols from digitizer libraries or from menus on screens.

ANSI prefers unidirectional dimensioning over the aligned system of placing dimensions on a drawing. Fractional inch dimensions are only recommended for specialized fields which do not require decimal accuracy such as furniture making, sheet metal fabrication, and forging and foundry areas.

ANSI standards call for threads to be drawn using the simplified method.

The part's material should be described in the Bill of Materials. Thus, section lines should be constructed at a 45 degree angle regardless of the part's material.

Drawings done to ANSI standards should contain a note "Drawn in accordance to ANSI Y14.5M."

Major symbols and their meanings are shown in the chart. Changes to be noted include:
   The diameter symbol appears before each diameter value.
   The radius symbol appears before the radius value.
   New symbols are used to describe holes.

This is a very brief description of standards which industry used for geometric dimensioning and tolerancing. The new symbols will improve the international understanding of drafting. Additional information can be obtained from the American National Standards Institute, 1430 Broadway, New York, New York, 10018 or from the American Society of Mechanical Engineers, 345 East 47th Street, New York, New York, 10017.

Meaning	ANSI Standards	Old Method
Not to scale	12	12 or NTS
Diameter	Ø26	26 DIA
Radius	R13	13 R
Reference dimension	(14)	14 REF
Times/places	6X	6 PLACES
Counterbore	⌴Ø7	7 CBORE
Countersink	∨Ø3	3 CSK
Deep	T14	14 DP
Square	□ 14	14 SQ
	4X 1/2−12 UNC−3B Ø.4062 T1.25 ⌴Ø.50 T.062 —→	1/2−12 UNC−3B−4 HOLES 13/32 DRILL−1 1/4 DP 1/2 C BORE−1/16 DP —→

MEASUREMENT SYSTEMS

ENGLISH SYSTEM

MEASURES OF TIME
60 sec. = 1 min.
60 min. = 1 hr.
24 hr. = 1 day
365 dy. = 1 common yr.
366 dy. = 1 leap yr.

DRY MEASURES
2 pt.   = 1 qt.
8 qt.   = 1 pk.
4 pk.   = 1 bu.
2150.42 cu. in. = 1 bu.

MEASURES OF LENGTH
12 in.   = 1 ft.
3 ft.   = 1 yd.
5 1/2 yd. = 1 rod
320 rods   = 1 mile
5,280 ft. = 1 mile
1,760 yd. = 1 mile
6,080 ft. = 1 knot

LIQUID MEASURES
16 fluid oz. = 1 pt.
2 pt.   = 1 qt.
32 fl. oz.   = 1 qt.
4 qt.   = 1 gal.
31 1/2 gal. = 1 bbl.
231 cu. in. = 1 gal.
7 1/2 gal.   = 1 cu. ft.

MEASURES OF AREA

144 sq. in.   = 1 sq. ft.
9 sq. ft.   = 1 sq. yd.
30 1/4 sq. yd. = 1 sq. rod
160 sq. rods   = 1 acre
640 acres   = 1 sq. mile

MEASURES OF WEIGHT (Avoirdupois)
7,000 grains (gr.) = 1 lb.
16 oz.   = 1 lb.
100 lb.   = 1 cwt.
2,000 lb.   = 1 short ton
2,240 lb.   = 1 long ton

MEASURES OF VOLUME
1,728 cu. in. = 1 cu. ft.
27 cu. ft.   = 1 cu. yd.
128 cu. ft.   = 1 cord

METRIC SYSTEM

The basic unit of the metric system is the meter (m). The meter is exactly 39.37 in. long. This is 3.37 in. longer than the English yard. Units that are multiples or fractional parts of the meter are designated as such by prefixes to the word "meter". For example:

1 millimeter (mm.) = 0.001 meter or 1/1000 meter
1 centimeter (cm.) = 0.01 meter or 1/100 meter
1 decimeter (dm.) = 0.1 meter or 1/10 meter
            1 meter (m.)
1 decameter (dkm.) = 10 meters
1 hectometer (hm.) = 100 meters
1 kilometer (km.) = 1000 meters

These prefixes may be applied to any unit of length, weight, volume, etc. The meter is adopted as the basic unit of length, the gram for mass, and the liter for volume.

In the metric system, area is measured in square kilometers (sq. km. or km.^{2}), square centimeters (sq. cm. or cm.^{2}), etc. Volume is commonly measured in cubic centimeters, etc. One liter (1) is equal to 1,000 cubic centimeters.

The metric measurements in most common use are shown in the following tables:

MEASURES OF LENGTH
10 millimeters = 1 centimeter
10 centimeters = 1 decimeter
10 decimeters = 1 meter
1000 meters   = 1 kilometer

MEASURES OF WEIGHT
100 milligrams = 1 gram
1000 grams   = 1 kilogram
1000 kilograms = 1 metric ton

MEASURES OF VOLUME
1000 cubic centimeters = 1 liter
100 liters   = 1 hectoliter

METRIC PREFIXES, EXPONENTS, AND SYMBOLS

DECIMAL FORM	EXPONENT OR POWER	PREFIX	PRONUNCIATION	SYMBOL	MEANING
1 000 000 000 000 000 000	$= 10^{18}$	exa	ex'a	E	quintillion
1 000 000 000 000 000	$= 10^{15}$	peta	pet'a	P	quadrillion
1 000 000 000 000	$= 10^{12}$	tera	tĕr'ȧ	T	trillion
1 000 000 000	$= 10^{9}$	giga	ji'gȧ	G	billion
1 000 000	$= 10^{6}$	mega	mĕg'ȧ	M	million
1 000	$= 10^{3}$	kilo	kĭl'ō	k	thousand
100	$= 10^{2}$	hecto	hĕk'to	h	hundred
10	$= 10^{1}$	deka	dĕk'a	da	ten
1					base unit
0.1	$= 10^{-1}$	deci	dĕs'ĭ	d	tenth
0.01	$= 10^{-2}$	centi	sĕn'tĭ	c	hundredth
0.001	$= 10^{-3}$	milli	mĭl'ĭ	m	thousandths
0.000 001	$= 10^{-6}$	micro	mi'krō	μ	millionth
0.000 000 001	$= 10^{-9}$	nano	năn'ō	n	billionth
0.000 000 000 001	$= 10^{-12}$	pico	pēc'o	p	trillionth
0.000 000 000 000 001	$= 10^{-15}$	femto	fĕm'to	f	quadrillionth
0.000 000 000 000 000 001	$= 10^{-18}$	atto	ăt'to	a	quintillionth

Most commonly used

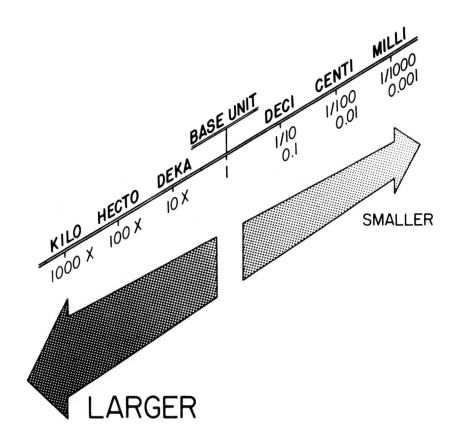

SI UNITS & CONVERSIONS

PROPERTY	UNIT	SYMBOL	EXACT CONVERSION			APPROXIMATE EQUIVALENCY
			FROM	TO	MULTIPLY BY	
length	meter	m	inch	mm	2.540×10	25mm = 1 in.
	centimeter	cm	inch	cm	2.540	300mm = 1 ft.
	millimeter	mm	foot	mm	3.048×10^{-4}	
mass	kilogram	kg	ounce	g	2.835×10	2.8g - 1 oz.
	gram	g	pound	kg	4.536×10^{-1}	kg = 2.2 lbs. = 35 oz.
	tonne (megagram)	t	ton (2000 lb)	kg	9.072×10^{2}	1t = 2200 lbs.
density	kilogram per cub. meter	kg/m^3	pounds per cu. ft.	kg/m^3	1.602×10	16kg/M^3 = 1 lb./ft^3
temperature	deg. Celsius	°C	deg. Fahr.	°C	$(^{\circ}F-32) \times 5/9$	0°C = 32°F
100°C = 212°F
area	square meter	m^2	sq. inch	mm^2	6.452×10^{2}	645mm^2 = 1 in.^{2}
	square millimeter	mm^2	sq. ft.	m^2	9.290×10^{-2}	1m^2 = 11 ft.^{2}
volume	cubic meter	m^3	cu. in.	mm^3	1.639×10^{4}	16400mm^3 = 1 in.^{3}
	cubic centimeter	cm^3	cu. ft.	m^3	2.832×10^{-2}	1m^3 = 35 ft.^{3}
	cubic millimeter	mm^3	cu. yd.	m^3	7.645×10^{-1}	1m^3 = 1.3 yd.^{3}
force	newton	N	ounce (Force)	N	2.780×10^{-1}	1N = 3.6 oz.
	kilonewton	kN	pound (Force)	kN	4.448×10^{-3}	4.4N = 1 lb.
	meganewton	MN	Kip	MN	4.448	1kN = 225 lb.
stress	megapascal	MPa	pound/in^2 (psi)	MPa	6.895×10^{-3}	1MPa = 145 psi
			Kip/in^2 (ksi)	MPa	6.895	7MPa = 1 ksi
torque	newton-meters	N.m	in-ounce	N.m	7.062×10^{3}	1N.m = 140 in.oz.
			in.pound	N.m	1.130×10^{-1}	1N.m = 9 in.lb.
			ft pound	N.m	1.356	1N.m = .75 ft.lb.
1.4N.m = 1 ft.lb.

FRACTIONAL INCHES
INTO DECIMALS AND MILLIMETERS

INCH	DECIMAL INCH	MILLIMETER	INCH	DECIMAL INCH	MILLIMETER
1/64	0.0156	0.3967	33/64	0.5162	13.0968
1/32	0.0312	0.7937	17/32	0.5312	13.4937
3/64	0.0468	1.1906	35/64	0.5468	13.8906
1/16	0.0625	1.5875	9/16	0.5625	14.2875
5/64	0.0781	1.9843	37/64	0.5781	14.6843
3/32	0.0937	2.3812	19/32	0.5937	15.0812
7/64	0.1093	2.7781	39/64	0.6093	15.4781
1/8	0.125	3.175	5/8	0.625	15.875
9/64	0.1406	3.5718	41/64	0.6406	16.2718
5/32	0.1562	3.9687	21/32	0.6562	16.6687
11/64	0.1718	4.3656	43/64	0.6718	17.0656
3/16	0.1875	4.7625	11/16	0.6875	17.4625
13/64	0.2031	5.1593	45/64	0.7031	17.8593
7/32	0.2187	5.5562	23/32	0.7187	18.2562
15/64	0.2343	5.9531	47/64	0.7343	18.6531
1/4	0.25	6.5	3/4	0.75	19.05
17/64	0.2656	6.7468	49/64	0.7656	19.4468
9/32	0.2812	7.1437	25/32	0.7812	19.8437
19/64	0.2968	7.5406	51/64	0.7968	20.2406
5/16	0.3125	7.9375	13/16	0.8125	20.6375
21/64	0.3281	8.3343	53/64	0.8281	21.0343
11/32	0.3437	8.7312	27/32	0.8437	21.4312
23/64	0.3593	9.1281	55/64	0.8593	21.8281
3/8	0.375	9.525	7/8	0.875	22.225
25/64	0.3906	9.9218	57/64	0.8906	22.6218
13/32	0.4062	10.3187	29/32	0.9062	23.0187
27/64	0.4218	10.7156	59/64	0.9218	23.4156
7/16	0.4375	11.1125	15/16	0.9375	23.8125
29/64	0.4531	11.5093	61/64	0.9531	24.2093
15/32	0.4687	11.9062	31/32	0.9687	24.6062
31/64	0.4843	12.3031	63/64	0.9843	25.0031
1/2	0.50	12.7	1	1.0000	25.4

Exploring Drafting

PREFERRED METRIC SIZES FOR ENGINEERS

SIZES 1 mm to 10 mm			SIZES 10 mm to 100 mm			SIZES 100 mm to 1000 mm		
1st	2nd	3rd	1st	2nd	3rd	1st	2nd	3rd
1			10			100		
	1.1			11		110		105
1.2			12			120		115
		1.3			13		130	125
	1.4			14		140		135
		1.5			15		150	145
1.6			16			160		155
		1.7			17		170	165
	1.8			18		180		175
		1.9			19		190	185
2			20			200		195
	2.1				21		210	
	2.2			22		220		
		2.4			23		230	
2.5			25		24	250	240	
		2.6			26		260	
	2.8			28		280		270
3			30			300		290
		3.2		32			320	
	3.5			35	34	350		340
		3.8		38	36	380		360
4			40			400		
	4.2			42		420		
					44			440
	4.5			45		450		
					46			460
		4.8		48		480		
5			50			500		
	5.2			52		520		
					54			540
	5.5			55		550		
					56			560
	5.8			58		580		
6			60			600		
					62		620	
								640
	6.5			65		650		
								660
					68		680	
	7			70		700		
					72		720	
								740
	7.5			75		750		
					78		780	760
8			80			800		
					82			820
	8.5			85		850		
					88			880
	9			90		900		
					92			920
	9.5			95		950		
					98			980
10			100			1000		

A COMPARISON OF INCH SERIES AND METRIC THREADS

INCH SERIES			METRIC			
Size	Dia. in.	TPI	Size	Dia. in.	Pitch (mm)	TPI (Approx.)
No. 0	.060	80	M1.4	.055	0.3	85
No. 1	.073	64 / 72	M1.6	.063	0.35(a)	74
No. 2	0.86	56 / 64	M2	.079	0.4(a)	64
No. 3	.099	48 / 56	M2.5	.098	0.45(a)	56
No. 4	.112	40 / 48				
No. 5	.125	40 / 44	M3	.118	0.5(a)	51
No. 6	.138	32 / 40	M3.5	.138	0.6(a)	42
No. 8	.164	32 / 36	M4	.158	0.7(a)	36
No. 10	.190	24 / 32				
			M5	.197	0.8(a)	32
1/4	.250	20 / 28	M6	.236	1.0	25
5/16	.312	18 / 24	M6.3	.248	1.0(a)	25
			M7	.276	1.0	25
3/8	.375	16 / 24	M8	.315	1.25(a) / 1.0	20 / 25
7/16	.437	14 / 20	M10	.394	1.5(a) / 1.25	18 / 20
1/2	.500	13 / 20	M12	.472	1.75(a) / 1.25	14.5 / 20
9/16	.562	12 / 18	M14	.551	2.0(a) / 1.5	12.5 / 17
5/8	.625	11 / 18	M16	.630	2.0(a) / 1.5	12.5 / 17
			M18	.709	2.5 / 1.5	10 / 17
3/4	.750	10 / 16	M20	.787	2.5(a) / 1.5	10 / 17
			M22	.866	2.5 / 1.5	10 / 17
7/8	.875	9 / 14	M24	.945	3.0(a) / 2.0	8.5 / 12.5
1	1.000	8 / 14	M27	1.063	3.0 / 2.0	8.5 / 12.5

(a)Preferred series for use in the United States

A comparison of inch series and metric threads shows that even though many of the threads are similar in size and pitch they are not interchangeable. Extreme care must be observed to keep the two series separate.

Useful Information

CONVERSION TABLES

TO REDUCE	MULTIPLY BY	TO REDUCE	MULTIPLY BY
LENGTH			
miles to km	1.61	km to miles	0.62
miles to m	1609.35	m to miles	0.00062
yd. to m	0.9144	m to yd.	1.0936
in. to cm	2.54	cm to in.	0.3937
in. to mm	25.4	mm to in.	0.03937
VOLUME			
cu. in. to cm^3 or mL	16.387	cm^3 to cu. in.	0.061
cu. in. to L	0.0164	l to cu. in.	61.024
gal. to L	3.785	l to gal.	0.264
WEIGHT			
lb. to kg	0.4536	kg to lb.	2.2
oz. to gm	28.35	gm to oz.	0.0353
gr. to gm	0.0648	gm to gr.	15.432

METRIC SIZE DRAFTING SHEETS

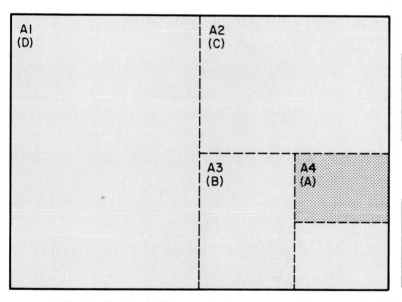

ISO STANDARD

SIZE	MILLIMETRES	INCHES
A0	841 x 1189	33.11 x 46.81
A1	594 x 841	23.39 x 33.11
A2	420 x 594	16.54 x 23.39
A3	297 x 420	11.69 x 16.54
A4	210 x 297	8.27 x 11.69

AMERICAN STANDARD

SIZE	INCHES	MILLIMETRES
E	34 x 44	863.6 x 1117.6
D	22 x 34	558.8 x 863.6
C	17 x 22	431.8 x 558.8
B	11 x 17	279.4 x 431.8
A	8 1/2 x 11	215.9 x 279.4

Left. Metric size drafting paper sheets are exactly proportional in size. Right. Charts show millimetre sizes and customary inch equivalent sizes of ISO papers.

ABBREVIATIONS

(SOME FORMER ABBREVIATIONS ARE NOW SYMBOLS)

Across flats	ACR FLT	Inside diameter	ID
Centers	CTR	Left hand	LH
Center line	CL	Material	MATL
Centimetre	cm	Metre	m
Chamfer	CHAM	Millimetre	mm
Counterbore	CBORE	Number	NO
Countersink	CSK	Outside diameter	OD
Countersunk head	CSK H	Pitch diameter	PD
Diameter (before dimension)	Ø	Radius	R
Diameter (in a note)	DIA	Right hand	RH
Drawing	DWG	Round	RD
Figure	FIG	Square (before dimension)	□
Hexagon	HEX	Square (in a note)	SQ
Hexagonal head	HEX HD	Thread	THD

FORMULAS

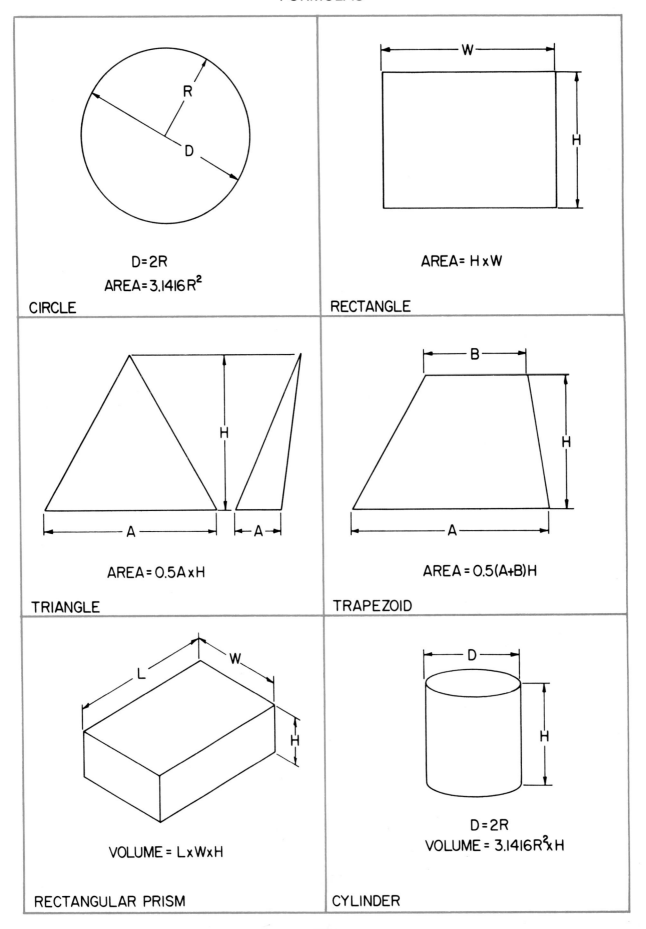

$D = 2R$
$AREA = 3.1416 R^2$

CIRCLE

$AREA = H \times W$

RECTANGLE

$AREA = 0.5A \times H$

TRIANGLE

$AREA = 0.5(A+B)H$

TRAPEZOID

$VOLUME = L \times W \times H$

RECTANGULAR PRISM

$D = 2R$
$VOLUME = 3.1416 R^2 \times H$

CYLINDER

FORMULAS

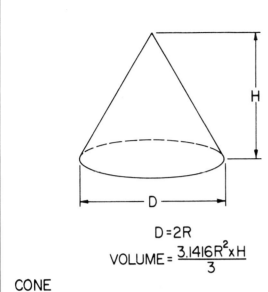

D = 2R

$$\text{VOLUME} = \frac{3.1416 R^2 \times H}{3}$$

CONE

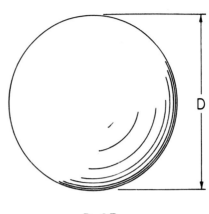

D = 2R

$$\text{VOLUME} = \frac{4 \times 3.1416 R^3}{3}$$

SPHERE

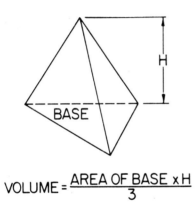

$$\text{VOLUME} = \frac{\text{AREA OF BASE} \times H}{3}$$

TRIANGULAR PYRAMID

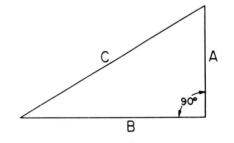

$$A = \sqrt{C^2 - B^2}$$

$$B = \sqrt{C^2 - A^2}$$

$$C = \sqrt{A^2 - B^2}$$

PYTHAGOREAN THEOREM

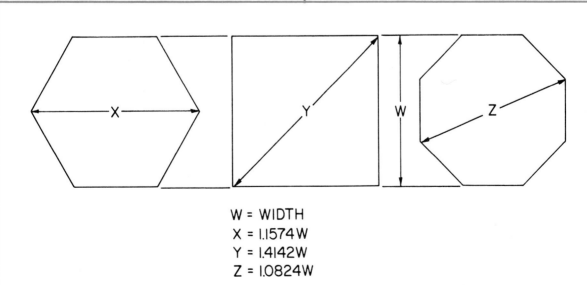

W = WIDTH
X = 1.1574 W
Y = 1.4142 W
Z = 1.0824 W

STANDARD ABBREVIATIONS
FOR USE ON DRAWINGS

A

Abrasive	ABRSV
Accessory	ACCESS
Accumulator	ACCUMR
Acetylene	ACET
Actual	ACT
Actuator	ACTR
Addendum	ADD
Adhesive	ADH
Adjust	ADJ
Advance	ADV
Aeronautic	AERO
Alclad	CLAD
Alignment	ALIGN
Allowance	ALLOW
Alloy	ALY
Alteration	ALT
Alternate	ALT
Alternating Current	AC
Aluminum	AL
American National Standards Institute	ANSI
American Wire Gage	AWG
Ammeter	AMM
Amplifier	AMPL
Anneal	ANL
Anodize	ANOD
Antenna	ANT
Approved	APPD
Approximate	APPROX
Arrangement	ARR
As Required	AR
Assemble	ASSEM
Assembly	ASSY
Automatic	AUTO
Auxiliary	AUX
Average	AVG

B

Babbit	BAB
Base Line	BL
Battery	BAT
Bearing	BRG
Bend Radius	BR
Bevel	BEV
Bill of Material	B/M
Blueprint	BP or B/P
Bolt Circle	BC
Bracket	BRKT
Brass	BRS
Brazing	BRZG

Brinnell Hardness Number	BHN
Bronze	BRZ
Brown & Sharpe (Gage)	B&S
Burnish	BNH
Bushing	BUSH

C

Cabinet	CAB
Calculated	CACL
Cancelled	CANC
Capacitor	CAP
Capacity	CAP
Carburize	CARB
Case Harden	CH
Casting	CSTG
Cast Iron	CI
Cathode-Ray Tube	CRT
Center	CTR
Center to Center	C to C
Centigrade	C
Centimeter	CM
Centrifugal	CENT
Chamfer	CHAM
Circuit	CKT
Circular	CIR
Circumference	CIRC
Clearance	CL
Clockwise	CW
Closure	CLOS
Coated	CTD
Cold-Drawn Steel	CDS
Cold-Rolled Steel	CRS
Color Code	CC
Commercial	COMM
Concentric	CONC
Condition	COND
Conductor	CNDCT
Contour	CTR
Control	CONT
Copper	COP
Counterbore	CBORE
Counterclockwise	CCW
Counter-Drill	CDRILL
Countersink	CSK
Cubic	CU
Cylinder	CYL

D

Datum	DAT
Decimal	DEC

Decrease	DECR
Degree	DEG
Detail	DET
Detector	DET
Developed Length	DL
Developed Width	DW
Deviation	DEV
Diagonal	DIAG
Diagram	DIAG
Diameter	DIA
Diameter Bolt Circle	DBC
Diametral Pitch	DP
Dimension	DIM
Direct Current	DC
Disconnect	DISC
Double-Pole Double-Throw	DPDT
Double-Pole Single-Throw	DPST
Dowel	DWL
Draft	DFT
Drafting Room Manual	DRM
Drawing	DWG
Drawing Change Notice	DCN
Drill	DR
Drop Forge	DF
Duplicate	DUP

E

Each	EA
Eccentric	ECC
Effective	EFF
Electric	ELEC
Enclosure	ENCL
Engine	ENG
Engineer	ENGR
Engineering	ENGRG
Engineering Change Order	ECO
Engineering Order	EO
Equal	EQ
Equivalent	EQUIV
Estimate	EST

F

Fabricate	FAB
Fillet	FIL
Finish	FIN
Finish All Over	FAO
Fitting	FTG
Fixed	FXD
Fixture	FIX
Flange	FLG

STANDARD ABBREVIATIONS
(Continued)

Flat Head	FHD	**J**		Modification	MOD	
Flat Pattern	F/P			Mold Line	ML	
Flexible	FLEX	Joggle	JOG	Motor	MOT	
Fluid	FL	Junction	JCT	Mounting	MTG	
Forged Steel	FST			Multiple	MULT	
Forging	FORG					
Furnish	FURN	**K**				
				N		
		Keyway	KWY			
G				Nickel Steel	NS	
				Nomenclature	NOM	
Gage	GA	**L**		Nominal	NOM	
Gallon	GAL			Normalize	NORM	
Galvanized	GALV	Laboratory	LAB	Not to Scale	NTS	
Gasket	GSKT	Lacquer	LAQ	Number	NO.	
Generator	GEN	Laminate	LAM			
Grind	GRD	Left Hand	LH			
Ground	GRD	Length	LG	**O**		
		Letter	LTR			
		Limited	LTD	Obsolete	OBS	
H		Limit Switch	LS	Opposite	OPP	
		Linear	LIN	Oscilloscope	SCOPE	
Half-Hard	1/2H	Liquid	LIQ	Ounce	OZ	
Handle	HDL	List of Material	L/M	Outside Diameter	OD	
Harden	HDN	Long	LG	Over-All	OA	
Head	HD	Low Carbon	LC			
Heat Treat	HT TR	Low Voltage	LV			
Hexagon	HEX	Lubricate	LUB	**P**		
High Carbon Steel	HCS					
High Frequency	HF			Package	PKG	
High Speed	HS	**M**		Parting Line (Castings)	PL	
Horizontal	HOR			Parts List	P/L	
Hot-Rolled Steel	HRS	Machine(ing)	MACH	Pattern	PATT	
Hour	HR	Magnaflux	M	Piece	PC	
Housing	HSG	Magnesium	MAG	Pilot	PLT	
Hydraulic	HYD	Maintenance	MAINT	Pitch	P	
		Major	MAJ	Pitch Circle	PC	
		Malleable	MALL	Pitch Diameter	PD	
I		Malleable Iron	MI	Plan View	PV	
		Manual	MAN	Plastic	PLSTC	
Identification	IDENT	Manufacturing (ed, er)	MFG	Plate	PL	
Inch	IN	Mark	MK	Pneumatic	PNEU	
Inclined	INCL	Master Switch	MS	Port	P	
Include, Including,	INCL	Material	MATL	Positive	POS	
Inclusive		Maximum	MAX	Potentiometer	POT	
Increase	INCR	Measure	MEAS	Pounds Per Square Inch	PSI	
Independent	INDEP	Mechanical	MECH	Pounds Per Square	PSIG	
Indicator	IND	Medium	MED	Inch Gage		
Information	INFO	Meter	MTR	Power Amplifier	PA	
Inside Diameter	ID	Middle	MID	Power Supply	PWR SPLY	
Installation	INSTL	Military	MIL	Pressure	PRESS	
International Standards	ISO	Millimeter	MM	Primary	PRI	
Organization		Minimum	MIN	Process, Procedure	PROC	
Interrupt	INTER	Miscellaneous	MISC	Product, Production	PROD	

STANDARD ABBREVIATIONS
(Continued)

Q

Quality	QUAL
Quantity	QTY
Quarter-Hard	1/4H

R

Radar	RDR
Radio	RAD
Radio Frequency	RF
Radius	RAD or R
Ream	RM
Receptacle	RECP
Reference	REF
Regular	REG
Regulator	REG
Release	REL
Required	REQD
Resistor	RES
Revision	REV
Revolutions Per Minute	RPM
Right Hand	RH
Rivet	RIV
Rockwell Hardness	RH
Round	RD

S

Schedule	SCH
Schematic	SCHEM
Screw	SCR
Screw Threads	
American National Coarse	NC
American National Fine	NF
American National Extra Fine	NEF
American National 8 Pitch	8N
American Standard Taper Pipe	NTP
American Standard Straight Pipe	NPSC
American Standard Taper (Dryseal)	NPTF
American Standard Straight (Dryseal)	NPSF
Unified Screw Thread Coarse	UNC
Unified Screw Thread Fine	UNF
Unified Screw Thread Extra Fine	UNEF
Unified Screw Thread 8 Thread	8UN
Section	SECT
Sequence	SEQ
Serial	SER
Serrate	SERR
Sheathing	SHTHG
Sheet	SH
Silver Solder	SILS
Single-Pole Double-Throw	SPDT
Single-Pole Single-Throw	SPST
Society of Automotive Engineers	SAE
Solder	SLD
Solenoid	SOL
Speaker	SPKR
Special	SPL
Specification	SPEC
Spot Face	SF
Spring	SPG
Square	SQ
Stainless Steel	SST
Standard	STD
Steel	STL
Stock	STK
Support	SUP
Switch	SW
Symbol	SYM
Symmetrical	SYM
System	SYS

T

Tabulate	TAB
Tangent	TAN
Tapping	TAP
Technical Manual	TM
Teeth	T
Television	TV
Temper	TEM
Temperature	TEM
Tensile Strength	TS
Thick	THK
Thread	THD
Through	THRU
Tolerance	TOL
Tool Steel	TS
Torque	TOR
Total Indicator Reading	TIR
Transformer	XFMR
Transistor	XSTR
Transmitter	XMTR
Tungsten	TU
Typical	TYP

U

Ultra-High Frequency	UHF
Unit	U
Universal	UNIV
Unless Otherwise Specified	UOS

V

Vacuum	VAC
Vacuum Tube	VT
Variable	VAR
Vernier	VER
Vertical	VERT
Very High Frequency	VHF
Vibrate	VIB
Video	VD
Void	VD
Volt	V
Volume	VOL

W

Washer	WASH
Watt	W
Weatherproof	WP
Weight	WT
Wide, Width	W
Wire Wound	WW
Wood	WD
Wrought Iron	WI

Y

Yield Point (PSI)	YP
Yield Strength (PSI)	YS

NATIONAL COARSE AND NATIONAL FINE THREADS AND TAP DRILLS

SIZE	THREADS PER INCH	MAJOR DIA.	MINOR DIA.	PITCH DIA.	TAP DRILL 75 PERCENT THREAD	DECIMAL EQUIVALENT	CLEARANCE DRILL	DECIMAL EQUIVALENT
2	56	.0860	.0628	.0744	50	.0700	42	.0935
	64	.0860	.0657	.0759	50	.0700	42	.0935
3	48	.099	.0719	.0855	47	.0785	36	.1065
	56	.099	.0758	.0874	45	.0820	36	.1065
4	40	.112	.0795	.0958	43	.0890	31	.1200
	48	.112	.0849	.0985	42	.0935	31	.1200
6	32	.138	.0974	.1177	36	.1065	26	.1470
	40	.138	.1055	.1218	33	.1130	26	.1470
8	32	.164	.1234	.1437	29	.1360	17	.1730
	36	.164	.1279	.1460	29	.1360	17	.1730
10	24	.190	.1359	.1629	25	.1495	8	.1990
	32	.190	.1494	.1697	21	.1590	8	.1990
12	24	.216	.1619	.1889	16	.1770	1	.2280
	28	.216	.1696	.1928	14	.1820	2	.2210
1/4	20	.250	.1850	.2175	7	.2010	G	.2610
	28	.250	.2036	.2268	3	.2130	G	.2610
5/16	18	.3125	.2403	.2764	F	.2570	21/64	.3281
	24	.3125	.2584	.2854	I	.2720	21/64	.3281
3/8	16	.3750	.2938	.3344	5/16	.3125	25/64	.3906
	24	.3750	.3209	.3479	Q	.3320	25/64	.3906
7/16	14	.4375	.3447	.3911	U	.3680	15/32	.4687
	20	.4375	.3725	.4050	25/64	.3906	29/64	.4531
1/2	13	.5000	.4001	.4500	27/64	.4219	17/32	.5312
	20	.5000	.4350	.4675	29/64	.4531	33/64	.5156
9/16	12	.5625	.4542	.5084	31/64	.4844	19/32	.5937
	18	.5625	.4903	.5264	33/64	.5156	37/64	.5781
5/8	11	.6250	.5069	.5660	17/32	.5312	21/32	.6562
	18	.6250	.5528	.5889	37/64	.5781	41/64	.6406
3/4	10	.7500	.6201	.6850	21/32	.6562	25/32	.7812
	16	.7500	.6688	.7094	11/16	.6875	49/64	.7656
7/8	9	.8750	.7307	.8028	49/64	.7656	29/32	.9062
	14	.8750	.7822	.8286	13/16	.8125	57/64	.8906
1	8	1.0000	.8376	.9188	7/8	.8750	1-1/32	1.0312
	14	1.0000	.9072	.9536	15/16	.9375	1-1/64	1.0156
1-1/8	7	1.1250	.9394	1.0322	63/64	.9844	1-5/32	1.1562
	12	1.1250	1.0167	1.0709	1-3/64	1.0469	1-5/32	1.1562
1-1/4	7	1.2500	1.0644	1.1572	1-7/64	1.1094	1-9/32	1.2812
	12	1.2500	1.1417	1.1959	1-11/64	1.1719	1-9/32	1.2812
1-1/2	6	1.5000	1.2835	1.3917	1-11/32	1.3437	1-17/32	1.5312
	12	1.5000	1.3917	1.4459	1-27/64	1.4219	1-17/32	1.5312

SCREW THREAD ELEMENTS FOR UNIFIED AND NATIONAL FORM OF THREAD

THREADS PER INCH (n)	PITCH (p) $p = \frac{1}{n}$	SINGLE HEIGHT SUBTRACT FROM BASIC MAJOR DIAMETER TO GET BASIC PITCH DIAMETER	DOUBLE HEIGHT SUBTRACT FROM BASIC MAJOR DIAMETER TO GET BASIC MINOR DIAMETER	83 1/3 PERCENT DOUBLE HEIGHT SUBTRACT FROM BASIC MAJOR DIAMETER TO GET MINOR DIAMETER OF RING GAGE	BASIC WIDTH OF CREST AND ROOT FLAT $\frac{p}{8}$	CONSTANT FOR BEST SIZE WIRE ALSO SINGLE HEIGHT OF 60 DEG. V–THREAD	DIAMETER OF BEST SIZE WIRE
3	.333333	.216506	.43301	.36084	.0417	.28868	.19245
3 1/4	.307692	.199852	.39970	.33309	.0385	.26647	.17765
3 1/2	.285714	.185577	.37115	.30929	.0357	.24744	.16496
4	.250000	.162379	.32476	.27063	.0312	.21651	.14434
4 1/2	.222222	.144337	.28867	.24056	.0278	.19245	.12830
5	.200000	.129903	.25981	.21650	.0250	.17321	.11547
5 1/2	.181818	.118093	.23619	.19682	.0227	.15746	.10497
6	.166666	.108253	.21651	.18042	.0208	.14434	.09623
7	.142857	.092788	.18558	.15465	.0179	.12372	.08248
8	.125000	.081189	.16238	.13531	.0156	.10825	.07217
9	.111111	.072168	.14434	.12028	.0139	.09623	.06415
10	.100000	.064952	.12990	.10825	.0125	.08660	.05774
11	.090909	.059046	.11809	.09841	.0114	.07873	.05249
11 1/2	.086956	.056480	.11296	.09413	.0109	.07531	.05020
12	.083333	.054127	.10826	.09021	.0104	.07217	.04811
13	.076923	.049963	.09993	.08327	.0096	.06662	.04441
14	.071428	.046394	.09279	.07732	.0089	.06186	.04124
16	.062500	.040595	.08119	.06766	.0078	.05413	.03608
18	.055555	.036086	.07217	.06014	.0069	.04811	.03208
20	.050000	.032475	.06495	.05412	.0062	.04330	.02887
22	.045454	.029523	.05905	.04920	.0057	.03936	.02624
24	.041666	.027063	.05413	.04510	.0052	.03608	.02406
27	.037037	.024056	.04811	.04009	.0046	.03208	.02138
28	.035714	.023197	.04639	.03866	.0045	.03093	.02062
30	.033333	.021651	.04330	.03608	.0042	.02887	.01925
32	.031250	.020297	.04059	.03383	.0039	.02706	.01804
36	.027777	.018042	.03608	.03007	.0035	.02406	.01604
40	.025000	.016237	.03247	.02706	.0031	.02165	.01443
44	.022727	.014761	.02952	.02460	.0028	.01968	.01312
48	.020833	.013531	.02706	.02255	.0026	.01804	.01203
50	.020000	.012990	.02598	.02165	.0025	.01732	.01155
56	.017857	.011598	.02320	.01933	.0022	.01546	.01031
60	.016666	.010825	.02165	.01804	.0021	.01443	.00962
64	.015625	.010148	.02030	.01691	.0020	.01353	.00902
72	.013888	.009021	.01804	.01503	.0017	.01203	.00802
80	.012500	.008118	.01624	.01353	.0016	.01083	.00722
90	.011111	.007217	.01443	.01202	.0014	.00962	.00642
96	.010417	.006766	.01353	.01127	.0013	.00902	.00601
100	.010000	.006495	.01299	.01082	.0012	.00866	.00577
120	.008333	.005413	.01083	.00902	.0010	.00722	.00481

Using the Best Size Wires, the measurement over three wires minus the Constant for Best Size Wire equals the Pitch Diameter.

MACHINE SCREW AND CAP SCREW HEADS

FILLISTER HEAD

SIZE	A	B	C	D
#8	.260	.141	.042	.060
#10	.302	.164	.048	.072
1/4	3/8	.205	.064	.087
5/16	7/16	.242	.077	.102
3/8	9/16	.300	.086	.125
1/2	3/4	.394	.102	.168
5/8	7/8	.500	.128	.215
3/4	1	.590	.144	.258
1	1 5/16	.774	.182	.352

FLAT HEAD

SIZE	A	B	C	D
#8	.320	.092	.043	.037
#10	.372	.107	.048	.044
1/4	1/2	.146	.064	.063
5/16	5/8	.183	.072	.078
3/8	3/4	.220	.081	.095
1/2	7/8	.220	.102	.090
5/8	1 1/8	.293	.128	.125
3/4	1 3/8	.366	.144	.153

ROUND HEAD

SIZE	A	B	C	D
#8	.297	.113	.044	.067
#10	.346	.130	.048	.073
1/4	7/16	.1831	.064	.107
5/16	9/16	.236	.072	.150
3/8	5/8	.262	.081	.160
1/2	13/16	.340	.102	.200
5/8	1	.422	.128	.255
3/4	1 1/4	.526	.144	.320

HEXAGON HEAD

SIZE	A	B	C
1/4	.494	.170	7/16
5/16	.564	.215	1/2
3/8	.635	.246	9/16
1/2	.846	.333	3/4
5/8	1.058	.411	15/16
3/4	1.270	.490	1 1/8
7/8	1.482	.566	1 5/16
1	1.693	.640	1 1/2

SOCKET HEAD

SIZE	A	B	C
#8	.265	.164	1/8
#10	5/16	.190	5/32
1/4	3/8	1/4	3/16
5/16	7/16	5/16	7/32
3/8	9/16	3/8	5/16
7/16	5/8	7/16	5/16
1/2	3/4	1/2	3/8
5/8	7/8	5/8	1/2
3/4	1	3/4	9/16
7/8	1 1/8	7/8	9/16
1	1 5/16	1	5/8

TWIST DRILL DATA

METRIC DRILL SIZES (mm)[1]		Decimal Equivalent in Inches (Ref)	METRIC DRILL SIZES (mm)[1]		Decimal Equivalent in Inches (Ref)
Preferred	Available		Preferred	Available	
	.40	.0157	1.70		.0669
	.42	.0165		1.75	.0689
	.45	.0177	1.80		.0709
	.48	.0189		1.85	.0728
.50		.0197	1.90		.0748
	.52	.0205		1.95	.0768
.55		.0217	2.00		.0787
	.58	.0228		2.05	.0807
.60		.0236	2.10		.0827
	.62	.0244		2.15	.0846
.65		.0256	2.20		.0866
	.68	.0268		2.30	.0906
.70		.0276	2.40		.0945
	.72	.0283	2.50		.0984
.75		.0295	2.60		.1024
	.78	.0307		2.70	.1063
.80		.0315	2.80		.1102
	.82	.0323		2.90	.1142
.85		.0335	3.00		.1181
	.88	.0346		3.10	.1220
.90		.0354	3.20		.1260
	.92	.0362		3.30	.1299
.95		.0374	3.40		.1339
	.98	.0386		3.50	.1378
1.00		.0394	3.60		.1417
	1.03	.0406		3.70	.1457
1.05		.0413	3.80		.1496
	1.08	.0425		3.90	.1535
1.10		.0433	4.00		.1575
	1.15	.0453		4.10	.1614
1.20		.0472	4.20		.1654
1.25		.0492		4.40	.1732
1.30		.0512	4.50		.1772
	1.35	.0531		4.60	.1811
1.40		.0551	4.80		.1890
	1.45	.0571	5.00		.1969
1.50		.0591		5.20	.2047
	1.55	.0610	5.30		.2087
1.60		.0630		5.40	.2126
	1.65	.0650	5.60		.2205
				5.80	.2283

1 Metric drill sizes listed in the "Preferred" column are based on the R′40 series of preferred numbers shown in the ISO Standard R497. Those listed in the "Available" column are based on the R80 series from the same document.

TWIST DRILL DATA (Continued)

METRIC DRILL SIZES (mm)[1]		Decimal Equivalent in Inches (Ref)	METRIC DRILL SIZES (mm)[1]		Decimal Equivalent in Inches (Ref)
Preferred	Available		Preferred	Available	
6.00		.2362		19.50	.7677
	6.20	.2441	20.00		.7874
6.30		.2480		20.50	.8071
	6.50	.2559	21.00		.8268
6.70		.2638		21.50	.8465
	6.80[2]	.2677	22.00		.8661
	6.90	.2717		23.00	.9055
7.10		.2795	24.00		.9449
	7.30	.2874	25.00		.9843
7.50		.2953	26.00		1.0236
	7.80	.3071		27.00	1.0630
8.00		.3150	28.00		1.1024
	8.20	.3228		29.00	1.1417
8.50		.3346	30.00		1.1811
	8.80	.3465		31.00	1.2205
9.00		.3543	32.00		1.2598
	9.20	.3622		33.00	1.2992
9.50		.3740	34.00		1.3386
	9.80	.3858		35.00	1.3780
10.00		.3937	36.00		1.4173
	10.30	.4055		37.00	1.4567
10.50		.4134	38.00		1.4961
	10.80	.4252		39.00	1.5354
11.00		.4331	40.00		1.5748
	11.50	.4528		41.00	1.6142
12.00		.4724	42.00		1.6535
12.50		.4921		43.50	1.7126
13.00		.5118	45.00		1.7717
	13.50	.5315		46.50	1.8307
14.00		.5512	48.00		1.8898
	14.50	.5709	50.00		1.9685
15.00		.5906		51.50	2.0276
	15.50	.6102	53.00		2.0866
16.00		.6299		54.00	2.1260
	16.50	.6496	56.00		2.2047
17.00		.6693		58.00	2.2835
	17.50	.6890	60.00		2.3622
18.00		.7087			
	18.50	.7283			
19.00		.7480			

1 Metric drill sizes listed in the "Preferred" column are based on the R'40 series of preferred numbers shown in the ISO Standard R497. Those listed in the "Available" column are based on the R80 series from the same document.

2 Recommended only for use as a tap drill size.

LETTER SIZE DRILLS

A	0.234	J	0.277	S	0.348
B	0.238	K	0.281	T	0.358
C	0.242	L	0.290	U	0.368
D	0.246	M	0.295	V	0.377
E	0.250	N	0.302	W	0.386
F	0.257	O	0.316	X	0.397
G	0.261	P	0.323	Y	0.404
H	0.266	Q	0.332	Z	0.413
I	0.272	R	0.339		

WOOD SCREW TABLE

LENGTH	GAUGE STEEL SCREW	GAUGE BRASS SCREW	GAUGE NO.	DECIMAL	APPROX. FRACT.	DRILL SIZE A	DRILL SIZE B	DRILL SIZE C
1/4	0 to 4	0 to 4	0	.060	1/16	1/16		
3/8	0 to 8	0 to 6	1	.073	5/64	3/32		
1/2	1 to 10	1 to 8	2	.086	5/64	3/32	1/16	3/16
5/8	2 to 12	2 to 10	3	.099	3/32	1/8	1/16	1/4
3/4	2 to 14	2 to 12	4	.112	7/64	1/8	1/16	1/4
7/8	3 to 14	4 to 12	5	.125	1/8	1/8	3/32	1/4
1	3 to 16	4 to 14	6	.138	9/64	5/32	3/32	5/16
1 1/4	4 to 18	6 to 14	7	.151	5/32	5/32	1/8	5/16
1 1/2	4 to 20	6 to 14	8	.164	5/32	3/16	1/8	3/8
1 3/4	6 to 20	8 to 14	9	.177	11/64	3/16	1/8	3/8
2	6 to 20	8 to 18	10	.190	3/16	3/16	1/8	3/8
2 1/4	6 to 20	10 to 18	11	.203	13/64	7/32	5/32	7/16
2 1/2	8 to 20	10 to 18	12	.216	7/32	7/32	5/32	7/16
2 3/4	8 to 20	8 to 20	14	.242	15/64	1/4	3/16	1/2
3	8 to 24	12 to 18	16	.268	17/64	9/32	7/32	9/16
3 1/2	10 to 24	12 to 18	18	.294	19/64	5/16	1/4	5/8
4	12 to 24	12 to 24	20	.320	21/64	11/32	9/32	11/16
4 1/2	14 to 24	14 to 24	24	.372	3/8	3/8	5/16	3/4
5	14 to 24	14 to 24						

SCREWS ALSO AVAILABLE WITH PHILLIPS HEAD

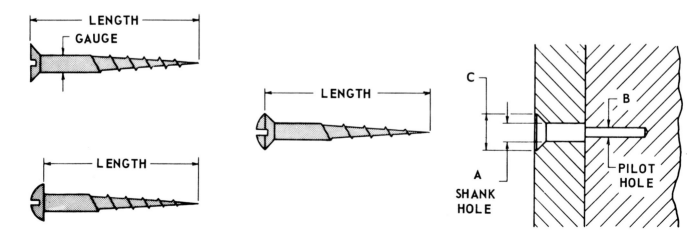

DIAMETER/THREAD PITCH COMPARISON

INCH SERIES			METRIC			
Size	Dia. (In.)	TPI	Size	Dia. (In.)	Pitch (mm)	TPI (Approx)
			M1.4	.055	.3 .2	85 127
#0	.060	80				
			M1.6	.063	.35 .2	74 127
#1	.073	64 72				
			M2	.079	.4 .25	64 101
#2	.086	56 64				
			M2.5	.098	.45 .35	56 74
#3	.099	48 56				
#4	.112	40 48				
			M3	.118	.5 .35	51 74
#5	.125	40 44				
#6	.138	32 40				
			M4	.157	.7 .5	36 51
#8	.164	32 36				
#10	.190	24 32				
			M5	.196	.8 .5	32 51
			M6	.236	1.0 .75	25 34
1/4	.250	20 28				
5/16	.312	18 24				
			M8	.315	1.25 1.0	20 25
3/8	.375	16 24				
			M10	.393	1.5 1.25	17 20
7/16	.437	14 20				
			M12	.472	1.75 1.25	14.5 20
1/2	.500	13 20				
			M14	.551	2 1.5	12.5 17
5/8	.625	11 18				
			M16	.630	2 1.5	12.5 17
			M18	.709	2.5 1.5	10 17
3/4	.750	10 16				
			M20	.787	2.5 1.5	10 17
			M22	.866	2.5 1.5	10 17
7/8	.875	9 14				
			M24	.945	3 2	8.5 12.5
1″	1.000	8 12				
			M27	1.063	3 2	8.5 12.5

CONVERSION CHART INCH/mm

Drill No. or Letter		Inch	mm
		.001	0.0254
		.002	0.0508
		.003	0.0762
		.004	0.1016
		.005	0.1270
		.006	0.1524
		.007	0.1778
		.008	0.2032
		.009	0.2286
		.010	0.2540
		.011	0.2794
		.012	0.3048
		.013	0.3302
80	.0135		
		.014	0.3556
79	.0145		
		.015	0.3810
	1/64	.0156	0.3969
78		.016	0.4064
		.017	0.4318
77		.018	0.4572
		.019	0.4826
76		.020	0.5080
75		.021	0.5334
		.022	0.5588
74	.0225		
		.023	0.5842
73		.024	0.6096
72		.025	0.6350
71		.026	0.6604
		.027	0.6858
70		.028	0.7112
		.029	0.7366
69	.0292		
		.030	0.7620
68		.031	0.7874
	1/32	.0312	0.7937
67		.032	0.8128
66		.033	0.8382
		.034	0.8636
65		.035	0.8890
64		.036	0.9144
63		.037	0.9398
62		.038	0.9652
61		.039	0.9906
		.0394	1.0000
60		.040	1.0160
59		.041	1.0414
58		.042	1.0668
57		.043	1.0922
		.044	1.1176
		.045	1.1430
56	.0465	.046	1.1684
	3/64	.0469	1.1906
		.047	1.1938
		.048	1.2192
		.049	1.2446
		.050	1.2700
		.051	1.2954
55		.052	1.3208
		.053	1.3462
		.054	1.3716
54		.055	1.3970
		.056	1.4224
		.057	1.4478
		.058	1.4732
		.059	1.4986
53	.0595		
		.060	1.5240
		.061	1.5494
		.062	1.5748
	1/16	.0625	1.5875
		.063	1.6002
52	.0635		
		.064	1.6256
		.065	1.6510
		.066	1.6764
51		.067	1.7018
		.068	1.7272
		.069	1.7526
50		.070	1.7780
		.071	1.8034
		.072	1.8288
49		.073	1.8542
		.074	1.8796
		.075	1.9050
48		.076	1.9304
		.077	1.9558
47	.0785	.078	1.9812
	5/64	.0781	1.9844
		.0787	2.0000
		.079	2.0066
		.080	2.0320
46		.081	2.0574
45		.082	2.0828
		.083	2.1082
		.084	2.1336
		.085	2.1590
44		.086	2.1844
		.087	2.2098
		.088	2.2352
43		.089	2.2606
		.090	2.2860
		.091	2.3114
		.092	2.3368
42	.0935	.093	2.3622
	3/32	.0937	2.3812
		.094	2.3876
		.095	2.4130
41		.096	2.4384
		.097	2.4638
40		.098	2.4892
		.099	2.5146
39	.0995		
		.100	2.5400

Drill No. or Letter		Inch	mm
		.101	2.5654
38	.1015	.102	2.5908
		.103	2.6162
37		.104	2.6416
		.105	2.6670
36	.1065	.106	2.6924
		.107	2.7178
		.108	2.7432
		.109	2.7686
	7/64	.1094	2.7781
35		.110	2.7490
34		.111	2.8194
		.112	2.8448
33		.113	2.8702
		.114	2.8956
		.115	2.9210
32		.116	2.9464
		.117	2.9718
		.118	2.9972
		.1181	3.0000
		.119	3.0226
31		.120	3.0480
		.121	3.0734
		.122	3.0988
		.123	3.1242
		.124	3.1496
	1/8	.125	3.1750
		.126	3.2004
		.127	3.2258
		.128	3.2512
30	.1285	.129	3.2766
		.130	3.3020
		.131	3.3274
		.132	3.3528
		.133	3.3782
		.134	3.4036
		.135	3.4290
29		.136	3.4544
		.137	3.4798
		.138	3.5052
		.139	3.5306
28	.1405	.140	3.5560
	9/64	.1406	3.5719
		.141	3.5814
		.142	3.6068
		.143	3.6322
27		.144	3.6576
		.145	3.6830
		.146	3.7084
26		.147	3.7338
		.148	3.7592
		.149	3.7846
25	.1495		
		.150	3.8100
		.151	3.8354
24		.152	3.8608
		.153	3.8862
23		.154	3.9116
		.155	3.9370
		.156	3.9624
	5/32	.1562	3.9687
22		.157	3.9878
		.1575	4.0000
		.158	4.0132
21		.159	4.0386
		.160	4.0640
20		.161	4.0894
		.162	4.1148
		.163	4.1402
		.164	4.1656
		.165	4.1910
19		.166	4.2164
		.167	4.2418
		.168	4.2672
		.169	4.2926
18	.1635	.170	4.3180
		.171	4.3434
	11/64	.1719	4.3656
		.172	4.3688
17		.173	4.3942
		.174	4.4196
		.175	4.4450
		.176	4.4704
16		.177	4.4958
		.178	4.5212
		.179	4.5466
15		.180	4.5720
		.181	4.5974
14		.182	4.6228
		.183	4.6482
		.184	4.6736
13		.185	4.6990
		.186	4.7244
		.187	4.7498
	3/16	.1875	4.7625
		.188	4.7752
12		.189	4.8006
		.190	4.8260
11		.191	4.8514
		.192	4.8768
		.193	4.9022
10	.1935	.194	4.9276
		.195	4.9530
9		.196	4.9784
		.1969	5.0000
		.197	5.0038
		.198	5.0292
		.199	5.0546
8		.200	5.0800

Drill No. or Letter		Inch	mm
7		.201	5.1054
		.202	5.1308
		.203	5.1562
	13/64	.2031	5.1594
6		.204	5.1816
5	.2055	.205	5.2070
		.206	5.2324
		.207	5.2578
		.208	5.2832
4		.209	5.3086
		.210	5.3340
		.211	5.3594
		.212	5.3848
3		.213	5.4102
		.214	5.4356
		.215	5.4610
		.216	5.4864
		.217	5.5118
		.218	5.5372
	7/32	.2187	5.5562
		.219	5.5626
		.220	5.5880
2		.221	5.6134
		.222	5.6388
		.223	5.6642
		.224	5.6896
		.225	5.7150
		.226	5.7404
		.227	5.7658
1		.228	5.7912
		.229	5.8166
		.230	5.8410
		.231	5.8674
		.232	5.8928
		.233	5.9182
A		.234	5.9436
	15/64	.2344	5.9531
		.235	5.9690
		.236	5.9944
		.2362	6.0000
		.237	6.0198
B		.238	6.0452
		.239	6.0706
		.240	6.0960
		.241	6.1214
C		.242	6.1468
		.243	6.1722
		.244	6.1976
		.245	6.2230
D		.246	6.2484
		.247	6.2738
		.248	6.2992
		.249	6.3246
E	1/4	.250	6.3500
		.251	6.3754
		.252	6.4008
		.253	6.4262
		.254	6.4516
		.255	6.4770
		.256	6.5024
F		.257	6.5278
		.258	6.5532
		.259	6.5786
		.260	6.6040
G		.261	6.6294
		.262	6.6548
		.263	6.6802
		.264	6.7056
		.265	6.7310
	17/64	.2656	6.7469
H		.266	6.7564
		.267	6.7818
		.268	6.8072
		.269	6.8326
		.270	6.8580
		.271	6.8834
I		.272	6.9088
		.273	6.9342
		.274	6.9596
		.275	6.9850
		.2756	7.0000
		.276	7.0104
J		.277	7.0358
		.278	7.0612
		.279	7.0866
		.280	7.1120
K		.281	7.1374
	9/32	.2812	7.1437
		.282	7.1628
		.283	7.1882
		.284	7.2136
		.285	7.2390
		.286	7.2644
		.287	7.2898
		.288	7.3152
		.289	7.3406
L		.290	7.3660
		.291	7.3914
		.292	7.4168
		.293	7.4422
		.294	7.4676
M		.295	7.4930
		.296	7.5184
	19/64	.2969	7.5406
		.297	7.5438
		.298	7.5692
		.299	7.5946
		.300	7.6200

Drill No. or Letter		Inch	mm
		.301	7.6454
N		.302	7.6708
		.303	7.6962
		.304	7.7216
		.305	7.7470
		.306	7.7724
		.307	7.7978
		.308	7.8232
		.309	7.8486
		.310	7.8740
		.311	7.8994
		.312	7.9248
	5/16	.3125	7.9375
		.313	7.9502
		.314	7.9756
		.3150	8.0000
		.315	8.0010
O		.316	8.0264
		.317	8.0518
		.318	8.0772
		.319	8.1026
		.320	8.1280
		.321	8.1534
		.322	8.1788
P		.323	8.2042
		.324	8.2296
		.325	8.2550
		.326	8.2804
		.327	8.3058
		.328	8.3312
	21/64	.3281	8.3344
		.329	8.3566
		.330	8.3820
		.331	8.4074
Q		.332	8.4328
		.333	8.4582
		.334	8.4836
		.335	8.5090
		.336	8.5344
		.337	8.5598
		.338	8.5852
R		.339	8.6106
		.340	8.6360
		.341	8.6614
		.342	8.6868
		.343	8.7122
	11/32	.3437	8.7312
		.344	8.7376
		.345	8.7630
		.346	8.7884
		.347	8.8138
S		.348	8.8392
		.349	8.8646
		.350	8.8900
		.351	8.9154
		.352	8.9408
		.353	8.9662
		.354	8.9916
		.3543	9.0000
		.355	9.0170
		.356	9.0424
		.357	9.0678
T		.358	9.0932
		.359	9.1186
	23/64	.3594	9.1281
		.360	9.1440
		.361	9.1694
		.362	9.1948
		.363	9.2202
		.364	9.2456
		.365	9.2710
		.366	9.2964
		.367	9.3218
U		.368	9.3472
		.369	9.3726
		.370	9.3980
		.371	9.4234
		.372	9.4488
		.373	9.4742
		.374	9.4996
	3/8	.375	9.5250
		.376	9.5504
V		.377	9.5758
		.378	9.6012
		.379	9.6266
		.380	9.6520
		.381	9.6774
		.382	9.7028
		.383	9.7282
		.384	9.7536
		.385	9.7790
W		.386	9.8044
		.387	9.8298
		.388	9.8552
		.389	9.8806
		.390	9.9060
	25/64	.3906	9.9219
		.391	9.9314
		.392	9.9568
		.393	9.9822
		.3937	10.0000
		.394	10.0076
		.395	10.0330
		.396	10.0584
X		.397	10.0838
		.398	10.1092
		.399	10.1346
		.400	10.1600

Drill No. or Letter		Inch	mm
		.401	10.1854
		.402	10.2108
		.403	10.2362
Y		.404	10.2616
		.405	10.2870
		.406	10.3124
	13/32	.4062	10.3187
		.407	10.3378
		.408	10.3632
		.409	10.3886
		.410	10.4140
		.411	10.4394
		.412	10.4648
Z		.413	10.4902
		.414	10.5156
		.415	10.5410
		.416	10.5664
		.417	10.5918
		.418	10.6172
		.419	10.6426
		.420	10.6680
		.421	10.6934
	27/64	.4219	10.7156
		.422	10.7188
		.423	10.7442
		.424	10.7696
		.425	10.7950
		.426	10.8204
		.427	10.8458
		.428	10.8712
		.429	10.8966
		.430	10.9220
		.431	10.9474
		.432	10.9728
		.433	10.9982
		.4331	11.0000
		.434	11.0236
		.435	11.0490
		.436	11.0744
		.437	11.0998
	7/16	.4375	11.1125
		.438	11.1252
		.439	11.1506
		.440	11.1760
		.441	11.2014
		.442	11.2268
		.443	11.2522
		.444	11.2776
		.445	11.3030
		.446	11.3284
		.447	11.3538
		.448	11.3792
		.449	11.4046
		.450	11.4300
		.451	11.4554
		.452	11.4808
		.453	11.5062
	29/64	.4531	11.5094
		.454	11.5316
		.455	11.5570
		.456	11.5824
		.457	11.6078
		.458	11.6332
		.459	11.6586
		.460	11.6840
		.461	11.7094
		.462	11.7348
		.463	11.7602
		.464	11.7856
		.465	11.8110
		.466	11.8364
		.467	11.8618
		.468	11.8872
	15/32	.4687	11.9062
		.469	11.9126
		.470	11.9380
		.471	11.9634
		.472	11.9888
		.4724	12.0000
		.473	12.0142
		.474	12.0396
		.475	12.0650
		.476	12.0904
		.477	12.1158
		.478	12.1412
		.479	12.1666
		.480	12.1920
		.481	12.2174
		.482	12.2428
		.483	12.2682
		.484	12.2936
	31/64	.4844	12.3031
		.485	12.3190
		.486	12.3444
		.487	12.3698
		.488	12.3952
		.489	12.4206
		.490	12.4460
		.491	12.4714
		.492	12.4968
		.493	12.5222
		.494	12.5476
		.495	12.5730
		.496	12.5984
		.497	12.6238
		.498	12.6492
		.499	12.6746
	1/2	.500	12.7000

PHYSICAL PROPERTIES OF METALS

| METAL | SYMBOL | SPECIFIC GRAVITY | SPECIFIC HEAT | MELTING POINT* | | LBS. PER CUBIC INCH |
				DEG. C	DEG. F.	
Aluminum (Cast)	Al	2.56	.2185	658	1217	.0924
Aluminum (Rolled).	Al	2.71	–	–	–	.0978
Antimony	Sb	6.71	.051	630	1166	.2424
Bismuth	Bi	9.80	.031	271	520	.3540
Boron.	B	2.30	.3091	2300	4172	.0831
Brass.	–	8.51	.094	–	–	.3075
Cadmium.	Cd	8.60	.057	321	610	.3107
Calcium	Ca	1.57	.170	810	1490	.0567
Carbon.	C	2.22	.165	–	–	.0802
Chromium	Cr	6.80	.120	1510	2750	.2457
Cobalt	Co	8.50	.110	1490	2714	.3071
Copper.	Cu	8.89	.094	1083	1982	.3212
Columbium . . .	Cb	8.57	–	1950	3542	.3096
Gold	Au	19.32	.032	1063	1945	.6979
Iridium.	Ir	22.42	.033	2300	4170	.8099
Iron.	Fe	7.86	.110	1520	2768	.2634
Iron (Cast) . . .	Fe	7.218	.1298	1375	2507	.2605
Iron (Wrought) .	Fe	7.70	.1138	1500–1600	2732–2912	.2779
Lead	Pb	11.37	.031	327	621	.4108
Lithium	Li	.057	.941	186	367	.0213
Magnesium . . .	Mg	1.74	.250	651	1204	.0629
Manganese . . .	Mn	8.00	.120	1225	2237	.2890
Mercury	Hg	13.59	.032	38.7	37.7	.4909
Molybdenum. . .	Mo	10.2	.0647	2620	4748	.368
Monel Metal. . .	–	8.87	.127	1360	2480	.320
Nickel	Ni	8.80	.130	1452	2646	.319
Phosphorus. . .	P	1.82	.177	43	111.4	.0657
Platinum.	Pt	21.50	.033	1755	3191	.7767
Potassium. . . .	K	0.87	.170	62	144	.0314
Selenium.	Se	4.81	.084	220	428	.174
Silicon.	Si	2.40	.1762	1427	2600	.087
Silver.	Ag	10.53	.056	961	1761	.3805
Sodium.	Na	0.97	.290	97	207	.0350
Steel	–	7.858	.1175	1330–1378	2372–2532	.2839
Strontium	Sr	2.54	.074	–	–	.0918
Sulphur.	S	2.07	.175	115	235.4	.075
Tantalum	Ta	10.80	–	2850	5160	.3902
Tin	Sn	7.29	.056	232	450	.2634
Titanium.	Ti	5.3	.130	1900	3450	.1915
Tungsten	W	19.10	.033	3000	5432	.6900
Uranium	U	18.70	–	–	–	.6755
Vanadium	V	5.50	–	1730	3146	.1987
Zinc	Zn	7.19	.094	419	786	.2598

* Circular of the Bureau of Standards No. 35, Department of Commerce and Labor.

GLOSSARY

ACCURATE: Made within the tolerances allowed.

ACOUSTIC TILE: Tile made of sound absorbing materials.

ACUTE ANGLE: An angle less than 90 deg.

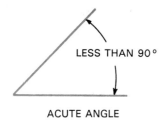

ADDRESS: A number, letter, or label specifying where a unit of information is stored in the computer's memory.

AI: Artificial intelligence. (Letters pronounced individually.)

AIR BRUSH: A device used to spray paint by means of compressed air.

AIR CONDITIONING: The control of the temperature, humidity, motion, dust, and distribution of the air within a structure.

ALLOY: A mixture of two or more metals fused or melted together to form a new metal.

ALLOWANCE: The limits permitted for satisfactory performance of machined parts.

ALPHANUMERIC: Pertaining to a set of characters that contains both letters and numbers.

ANALOG: Denotes the use of physical variables —distance, rotation, voltage, etc., to represent and correspond with numerical variables that occur in computation.

ANGLE: The figure formed by two lines coming together to a point.

ANNEALING: The process of heating metal to a given temperature (the exact temperature and period the temperature is held depends upon the composition of the metal being annealed) and cooling it slowly to remove stresses and induce softness.

ANSI: American National Standards Institute. A non-governmental organization that proposes, modifies, approves, and publishes drafting and manufacturing standards for voluntary use in the United States.

ARC: Portion of a circle.

ARTIFICIAL INTELLIGENCE: Computer techniques that mimic certain functions typically associated with human intelligence. Also refers to a computer or machine capable of improving their operation as a result of repeated experience.

ASPHALT ROOFING: A roofing material made by saturating felt with asphalt.

ASSEMBLY: A unit fitted together from manufactured parts.

ATTIC: The part of a building just below the roof.

AUTOMATIC DIMENSIONING: A form of dimensioning where a CAD program automatically measures distances, and assigns arrows and dimensions to the display and/or hard copy.

AXES: The plural of axis.

AXIS: The center line of a view or of a geometric figure.

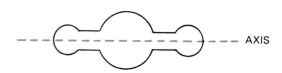

BASEMENT: The base story of a structure. It is usually underground.

BEAM: A term used to describe joists, rafters, and girders.

BEAM COMPASS: Compass used to draw large circles and arcs.

BEVEL: The angle that is formed by a line or a surface that is not at right angles to another line or surface.

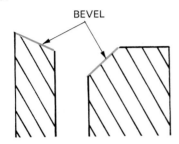

BEVEL

BINARY: A number system consisting of just two digits—zero (0) and one (1). These digits represent bits in computer language.

BIT: The smallest unit of information for a computer. It is represented by a zero (0) or a one (1) in a computer by the absence (0) or presence (1) of an electric current.

BLOWHOLE: A hole produced in a casting when gasses are entrapped during the pouring operation.

BLUEPRINT: A reproduction of a drawing that has a bright background with white lines.

BRAZING: Joining metals by the fusion of nonferrous alloys that have melting temperatures above 800 deg. F but lower than the metals being joined.

BUG: A mistake, malfunction, or defect that prevents a program from working properly.

BUSHING: A bearing for a revolving shaft. A hardened steel tube used on jigs to guide drills and reamers.

BYTE: A group of 8 bits treated as a unit and often used to represent one alphanumeric character.

CAD: Computer aided design and drafting. Sometimes listed as CADD. Computer automated design and drafting operations.

CASTING: An object made by pouring molten metal into a mold.

CATHODE RAY TUBE (CRT): The electronic tube used to show information as text or images on a video display terminal. Similar to the picture tube in a TV set.

CAULK: To fill cracks and seams with caulking material.

CASEHARDEN: A process of surface hardening iron base alloys so that the surface layer or case is made substantially harder than the interior or core of the metal.

HARDENED AREA OF METAL

CASEHARDENING

CENTER LINE SYMBOL:

CHAMFER: See BEVEL.

CHIP: Small piece of material, usually silicon, in which tiny amounts of other elements (with desired electronic properties) are deposited to form integrated circuits (IC). Also called a microchip.

CIRCUMFERENCE: The perimeter of a circle.

CIRCUMSCRIBE: To draw a line around.

CLEARANCE: The distance by which one part clears another part.

CLOCKWISE: From left to right in a circular motion. The direction clock hands move.

CLOCKWISE

CODE: The statements that make up a computer language.

COMPUTER: An electronic device capable of solving problems, accepting and processing data according to previous instructions of a program.

CONCAVE SURFACE: A curved depression in the surface of an object.

CONCAVE SURFACE

CONCENTRIC: Having a common center.

CONCENTRIC CIRCLES

CONCRETE: A mixture of portland cement, sand, gravel, and water.

CONICAL: Shaped like a cone.

CONICAL

CONTOUR: The outline of an object.

CONVENTIONAL: Not original; customary, or traditional.

CONVEX SURFACE: A rounded surface on an object.

CONVEX SURFACE

COORDINATES: The positions or locations of points on the x, y, and z planes.

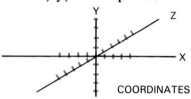

COORDINATES

CORE: A body of sand or other material that is formed to a desired shape and placed in a mold to produce a cavity or opening in a casting.

COUNTERBORE: To enlarge a hole to a given depth.

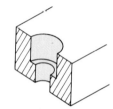

COUNTERBORED HOLE

COUNTERCLOCKWISE: From right to left in a circular motion.

COUNTERCLOCKWISE

COUNTERSINK: To chamfer a hole to receive a flat or fillister head fastener.

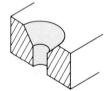

COUNTERSUNK HOLE

CPU: (Central Processing Unit). It manipulates data and processes instructions from a program or human operator. The "brain" of a computer, it retrieves data from memory storage.

CROSSHAIR: Crossed vertical and horizontal lines on a CAD graphic display that represents the cursor or current location. Also, the vertical and horizontal lines in an input device like a digitizing puck.

CURSOR: A special character on the video display that indicates the next position at which a character will be entered or deleted.

CURVED LINE: A line of which no part is straight.

CYLINDER: A geometric figure with a uniform circular cross-section through its entire length.

CYLINDER

DATA: Facts or information taken in and acted upon (processed) by a computer.

DEBUG: The process of detecting and correcting errors (bugs) in a computer program.

DIAGONAL: A line running across from corner to corner.

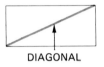

DIAGONAL

DIAMETER: The length of a straight line running through the center of a circle.

DIAMETER

DIE: The tool used to cut external threads. A tool used to shape materials.

DIE CASTING: A method of casting metal under pressure by injecting it into metal dies of a die casting machine.

DIGITIZE: To convert a drawing into digital form.

DIGITIZER: Device for inputting x and y coordinate data into a computer. Coordinate points are selected with a puck or stylus. Also known as a digitizer tablet or graphics tablet.

DISPLAY: Screen representation. Also the screen itself.

DRAFT: The clearance on a pattern or mold that allows easy withdrawal of the pattern from the mold.

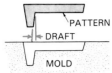

PATTERN
DRAFT
MOLD

DRILLING: Cutting round holes by use of a cutting tool called a drill.

DRILL ROD: Accurately ground and polished tool steel rods.

DRIVE FIT: Using force or pressure to fit two pieces together. One of several classes of fits.

DUCT: A sheet metal pipe or passageway used for conveying air.

ECCENTRIC: Not on a common center.

ECCENTRIC CIRCLES

ELIMINATE: To do away with.

ELLIPSE: A closed curve in the form of a symmetrical oval.

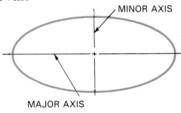

MINOR AXIS

MAJOR AXIS
ELLIPSE

ENGINEERING DRAWING: Also known as a technical drawing or drafting, is the graphic language of the engineer.

EQUAL: The same.

EQUILATERAL: A figure having equal length sides.

EXPANSION FIT: The reverse of shrink fit. The piece to be fitted is placed in liquid nitrogen or dry ice until it shrinks enough to fit into the mating piece. Interference develops between the fitted pieces as the cooled piece expands to normal size.

FERROUS METAL: A metal that contains iron as its major ingredient.

FILLET: The curved surface connecting two surfaces that form an angle.

FILLET

FIXTURE: A device for holding metal while it is being machined.

FLOPPY DISK: A flexible, flat circular plastic medium on which computer data and programs are recorded magnetically in a digital format.

FORGE: To form material using heat and pressure.

FLASH: A thin fin of metal formed at the parting line of a forging or casting where a small portion of metal is forced out between the edges of the dic.

FLASK: A frame of wood or metal consisting of a cope (the top portion) and a drag (the bottom portion) used to hold sand that forms the mold used in the foundry.

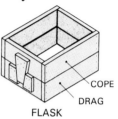

COPE
DRAG
FLASK

FLOOR PLAN: The drawing that shows the exact shape, dimensions, and arrangements of the rooms of a building.

FORCE FIT: The interference between two mating parts sufficient to require force to press the pieces together. The joined pieces are considered permanently assembled.

FREE FIT: Used when tolerances are liberal. Clearance is sufficient to permit a shaft to turn freely without binding or overheating when properly lubricated.

FRUSTUM: The figure formed by cutting off a portion of a cone or pyramid parallel to its base.

FRUSTUM

GATE: The opening that guides the molten metal into the cavity that forms the mold. See SPRUE.

GEARS: Toothed wheels that transmit rotary motion from one shaft without slippage.

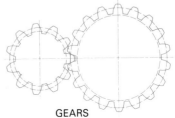

GEARS

GRAPHIC: Written or drawn.

GRID: A CAD drawing aid for determining distance. Formed by a network of uniformly spaced points or lines on a display device.

HARDCOPY: Any printed or plotter output.

HARDENING: The heating and quenching of certain iron-base alloys for the purpose of producing a hardness superior to that of the untreated metal.

HARDWARE: The components that make up a computer system.

HEAT TREATMENT: The careful application of a combination of heating and cooling cycles to a metal or alloy to bring about certain desirable conditions such as hardness and toughness.

HEXAGON: A six-sided figure with each side forming a 60 deg. angle.

HEXAGON

IC: Integrated circuit. (Electronic circuit on a silicon chip.)

INCLINED: Making an angle with another line or plane.

INCLINE

INSCRIBE: To draw one figure within another figure.

STAR INSCRIBED IN PENTAGON

INSPECTION: The measuring and checking of finished parts to determine whether they have been made to specifications.

INTERCHANGEABLE: Refers to a part that has been made to specific dimensions and tolerances and is capable of being fitted in a mechanism in place of a similarly made part.

INVESTMENT CASTING: A process that involves making a wax, plastic, or a frozen mercury pattern, surrounding it with a wet refractory material, melting or burning the pattern out after the investment material has dried and set, and finally pouring molten metal into the cavity.

JAMB: The vertical parts of a door frame.

JIG: A device that holds the work in position and guides the cutting tool.

JOYSTICK: Entry device with a vertical stick mounted on a circle of electrical contacts. It can be tilted in various directions to indicate direction of cursor movement on a screen.

KEY: A small piece of metal imbedded partially in the shaft and partially in the hub to prevent rotation of a gear or pulley on the shaft.

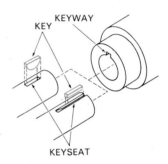

KEY   KEYWAY

KEYSEAT

KEYWAY: The slot or recess in the shaft that holds the key.

LAY OUT: To locate and scribe points for machining and forming operations.

LINE CONVENTIONS: Symbols that furnish a means of representing or describing some part of an object. It is expressed by a combination of line weight and appearance.

CONSTRUCTION AND GUIDE LINES (VERY THIN)

BORDER LINE (VERY THICK)

VISIBLE LINE (THICK)

8
DIMENSION LINE (THIN)

EXTENSION LINE (THIN)

HIDDEN LINE (MEDIUM)

CENTER LINE (THIN)

CUTTING PLANE LINE (THICK)

SECTION LINE (THIN)

PHANTOM LINE (THIN)

MACHINE TOOL: The name given to that class of machines which, taken as a group, can reproduce themselves.

MAJOR DIAMETER: The largest diameter of a thread measured perpendicular to the axis.

MEMORY: A device for storing information, usually in digital form. The device may be a magnetic tape, disk, or a silicon chip.

MENU: A program generated list of options, usually shown on the display screen. The user can refer to them and select in order to execute desired procedures.

MESH: To engage gears to a working contact.

MICROINCH: One-millionth of an inch.

MILL: To remove metal with a rotating cutter on a milling machine.

MINOR DIAMETER: The smallest diameter on a screw thread measured across the root of the thread and perpendicular to the axis. Also known as the "root diameter."

MIRROR: CAD function that allows a reverse graphic image to be formed around a selected axis.

MOLD: The material that forms the cavity into which molten metal is poured.

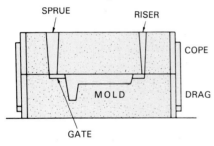

PARTS OF A MOLD

MOUSE: A movable input device for locating x and y coordinates on a surface.

OBTUSE ANGLE: An angle more than 90 deg.

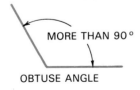

OBTUSE ANGLE

OCTAGON: An eight-sided geometric figure with each side forming a 45 deg. angle.

OCTAGON

OD: Abbreviation for outside diameter.

PENTAGON: A five-sided geometric figure with each side forming a 72 deg. angle.

PENTAGON

PERIMETER: The boundry of a geometric figure.

PERMANENT MOLD: Mold ordinarily made of metal that is used for the repeated production of similar castings.

PERPENDICULAR: A line at right angles to a given line.

PHOTO DRAWING: A drawing prepared using a photograph on which dimensions, notes, and specifications have been added.

PICK: To select an x, y coordinate point with an input device.

PITCH: The distance from a point on one thread to a corresponding point on the next thread.

PLATE: Another name given to a drawing.

PLOTTER: An output device for creating hard-copy of graphic images by controlling pens on a drawing media.

PRINTOUT: Hardcopy produced by a printer, plotter, or automated drafting machine.

PROFILE: The outline of an object.

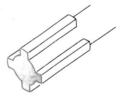

PROFILE

PROGRAM: A sequence of instructions given to a computer to perform specific functions or tasks. Also referred to as "software."

PROJECT: To extend from.

PUCK: An imput device used with a digitizing tablet to input x, y coordinates.

RACK: A flat strip with teeth designed to mesh with teeth on a gear. Used to change rotary motion to a reciprocating motion.

RADIUS: The length of a straight line running from the center of a circle to the perimeter of the circle.

RADIUS

RADII: The plural of radius.

RAM: (Random Access Memory). A type of computer memory where information can be stored and retrieved in miscellaneous order without disturbing adjacent memory cells.

REAL TIME: The time the computer needs to respond to a solution.

REAM: To finish a drilled hole to exact size with a reamer.

RECTANGLE: A geometric figure with opposite sides equal in length and each corner forming a 90 deg. angle.

RECTANGLE

RIGHT ANGLE: A 90 deg. angle. The angle that is formed by a line that is perpendicular to another line.

RIGHT ANGLE

RISER: An opening in the mold that permits the gasses to escape. The gasses are formed when molten metal is poured into the mold.

ROM: (Read Only Memory). A type of computer memory that contains information stored during processing. Usually permanent memory.

ROTATE: To turn or revolve around a point.

ROUGH LAYOUT: A rough pencil plan that arranges lines and symbols so that they have a pleasing relation to one another.

RUBBERBANDING: A CAD drawing aid to display a line with one end fixed while a second point is moved and fixed.

SEAM: The line formed where two edges are joined together.

SEGMENT: Any part of a divided line.

SKETCH: To draw without the aid of drafting instruments.

SNAP: A CAD drawing aid which allows placement of a coordinate point using a uniform grid as a guide.

SOFTWARE: The means of communicating with a computer. The program.

SPLINE: A series of grooves, cut lengthwise, around a shaft or hole.

SPOTFACE: To machine a circular spot on the surface of a casting to furnish a bearing surface for the head of a bolt or a nut.

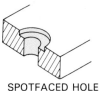

SPOTFACED HOLE

SPRUE: The opening in a mold that leads to the gate, which in turn leads to the cavity into which molten metal is poured.

SQUARE: To machine or cut at right angles. A geometric figure with four equal length sides and four right (90 deg.) angles.

STYLUS: A hand held input device used with a digitizing tablet.

SYMBOL: A figure or character used in place of a word or group of words.

TAP: The tool used to cut internal threads.

TAPER: A piece that increases or decreases in size at a uniform rate to assume a wedge or conical shape.

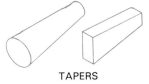

TAPERS

TEMPLATE: A pattern or guide.

THREAD: The act of cutting a screw thread.

TRAIN: A series of meshed gears.

TRIANGLE: A three-sided geometric figure.

EQUILATERAL TRIANGLE   ISOSCELES TRIANGLE   IRREGULAR TRIANGLE

TRUNCATE: To cut off a geometric solid at an angle to its base.

TRUNCATED PRISM

UNIFIED COARSE (UNC): Coarse series of American Standard for Unified Screw Threads.

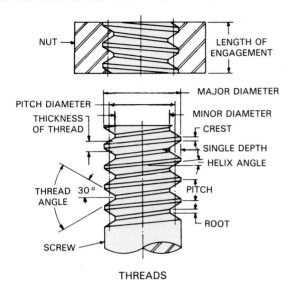

THREADS

UNIFIED FINE (UNF): Fine series of American Standard for Unified Screw Threads.

UNC: Abbreviation for coarse series of American Standard for Unified Screw Threads.

UNF: Abbreviation for fine series of American Standard for Unified Screw Threads.

UNIFIED THREADS: A series of screw threads that have been adopted by the United States, Canada, and Great Britain to attain interchangeability of certain screw threads.

VERTICAL: At right angles to a horizontal line or plane.

WORKING DRAWING: A drawing that gives the craftworker the necessary information to make and assemble a product.

ZOOM: To enlarge or decrease the displayed image in a CAD program.

ACKNOWLEDGEMENTS

While it would be a most pleasant task, it would be impossible for one person to develop the material included in this text by visiting the various industries represented and observing, studying and taking photos first hand.

My sincere thanks to those in industry who helped in gathering the necessary material, information, and photographs. Their cooperation was most appreciated.

I would also like to thank the teachers using the text who were kind enough to offer suggestions for improving EXPLORING DRAFTING.

John R. Walker
Charlottesville, VA

INDEX